D1084843

THE ILLUSTRATED FLORA OF ILLINOIS

The Illustrated Flora of Illinois

ROBERT H. MOHLENBROCK, General Editor

ADVISORY BOARD:

Constantine J. Alexopoulos, *University of Texas*
Gerald W. Prescott, *University of Montana*
Aaron J. Sharp, *University of Tennessee*
Robert F. Thorne, *Rancho Santa Ana Botanical Garden*
Rolla M. Tryon, Jr., *The Gray Herbarium*

THE ILLUSTRATED FLORA OF ILLINOIS

DIATOMS

John Jeffrey Dodd
With a Foreword by Robert H. Mohlenbrock

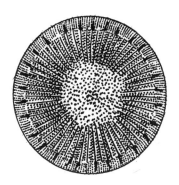

SOUTHERN ILLINOIS UNIVERSITY PRESS
Carbondale and Edwardsville

Copyright © 1987 by the Board of Trustees, Southern Illinois
University
All rights reserved
Printed in the United States of America
Edited by Stephen W. Smith
Designed by Andor Braun
Production supervised by Natalia Nadraga
90 89 88 87   4 3 2 1

Library of Congress Cataloging-in-Publication Data
Dodd, John Jeffrey, 1947–
    Diatoms.

    (The Illustrated flora of Illinois)
    Bibliography: p.
    Includes index.
    1. Diatoms—Illinois. I. Title. II. Series.
QK569.D54D6 1987   589.4'81'09771   86-17696
ISBN 0-8093-1154-2

This book is dedicated to my parents,
John D. and Jeanne N. Dodd.

CONTENTS

FOREWORD

When The Illustrated Flora of Illinois project was conceived in 1963, it was my intention for the project to cover every group of plants in the state of Illinois, from algae and fungi through bryophytes to ferns and flowering plants. To date, the volume on ferns and nine on flowering plants have been published, and, by the time the present volume appears, a tenth volume on flowering plants will have been published. This work on diatoms is the twelfth volume in The Illustrated Flora of Illinois.

I am now happy to introduce this first work in the series on a nonvascular plant group. Dr. John Jeffrey Dodd has prepared the fully illustrated volume *Diatoms*. This book is intended for the person who wishes to learn the basic diatom flora of the state of Illinois. Dr. Dodd has provided the information required for someone beginning a study of diatoms. The book, however, will also prove useful to the professional diatomologist, who can utilize the work in compiling distributional data.

Since many freshwater diatoms have broad geographical ranges, *Diatoms* should prove useful in many parts of North America. State floras are virtually nonexistent, and Dr. Dodd's book should fill a gap in our knowledge of North American algae.

I am pleased to report that work is under way on the green algae, euglenophytes, liverworts, and mosses of Illinois, as well as several additional volumes on the flowering plants.

I wish to acknowledge the Donnelley Foundation for its continued support, which has made the publication of this volume possible.

Robert H. Mohlenbrock

Carbondale, Illinois
January 30, 1987

PREFACE

The study of diatoms in the United States has grown phenomenally in the last thirty years. Interest in these organisms has spread to all parts of the country, at least in part because of the usefulness of diatoms as biological indicators in water-quality work.

One of the stumbling blocks the beginning student meets almost at once is the problem of finding suitable sources of information on the identification of diatoms beyond genus. It is not that there is a lack of such material—there are hundreds of useful books and papers—but its nature. Many of the more important books are out of print, written in languages other than English, or both. Few of those that are still available represent complete coverage of all the freshwater genera, to say nothing of the marine genera.

The present book has been written as a basic guide to the diatom-flora of a comparatively small geographical area—Illinois—but, due to the widespread occurrence of many diatom species, should serve the same function for many other areas as well. I have sought to provide descriptions and illustrations for all of the diatom taxa reported for the state—both those I found and those reported by others—and identification keys for most of them. In addition, the Supplementary Section at the end of the text provides an introduction to diatom biology and gives instructions for the collection and preparation of diatoms for study.

Diatoms have proven fascinating for many people. It is my hope that the present work will encourage students to give diatom study a try. For those who find this pursuit interesting, the bibliographies will perhaps be useful as an entry into the literature.

Few works are the unaided effort of one person and this is no exception.

I have based the text largely on a survey of the diatom-flora of Illinois that appeared as my doctoral dissertation at Southern Illinois University, Carbondale. It is appropriate therefore to acknowledge first the assistance granted by the Botany Department of that institution and particularly the help (and patience!) of Dr. Donald Tindall and Dr. Robert Mohlenbrock. The work at SIU-C was supported in part by a grant from the National Science Foundation (N.S.F. DEB77–1206).

In addition, some of this work was carried out at Iowa State University. I would like to thank in this connection Dr. Lois Tiffany and Dr. John D. Dodd. The Iowa State University Library was an invaluable source of reference materials. Several summers were spent in residence at the Iowa Lakeside Laboratory and I would like to thank the director, Dr. Richard Bovbjerg, for courtesies extended me there. Dr. Ruth Patrick and Dr. Charles W. Reimer of the Academy of Natural Sciences of Philadelphia provided scientific advice at various times for which I am especially grateful. Their published works have been frequently cited as "critical references" (CR) in the descriptions and elsewhere. Dr. Marcia Grady and Dr. Douglas Zehr supplied me with samples from various sites that I could not visit in person. I also acknowledge a considerable debt to Dr. F. M. Begres, who guided my first steps in diatom study and whose teaching and advice have made much of this work possible.

Last in this list (but first in my thoughts), I thank my parents, John D. and Jeanne N. Dodd, without whose continuous encouragement this project would surely never have been finished.

John Jeffrey Dodd

Iowa City, Iowa
January 16, 1987

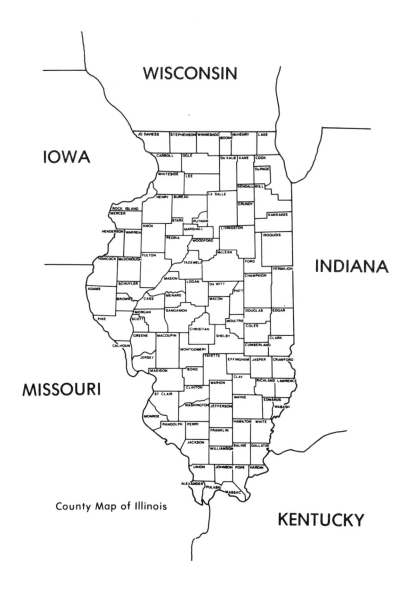

WISCONSIN

IOWA

INDIANA

MISSOURI

KENTUCKY

County Map of Illinois

JO DAVIESS | STEPHENSON | WINNEBAGO | McHENRY | LAKE
BOONE
CARROLL | OGLE | De KALB | KANE | COOK
DuPAGE
WHITESIDE | LEE
KENDALL | WILL
HENRY | BUREAU | La SALLE
GRUNDY
ROCK ISLAND | KANKAKEE
MERCER | KNOX | STARK | PUTNAM | LIVINGSTON
MARSHALL | IROQUOIS
HENDERSON | WARREN | PEORIA | WOODFORD
FULTON | McLEAN | FORD
HANCOCK | McDONOUGH | TAZEWELL | CHAMPAIGN | VERMILION
MASON | LOGAN | De WITT
SCHUYLER | PIATT
ADAMS | MENARD | MACON | DOUGLAS | EDGAR
BROWN | CASS | SANGAMON
MORGAN | CHRISTIAN | MOULTRIE | COLES
PIKE | SCOTT | SHELBY | CLARK
GREENE | MACOUPIN | CUMBERLAND
CALHOUN | MONTGOMERY | FAYETTE | EFFINGHAM | JASPER | CRAWFORD
JERSEY | BOND
MADISON | CLAY | RICHLAND | LAWRENCE
CLINTON | MARION | WAYNE | EDWARDS
ST CLAIR | WASHINGTON | JEFFERSON | WABASH
MONROE | HAMILTON | WHITE
RANDOLPH | PERRY | FRANKLIN
JACKSON | WILLIAMSON | SALINE | GALLATIN
UNION | JOHNSON | POPE | HARDIN
ALEXANDER | PULASKI | MASSAC

SYSTEMATIC SECTION

Introduction

The text contains two major sections: the Systematic Section and the Supplementary Section. Students unfamiliar with the diatoms should begin by reading the Supplementary, while more experienced workers will probably wish to go directly to the Systematic.

The Systematic Section contains the keys, descriptions and illustrations for the genera and species of diatoms that I found in Illinois, that others have reported, and, in addition, some genera that probably occur in the state but have not as yet been reported. I have provided also an illustrated glossary, a list of synonyms and names of uncertain application, and a list of the authorities for cited taxa.

The Supplementary Section contains remarks on diatom biology and classification and methods for collection and preparation. The bibliographies and index conclude the work.

Systematic Section

Unlike identification of higher plants, diatom identifications usually begin directly with the genus. In this text there is a key to genera, but if you are just beginning diatom study it might be easier if you scan the illustrations for the various genera before starting to use the key. The Glossary contains definitions or illustrations for all terms used.

If you are fairly certain of the genus, either use the key to the species of that genus or look for a likely identification by examining the illustrations. Keep in mind that the keys to species refer only to those taxa that I was able to confirm for the state. Other species may be found in Valid Taxa Reported by Other Authors and Genera Expected in Illinois but not yet Reported. If you are working from the illustrations in the first part of the Systematic Section, remember that there are illustrations associated with the chapters at the end as well. It almost goes without saying that you should read the description before deciding on an identification.

General Classification Scheme Used in this Book

I have provided a chart that shows the classification I have followed. With a few exceptions the genera are presented in text in the same order as in the chart.

Keys

I have provided dichotomous keys to the genera reported in Illinois and keys to the taxa I found in each genus if there is more than one species. The species reported by other authors are not included in my keys to species. For those who may be unfamiliar with dichotomous keys, there are brief instructions before the Key to Genera. The keys to species are constructed on the same principles.

Descriptions

Each description for a species, variety, or form contains the following information:

1. Name in trinomial form with the authority for the lowest level of classification in the name. (Trinomial nomenclature is explained in the Supplementary Section. A list of authorities can be found at the end of the Systematic Section.)

2. Number of the figure in the present text that contains the illustration(s).

3. Critical reference (CR). This is the source of information I used in making the identification.

4. Description. This is a verbal account of the important features of the taxon.

5. Dimensions. The dimensions listed are from the critical reference modified by any extensions that I encountered during the study. Lengths and widths are given in micrometers (μm). Densities are given as units (striae, puncta, etc.) in 10 micrometers (/10μm).

6. Map showing counties in which I found the taxon. If there is no map, the locations are given in text. When interpreting the maps, keep in mind that I took samples from only about half the counties in Illinois. Maps show where I found various species, but should not be interpreted as meaning that the species cannot be found elsewhere.

Illustrations

I have drawn most of the illustrations from nature, but those in the Glossary are diagrams that do not necessarily depict any known species, and illustrations for taxa that I did not find are stylized sketches based on the species concept of the author cited as the critical reference (CR) for the taxon in question. All scale lines on the figures represent 10 micrometers (10μm).

Classification of Diatom Genera Found or Expected In Illinois

The genera of diatoms found or expected in Illinois are arranged according to the following scheme based on Patrick and Reimer (1966, 1975); Hustedt (1930); and Silva (1962) as cited in Patrick and Reimer (1966).

Genera marked with an asterisk* have not been reported for Illinois but may be found here. Descriptions and illustrations for these genera are found in Genera Expected in Illinois but not yet Reported. These genera have been included, however, in the main Key to Genera.

Genera marked with a dagger[†] have been reported for Illinois by other authors but were not found during the present study. Descriptions and illustrations are found in Valid Taxa Reported by Other Authors. These genera also are included in the main Key to Genera.

Division Bacillariophyta
  Class Centrobacillariophyceae
    Order Eupodiscales
      Family Coscinodiscaceae
        Genus *Melosira*
            Thalassiosira
            Stephanodiscus
            Cyclotella
            Coscinodiscus
    Order Biddulphiales
      Family Biddulphiaceae
        Genus *Biddulphia*
            *Attheya**
    Order Rhizosoleniales
      Family Rhizosoleniaceae
        Genus *Rhizosolenia*[†]
  Class Pennatibacillariophyceae
    Order Fragilariales
      Family Fragilariaceae
        Genus *Tabellaria*
            Meridion

Diatoma
Asterionella
Fragilaria
Synedra
Opephora*
Order Eunotiales
Family Eunotiaceae
Genus *Eunotia*
Order Achnanthales
Family Achnanthaceae
Genus *Achnanthes*
Cocconeis
Rhoicosphenia
Order Naviculales
Family Naviculaceae
Genus *Mastogloia*
Amphipleura
Frustulia
Gyrosigma
Pleurosigma
Stauroneis
Capartogramma
Anomoeoneis
Neidium
Diploneis
Navicula
Caloneis
Pinnularia
Family *Entomoneidaceae*
Genus *Plagiotropis*
Entomoneis
Family Cymbellaceae
Genus *Cymbella*
Amphora
Family Gomphonemaceae
Genus *Gomphonema*
Gomphoneis[†]
Didymosphenia[†]
Order Epithemiales
Family Epithemiaceae

Genus *Denticula*
    Epithemia
    Rhopalodia
Order Bacillariales
  Family Bacillariaceae
    Genus *Bacillaria*
    Hantzschia
    Nitzschia
    Cylindrotheca*
Order Surirellales
  Family Surirellaceae
    Genus *Cymatopleura*
    Surirella
    Campylodiscus[†]

Key to the Genera of Diatoms Reported or Expected for Illinois

Those who are familiar with dichotomous keys should proceed directly to the key. All terms used in the key have been defined in the illustrated glossary at the end of the Systematic Section. Those genera marked with an asterisk* have not yet been reported for Illinois and will be found in Genera Expected in Illinois but not yet Reported. Genera marked with a dagger† were not found in my study but have been reported for the state by other authors and can be found in Valid Taxa Reported by Other Authors.

This key and the others in the book are dichotomous, that is, at each step you are asked to make a choice between mutually exclusive alternatives. You must begin with the first pair of numbers and proceed as directed. At the end of each choice will be either a number for the next pair of choices or the name of a genus (in other keys, name of a species). Be sure to read both choices for each number before deciding. Note any places where you feel that you are guessing. You will frequently find that if you make a wrong choice in the key, subsequent choices become more and more difficult and may often make no sense at all. If this happens, you may wish to go back to the point at which you made a guess and try the other path. Certainly you should do so if you reach an end point and the genus (or species) is clearly not what you have. Keep in mind too that every key is written for a limited selection of taxa and that the one you are looking for may not be in the key at all, especially in the case of species.

A classic remark on keys is that they are made by people who don't need them for people who can't use them. The problem arises for both parties because we rarely learn to recognize or describe things in the stepwise analytical fashion that a key requires. You will, however, soon recognize all the major genera by sight on the basis of overall appearance and will know many of the common species. The more you know, the easier the keys will be to use.

KEY TO GENERA

1. Both valves of each cell with a raphe, rudimentary raphe, or pseudoraphe. Valves distinctly longer than broad, or, if nearly circular, then

raphe or pseudoraphe distinct. Valve markings typically bilaterally symmetric to the longitudinal valve axis. Cells of some species united in nature into flat, ribbonlike filaments _____ 9

1. Neither valve with a raphe, rudimentary raphe or pseudoraphe. Valve outline typically circular or subcircular. Valve markings, if present, typically radially symmetric. Cells of some species united in nature into cylindrical filaments or into amorphous colonial masses _____ 2

    2. Cells lightly silicified and nearly invisible. Valves devoid of markings but bearing one or two long spines. Valve outline elliptic, so cells do not appear to be cylindrical. Numerous imbricate intercalary bands present. Cells always seen in girdle view. Planktonic only. _____ 3

    2. Cells typically clearly visible even when almost devoid of markings. Valves with or without spines, but, if spines present, then valves typically distinctly marked with rows of fine to coarse puncta. Valve outline circular, so cells in girdle view will appear cylindrical or drum-shaped _____ 4

3. Each valve with one long spine _____ *Rhizosolenia*†

3. Each valve with two long spines _____ *Attheya**

    4. Cells typically united valve-to-valve in cylindrical filaments (this colony structure may be lost during chemical cleaning). Valve margin with or without distinct spines. Valve mantle elongate and frequently marked with distinct rows of puncta. Cells and valves typically seen in girdle view _____ 1. *Melosira*

    4. Cells typically solitary, sometimes in groups in amorphous matrices, sometimes in short chains. Spines present or absent. Valve mantle short so valves almost always lie in valve view. Valve face marked with striae, costae, or puncta _____ 5

5. Each valve with two large oval protuberances. Cell wall heavily silicified and coarsely punctate-striate. Very rare _____ 6. *Biddulphia*

5. Oval protuberances lacking, but spines or dotlike warts often present --6

    6. Valve surface faintly and finely punctate-striate. Valve margin marked by a ring of dotlike warts. A single dashlike mark present near valve margin _____ 2. *Thalassiosira*

    6. Valve surface typically distinctly punctate but marginal warts and dashlike mark absent _____ 7

7. Valve markings divided into concentric zones, the outer zone costate, the inner zone variously marked _____ 4. *Cyclotella*

7. Valves marked by punctate striae that typically reach the center of the valve _____ 8

8. All striae follow radii of the valve. Marginal spines typically distinct _____ 3. *Stephanodiscus* (See also *Melosira roeseana*.)

8. Valve surface divided into wedge-shaped zones within which the striae are parallel to each other. Spines absent _ 5. *Coscinodiscus*

9. Pseudoraphe present on both valves or if not then neither valve with a true raphe or rudimentary raphe _____ 10

9. True raphe or rudimentary raphe present on at least one valve. (Be sure to check the ends of the cell carefully. I suggest that you examine illustrations of *Eunotia* and *Rhoicosphenia* for examples of valves with rudimentary raphes.) _____ 16

10. Valves marked with both striae and costae _____ 11

10. Valves marked with striae only _____ 12

11. Valves asymmetric to transverse axis _____ 8. *Meridion*

11. Valves symmetric to transverse axis _____ 9. *Diatoma*

12. Cells with internal septa associated with each valve _____ _____ 7. *Tabellaria*

12. Cells lacking internal septa_____ 13

13. Valves asymmetric to the transverse axis _____ 14

13. Valves symmetric to the transverse axis _____ 15

14. Valves elongate with somewhat bulbous ends, the one end broader than the other. Cells united in nature into star-shaped colonies __ _____ 10. *Asterionella*

14. Valves distinctly club-shaped, never forming star-shaped colonies ------ *Opephora** (See also *Fragilaria pinnata*, which has abnormal valves of similar shape.)

15. Valves typically under 50 μm length and united valve-to-valve in band-like or ribbonlike colonies (colonial form frequently destroyed by chemical "cleaning") _____ 11. *Fragilaria*

15. Valves often well in excess of 50 μm length. Cells not united into band-like colonies in nature _____ 12. *Synedra*

16. Both valves of each cell with short rudimentary raphes associated with enlarged polar nodules. Valves typically arc-shaped _____ _____ 13. *Eunotia*

16. One valve of each cell with a fully developed true raphe, the other with a similar true raphe or a rudimentary raphe or pseudoraphe _____ 17

17. Both valves with a fully developed true raphe _____ 20

17. Valve opposite the true-raphe valve with either a pseudoraphe or with short rudimentary raphe branches separated by a narrow clear space _____ 18

18. Pseudoraphe present on valve opposite true-raphe valve ____ 19

18. Rudimentary raphe present on valve opposite true-raphe valve. Valves club-shaped in outline _____ 16. *Rhoicosphenia*
19. Valves elliptical in outline and lacking distinctive ends. Valves arched around the longitudinal axis (Axial area and valve margins will usually not be simultaneously in focus.) Central area of pseudoraphe-valve never bearing a horseshoe-shaped mark _____ 15. *Cocconeis*
19. Valves typically not elliptical or if elliptical then central area of pseudoraphe valve bearing a horseshoe-shaped mark. Valve ends frequently distinctive. Valves often distinctly arched around the transverse axis so the cells appear bent if seen in girdle view _____ 14. *Achnanthes*.
   20. Each valve with a simple two-branch raphe structure not located in a canal or on a raised wing or keel. (See Glossary definitions.) Raphe typically follows the longitudinal axis of the valve ____ 32
   20. Each valve with a raphe structure found in a central or marginal canal or atop one or more raised wings or keels _____ 21
21. Raphe located in a central or marginal canal. Valve surface always marked by distinct costae alternating with groups of striae that may be faint or coarse and that may be clearly punctate or apparently nonpunctate. Some species have distinct multiloculate internal septa __22
21. Raphe(s) located atop one or more wings or keels. (See Glossary definitions.) Internal septa lacking _____ 24
   22. Valves symmetric to longitudinal and transverse axes _____
      _____ 35. *Denticula*
   22. Valves symmetric only to the transverse axis _____ 23
23. Valves typically have distinctly punctate striae. Canal clearly visible, especially at midvalve. Internal septa may be present _____
      _____ 36. *Epithemia*
23. Valves with faint striae that do not appear to be punctate under the light microscope. Canal closely following the apparent dorsal margin and therefore hard to distinguish. Internal septa absent _____
      _____ 37. *Rhopalodia*
   24. Raphe atop a single tall bilobed wing. Wing base is either sigmoid or straight and if straight lies along the longitudinal axis of the valve
      _____ 25
   24. Raphe(s) located atop one or two short wings or keels located usually, but not always, near the valve margin(s) _____ 26
25. Wing base sigmoid. Wing lobes so tall with respect to valve width that valves and cells almost always are seen in girdle view, one wing lobe of each wing out of focus due to the sigmoid base _____
      _____ 31. *Entomoneis* (rare)
25. Wing base straight. Cells and valves normally seen in girdle view but

occasionally in valve view _____ 30. *Plagiotropis* (very rare)
26. One keel per valve _____ 27
26. Two keels per valve, one along each valve margin _____ 30
27. Cells so lightly silicified that only the keels are easily visible. Cells twisted around the long axis so that the keels appear as intersecting helices _____ *Cylindrotheca**
27. Cells with outline distinctly visible, not twisted around the long axis. Keels more or less straight _____ 28
    28. Keel central or subcentral. Striae distinct and punctate. Cells united in nature into bandlike colonies that change shape constantly due to the individual motion of the cells __ 38. *Bacillaria*
    28. Keel marginal or submarginal or if nearly central then striae not visible except under strongly obliqued light. Cells not united in nature into colonies that change shape _____ 29
29. Keel located near the valve margin. Keels of both valves on the same side of the cell _____ 39. *Hantzschia*
29. Keel of both valves marginal. Keel of one valve on opposite side of cell from the other. In a few species, keels subcentral in position _____ _____ 40. *Nitzschia*
    30. Valve outline nearly circular, but cells strongly bent into a saddle-shape and therefore almost never seen in full valve view _____ _____ *Campylodiscus*†
    30. Valves typically elongate. Valves may have undulations or twists but are not bent into a saddle-shape _____ 31
31. Valve surface with undulations, the crests of which are more or less transverse _____ 41. *Cymatopleura*
31. Valve surface flat or arched, but not undulate _____ 42. *Surirella*
    32. Valves with sigmoid longitudinal axes and corresponding sigmoid outline _____ 33
    32. Valves with a straight longitudinal axis or if axis is curved then neither the axis nor the valve outline are sigmoid _____ 34
33. Striae in two diagonally intersecting systems _____ 21. *Pleurosigma*
33. Striae in two more or less perpendicularly intersecting systems or transverse striae alone are apparent _____ 20. *Gyrosigma*
    34. Valves symmetric to both transverse and longitudinal axes ___ 39
    34. Valves symmetric to transverse or longitudinal axes but not both _____ 35
35. Valves symmetric to transverse axis _____ 36
35. Valves symmetric to the longitudinal axis _____ 37
    36. Dorsal and ventral valve mantles of unequal height, the valve therefore "gabled" and the two valve planes not parallel _____ _____ 33. *Amphora*

36. Dorsal and ventral valve mantles of more or less equal height, the valve planes therefore more or less parallel _ _ _ _ _ 32. *Cymbella*
37. Striae composed of double rows of very fine puncta. Longitudinal lines cross the striae on both sides of the axial area _ _ _ _ _ _ _ *Gomphoneis†*
37. Striae composed of single rows of fine to coarse puncta. Longitudinal lines absent _ 38
    38. Valves large (often around 100 μm in length). Striae formed of single rows of very coarse "puncta" and the central area marked with 2–5 large stigmata _ _ _ _ _ _ _ _ _ _ _ _ _ _ _ _ _ *Didymosphenia†*
    38. Valves typically shorter than 100 μm. Isolated puncta sometimes present near the ends of some striae on one side of the central area, but these have the same form as normal puncta, though they may appear somewhat brighter or larger_ _ 34. *Gomphonema* (common)
39. Raphe branches short, not reaching midvalve. Each branch in a thickened silica rib, the two ribs connected by a narrow clear space. Striae not visible except with an excellent lens and strongly obliqued light _ 18. *Amphipleura*
39. Raphe branches nearly reaching midvalve, or, if moderately short, then the valve seen as distinctly striate without using obliqued light _ 40
    40. Central area reaching both margins and thickened to form a stauros. (The stauros will have a different luminosity from adjacent areas of the valve.) _ 41
    40. Central area not reaching both margins, or, if it does, then it is not thicker than the adjacent parts of the valve _ _ _ _ _ _ _ _ _ _ _ _ _ 42
41. Stauros butterfly-shaped to rectangular _ _ 22. *Stauroneis* (common)
41. Stauros X-shaped _ _ _ _ _ _ _ _ _ _ _ _ _ _ 23. *Capartogramma* (very rare)
    42. Each valve associated with an internal septum that has a large central locule with spoon-shaped ends that has along its margins a series of small locules _ 17. *Mastogloia*
    42. Valves not associated with any septa or if septum (craticular plate) present, then the locules irregular in appearance. (*Please Note:* Beyond this point in the key the separations are more uncertain due to the presence in the genus *Navicula* of some species, mostly rare, that superficially resemble members of the other genera.) _ _ _ 43
43. Striae arranged in two more or less perpendicularly intersecting systems (strongly obliqued light may be needed). Raphe branches lie within thickened silica ribs _ 19. *Frustulia*
43. Only transverse striae typically visible or no striae visible. If raphe branches lie within thickened ribs, then striae do not seem to form two intersecting systems _ 44
    44. Raphe branches lie within thickened silica ribs. Valves elliptic to

linear-elliptic in outline. Striae composed of one or two rows of puncta between thickened costae. Striae typically, but not always, distinctly punctate _____ 26. *Diploneis*

44. Raphe branches typically not enclosed in silica ribs, or, if so then spaces between striae not thickened into costae _____ 45

45. Striae typically distinctly punctate and crossed by at least one broad longitudinal band on each side of the valve. Valves linear to linear-elliptic _____ 25. *Neidium*

45. Striae visible or invisible, but if punctate then not crossed by broad longitudinal bands (though narrow longitudinal lines or clear spaces may be present) _____ 46

46. Striae punctate, the puncta so arranged that the gaps between puncta or groups of puncta form several undulating longitudinal spaces (not bands) on both sides of the valve __ 24. *Anomoeoneis*

46. Striae configured otherwise _____ 47

47. Striae distinct but cannot be resolved into puncta or crosslines even under very strongly obliqued light. Many species have a single open pore (foramen) in each stria, the pores so aligned as to form a single longitudinal line or band on each side of the valve. This band is not always distinct due to the size and position of the pores _____ 48

47. Striae distinct or invisible but frequently resolvable (under moderate to strongly obliqued light) into crosslines or puncta. The majority of species do not exhibit any longitudinal lines or bands __ 27. *Navicula*

48. Striae very thin and linelike _____ 28. *Caloneis*

48. Striae generally thick in appearance, frequently appearing thicker than the spaces between them _____ 29. *Pinnularia*

Genera and Species of Diatoms Found during the Present Study

1. *Melosira* C. A. Agardh 1824

Melosira belongs to the Centrobacillariophyceae and, like all members of the class, is characterized by radial symmetry of markings on the valve face and by the absence of any sort of raphe. The margin of the valve face is frequently equipped with a ring of short spines, and some species have long spines as well. One very common species lacks spines and, indeed, has no markings at all on the valve.

Most species of *Melosira* have elongate valve mantles. As a result, cells and valves are most often seen in girdle view. The valve mantle of most species is distinctly marked with punctate striae. The valve face is rarely seen in most species. Valve face markings vary from absent to scattered puncta to distinct punctate striae.

Cells of *Melosira* are united by the valve faces into short or long filaments. Most species are planktonic and some of them are among the most important components of the plankton of lakes. The least typical species, *M. roeseana* (which has punctate striae on the valve face), is associated with subaerial habitats.

KEY TO THE TAXA OF Melosira IN ILLINOIS

1. Valve margins lacking spines; valve face and mantle devoid of markings, except for a few ghostlike speckles _____ 12. *M. varians*
1. Valve margins with spines; valve mantle with regularly arranged puncta (sometime very fine); valve face variously marked _____ 2
   2. Terminal cells of a chain and some intermediate cells bearing long spines _____ 3
   2. All cells with short spines only _____ 7
3. Valve mantles with very fine striae (usually visible only with strongly oblique light) _____ 7. *M. herzogii*
3. Valve mantles with coarse, distinctly punctate striae _____ 4
   4. Pervalvar axis curved; filaments often helical _____
   _____ 5. *M. granulata* var. *angustissima* f. *spiralis*
   4. Pervalvar axis straight; filaments straight _____ 5
5. Length of a complete cell ten or more times the diameter _____
_____ 4. *M. granulata* var. *angustissima*
5. Length of a complete cell less than ten times the diameter _____ 6

6. Valve diameter greater than the length of the valve mantle _____
_____ 6. *M. granulata* var. *muzzanensis*
6. Valve diameter less than or equal to the length of the valve mantle
_____ 3. *M. granulata* var. *granulata*
7. Cells solitary or in very short chains; valve face striate, the striae distinctly punctate; valve margin with short spines which sometimes appear bifurcate; girdle with numerous overlapping intercalary bands; valve mantle punctate-striate _____ 11. *M. roeseana*
7. Cells usually in fairly long chains; spines not bifurcate; girdle not formed of numerous intercalary and girdle bands _____ 8
8. Sulcus distinct, square in cross section _____ 1. *M. ambigua*
8. Sulcus V-shaped in cross section _____ 9
9. Length of valve mantle usually greater than the diameter; cells rarely seen in valve view _____ 10
9. Length of valve mantle usually less than the diameter; cells often seen in valve view _____ 2. *M. distans* v. *alpigena*
10. Striae on valve mantle parallel to pervalvar axis _ 8. *M. islandica*
10. Striae on valve mantle helically arranged _____ 11
11. Valve diameter 3–5 μm _____ 10. *M. italica* v. *tenuissima*
11. Valve diameter greater than 5 μm _____ 9. *M. italica* v. *italica*

1. Melosira ambigua (Grun.) O. Muell. var. **ambigua** *Fig. 2f, g.*

CR: Hustedt (1930).

DESCRIPTION: Cells united into chains of moderate length; valve face not usually seen; valve margin ringed with short spines; long spines lacking; valve mantle ornamented with numerous rows of fine pores arranged helically; sulcus distinct, square in cross section.

DIMENSIONS: Height, 3.5–13.0 μm; diameter, 4–15 μm; striae, 16–24/10 μm.

2. Melosira distans var. **alpigena** Grun. *Fig. 2e*.

CR: Hustedt (1930).

DESCRIPTION: Cells united into short chains; valve face often seen; valve face marked with evenly scattered puncta; valve margin ringed with short spines; valve mantle short and ornamented with rows of fine pores arranged helically; sulcus present and angular in section.

DIMENSIONS: Height, about 4 μm; diameter, 3–8 μm; striae, 16–20/10 μm.

3. Melosira granulata (Ehr.) Ralfs var. **granulata** *Fig. 1f*

CR: Hustedt (1930).

DESCRIPTION: Cells united into chains of moderate to considerable length; valve face almost never seen; valve margin ringed with short spines; terminal cells of chains and some intercalary cells equipped with two or more long spines (the presence of spines is usually required for confirmation of this species); valve mantle ornamented by rows of very coarse puncta or rows of moderately fine puncta (both types of cells may be present in the same chain); arrangement of rows of puncta helical; sulcus present and of angular section.

DIMENSIONS: Height, 5–18 μm; diameter, 5–21 μm; striae, (coarse form) 8–9/10 μm; striae (fine form) 10–15/10 μm; puncta, 10–12/10 μm.

4. Melosira granulata var. **angustissima** O. Muell. *Fig. 1g*.

CR: Hustedt (1930).

DESCRIPTION: Cells united into chains of moderate to considerable length; valve face never seen; valve margin ringed with short spines; terminal cells of chains equipped with two or more long spines (the presence of spines is necessary for confirmation of identification); valve mantle ornamented with several rows of moderately fine puncta, the rows arranged helically on the mantle; sulcus present and angular in section. This taxon is separated from the nominate variety by the ratio of the length of the pervalvar axis to the cell diameter, which is ten or more times for the present taxon and less, usually considerably less, for the nominate.

DIMENSIONS: Length-to-breadth ratio as noted above; diameter, 3–5 μm; striae, as for the nominate variety.

5. Melosira granulata var. angustissima f. spiralis Hust. *Fig. 1h.*

CR: Hustedt (1930).

DESCRIPTION: as for var. *angustissima*, with the exception that the pervalvar axis of the cell is curved so that the chains of cells are helical rather than straight.

DIMENSIONS: As for var. *angustissima*.

6. Melosira granulata var. muzzanensis Meister *Fig. 1d.*

CR: Hustedt (1930).

DESCRIPTION: Cells similar in appearance to var. *granulata*, with the exception that the valve mantles are very short. This variety probably intergrades with the nominate but could not be shown to do so in the Illinois material

DIMENSIONS: Height, 3.5–8.0 μm; diameter, 8–25 μm; striae and puncta, as in the nominate variety.

7. Melosira herzogii Lemm. var. **herzogii** *Fig. 1e*.

CR: Bourelly (1968).

DESCRIPTION: Cells united into chains; cells almost never seen in valve view; valve margin bearing two long spines but no short spines. The third "spine" shown on the side of the valve mantle is in fact a groove. Valve mantle ornamented with numerous rows of very fine puncta (Bourelly did not report any visible striae at all); sulcus present and angular in section.

DIMENSIONS (my specimens): Height, around 17 μm; diameter, around 6.5 μm; striae, around 32/10 μm.

8. **Melosira islandica** O. Muell. var. **islandica** *Fig. 2h*.

CR: Hustedt (1930).

DESCRIPTION: Cells united into chains of moderate length; cells rarely seen in valve view; junction of valve face and valve mantle almost a right angle; pseudosulcus not developed; sulcus V-shaped; valve mantle ornamented with punctate striae that are oriented parallel to the pervalvar axis. Some cells may have coarser striae than others. Valve face ringed with very short spines.

DIMENSIONS: Height, 4–21 μm; diameter, 7–21 μm; striae (coarse form), 11–12/10 μm; striae (fine form), 14–16/10 μm, perhaps 18/10 μm; puncta (coarse form), 12–13/10 μm; puncta (fine form), 16–18/10 μm, perhaps 20/10 μm.

9. **Melosira italica** (Ehr.) Kuetz. var. **italica** *Fig. 2a–c*.

CR: Hustedt (1930).

DESCRIPTION: Cells united into chains of moderate length; valve ringed with short spines; long spines absent; valve mantle ornamented with rows of moderately fine pores arranged helically on the mantle; sulcus present and angular in section. I have illustrated (*Fig. 2c*) a resting spore of *M. italica*. This spore has very thick walls (especially near the sulcus) and a dome-shaped valve face. The resting spore was previously described as *M. laevis* Ehr. The forms with the striae nearly perpendicular to the long axis of the mantle and somewhat undulate have been described in error as the species *M. crenulata* Kuetz.

DIMENSIONS: Height, 8–21 μm; diameter, 5–28 μm; striae, 12–20/ 10 μm (mine occasionally reached 24–26/10 μm).

10. **Melosira italica** var. **tenuissima** (Grun.) O. Muell. *Fig. 2d.*

CR: Hustedt (1930).

DESCRIPTION: as for the nominate variety with the exception of diameter.

DIMENSIONS: Diameter, 3–5 μm; all other dimensions as for the nominate variety.

11. **Melosira roeseana** Rabh. var. **roeseana** *Fig. 1a, b.*

CR: Hustedt (1930).

DESCRIPTION: Cells united in short chains, occasionally solitary; valve face frequently seen. (The valve face appears much like a *Cyclotella* in having two distinct regions of marking. The outer, or marginal, ring is punctate-striate. The inner field is free of markings, except for two or three large puncta. The distinct bifurcate marginal spines and punctate striae distinguish it from *Cyclotella* species found in Illinois.) Valve mantle ornamented with moderately fine, punctate striae that are more or less parallel to the pervalvar axis; intercalary bands numerous and overlapping.

DIMENSIONS: Height, 14–25 μm; diameter, 14–25 μm; striae on valve mantle, about 12/10 μm.

12. Melosira varians C. A. Ag. var. **varians** *Fig. 1c.*

CR: Hustedt (1930).

DESCRIPTION: Cells united into chains of variable length in nature but usually seen isolated in chemically "cleaned" preparations; isolated valves often seen in valve view; valve face and valve mantle devoid of ornamentation except for a few ghostlike speckles; spines of all types lacking; sulcus not evident.

DIMENSIONS: Height, 9–13 μm; diameter, 8–35 μm.

2. *Thalassiosira* Cleve 1873

Thalassiosira belongs to the Centrobacillariophyceae and, as with other members of the class, is characterized by radial symmetry of markings on the valve face. The valve outline is circular. The valve margin is marked by a ring of low dotlike warts that are apparently short modified spines. The valve surface is striate, though the striae are so fine that they may not be noticed. A character which is most distinctive in the one species reported for Illinois is a dashlike mark near the margin of the valve. The long axis of the mark is aligned on a radius of the valve. The cells are usually seen in valve view since the pervalvar axis is fairly short. Hustedt (1930) noted that intercalary bands are present in the girdle.

Only the following taxon occurs in Illinois.

1. Thalassiosira fluviatilis Hust. var. **fluviatilis** *Fig. 3a.*

CR: Hustedt (1930).

DESCRIPTION: characters as for the genus, with these particulars: cells usually solitary.

DIMENSIONS: Diameter, 15–23 μm; marginal dots, 10–13/10 μm measured around the circumference; striae, very fine and finely punctate.

3. *Stephanodiscus* Ehrenberg 1845

This genus contains some of the more important planktonic diatoms. Members of the genus have been used by some as indicators of the beginning stages of eutrophication among other things and care should be taken in making identifications. Unfortunately, there appears to be considerable disagreement among the authorities about the taxonomy of the genus. The matter is under investigation currently in a number of laboratories, both in the United States and abroad. Those interested in diatom ecology would do well to follow the literature on this subject. The present work uses a fairly classical approach to taxonomy, which may ultimately require modification.

Stephanodiscus is a member of the Centrobacillariophyceae and, like other genera in that class, is characterized by radial symmetry of the markings of the valve face. Valve markings consist of coarse or fine puncta (which, by the way, are often not simple pores) arranged in striae. The striae consist of several to many rows of puncta and are separated from one another by thickened radial costae. Many or all of the costae are terminated by spines near the valve margin. In the smaller and more lightly silicified species, it sometimes appears that the striae, rather than the costae, are terminated by spines. This is apparently an illusion that results from the fact that only the rows of pores immediately adjacent to the very thin costae are visible. Thus each "stria" (more properly, pseudostria) consists of one costa plus the part of each adjacent true stria that is in contact with the costa. The spine that appears to terminate a stria thus in fact terminates a costa. I have chosen to illustrate the apparent, rather than the actual, state of affairs in some cases since it more accurately reflects the appearance under most light microscopes.

Valve face markings in *Stephanodiscus* are not divided into concentric zones in the same way as *Cyclotella*. Although the center part of the valve is often different in appearance from the outer part, it can be seen that the type of marking, that is, the puncta, is the same in both parts. In *Cyclotella*, the outer group of markings is typically not resolvable into puncta with the light microscope.

Other genera of centric diatoms can also be fairly easily distinguished from *Stephanodiscus*: *Stephanodiscus* lacks the dashlike marginal mark of *Thalassiosira*, the division of the surface into wedge-shaped groups of striae of *Coscinodiscus*, and the pair of large oval projections characteristic of *Biddulphia*. *Stephanodiscus* valves typically do not have elongate valve mantles as do those of

Melosira. One species of *Melosira*, *M*. *roeseana*, may cause confusion at first, but the presence of several very large, punctumlike marks near valve center in that species is a character not found in any of the *Stephanodiscus* species in the state. (Also, it should be noted that *M*. *roeseana* is rare and found in subaerial habitats, while *Stephanodiscus* is common and typically planktonic.)

All the Illinois species of *Stephanodiscus* have circular valve outlines and occur as solitary cells or rarely in chains of a very few cells.

KEY TO THE SPECIES OF Stephanodiscus IN ILLINOIS

1. Spines submarginal and very stout; a marginal zone of finely punctate striae usually evident outside the ring of spines; usually several striae between each pair of spines _____ 3. *S*. *niagarae*
1. Spines marginal (no outer striated zone present); each "stria" apparently terminated with a spine; spines not stout _____ 2
  2. Valve surface flat; striae faint, reaching the valve center _____ _____ 1. *S*. *hantzschii*
  2. Valve surface concentrically undulate; striae apparently becoming disorganized near valve center in many cases _____ 2. *S*. *minutus*

1. **Stephanodiscus hantzschii** Grun. var. **hantzschii** *Fig. 4c, e*.

CR: Cleve & Grunow (1880); Huber-Pestalozzi (1942); Håkansson (1976).

DESCRIPTION: Valves circular in outline; valve face flat (rather than undulate as for *S*. *minutus* var. *minutus*); each costa terminated with a spine, but as noted in the introduction to the genus, the spine may appear to be at the end of a pseudostria composed of a costa and the rows of puncta adjacent to it. The present taxon is very similar in appearance to *S*. *tenuis* Hustedt, which differs in certain ultrastructural characters and in having 4–6 striae/10 μm near the margin (as opposed to 9–12/ 10 μm in the present taxon). More important perhaps are the number of rows of puncta in each stria. Unfortunately this character is difficult or impossible to use in light microscopy. Grunow notes that *S*. *hantzschii* has two rows of puncta per stria near the margin, becoming a single row near the center of the valve, while the description of *S*. *tenuis* in Huber-Pestalozzi gives the number of rows of puncta as 3–5 near the margin and fewer near the center. On the basis of those characters that I could observe, the present specimens belong to *S*. *hantzschii*. The reader is cautioned to observe carefully any small *Stephanodiscus* species.

DIMENSIONS: Diameter, 8–20 μm; striae, 9–12/10 μm measured around the circumference of the valve.

2. **Stephanodiscus minutus** H. L. Smith var. **minutus** *Fig. 4b.*

CR: Huber-Pestalozzi (1942), as *S. astraea* var. *minutula* (Kütz.) Grun.

DESCRIPTION: Valves circular in outline; marginal spines usually very distinct; each costa terminated with a spine; puncta usually easily resolved; valve surface concentrically undulate (this focus feature is important in separating this taxon from *S. hantzschii*); central field rather broad and apparently irregularly punctate.

DIMENSIONS: Diameter, 8–30 μm; "striae," 9–12/10 μm measured around the circumference of the valve.

3. **Stephanodiscus niagarae** Ehr. var. **niagarae** *Fig. 4a.*

CR: Huber-Pestalozzi (1942).

DESCRIPTION: Valves circular in outline; valve face concentrically undulate; valve markings coarse and distinct, consisting of rows of coarse puncta that are arranged into distinct radial striae in the outer parts of the valve but which appear to become unordered and more widely spaced in a central zone that varies in diameter from population to population; each stria consisting of two to several rows of puncta in its outer part, becoming a single row nearer the valve center; striae alternate with costae; many but not all costae terminated at their outer end by stout spines; outside the ring of spines is a wider or narrower marginal band on which there are numerous radial rows of fine puncta. This taxon is similar in appearance to *S. astraea* (Ehr.) Grun., which, however, appears to lack the marginal band and which has nearly twice the number of striae in 10 μm.

DIMENSIONS: Diameter, 30–135 μm; striae, 4–5/10 μm; puncta, 12–14/10 μm measured along the striae.

4. *Cyclotella* Kützing 1834

Cyclotella belongs to class Centrobacillariophyceae and, like other genera in that class, has radical symmetry of markings on the valve face. The valve face is divided into an outer costate-striate zone and a central field having a structure that varies with the species but that is always different from that of the outer zone. Spinelike marginal structures are found in some species, but as these cannot easily be discerned with the light microscope, I have omitted them from the drawings. Many of the "isolated puncta" and other markings have been shown to be processes of various sorts. As with the marginal spines, however, their true nature is not evident under the light microscope, though it is easy to see with the electron microscope. In the costate-striate outer zone, the "striae" have been shown with the electron microscope to be variously perforate, while the spaces between them are indeed thickened into costae. The entire valve face may be flat or variously undulate.

Most species are observed as isolated cells, but some species form short filaments or colonies within an amorphous matrix. In chemically "cleaned" preparations colony structure is almost always lost. Occasionally, pairs of recently divided cells may remain connected even in "cleaned" preparations and in these cases the cell pairs are seen in girdle view. The very short pervalvar axis and the prominent appearance of the ends of the costae at the junction of the valve face and valve mantle are usually sufficient to distinguish such pairs of cells from short chains of *Melosira* and other small centric diatoms.

KEY TO THE TAXA OF Cyclotella CONFIRMED FOR ILLINOIS

1. Valves with numerous distinct markings in the central field _____ 2
1. Central field of valve marked only by a few "isolated puncta," which are often indistinct _____ 5
   2. A star-shaped cluster of marks in the central area __ 5. *C. stelligera*
   2. Markings in central area not forming a star-shaped cluster _____ 3
3. Small valves; markings in central field a tight cluster of dots _____ _____ 6. *C. stelligera* var. *tenuis*
3. Large valves; markings in central field radiate or scattered _____ 4
   4. "Striae" thick and wedge-shaped _____ 3. *C. meneghiniana*
   4. "Striae" thin, often undulate, and alternating longer and shorter __ _____ 2. *C. comta*
5. Central field with a single "isolated punctum"; a variable number of costae darker than the rest _____ 1. *C. atomus*
5. Central field with more than one "isolated punctum," or, if only one, then costae of equal brightness _____ 6
   6. Costae apparently thicker toward the outside of the valve than to the inside, the line of demarcation sharp between thin and thick portions; central field devoid of markings ____ 4. *C. michiganiana*
   6. Costae not suddenly changing thickness; central field with one or more large "isolated puncta" _____ 3. *C. meneghiniana*

1. Cyclotella atomus Hust. var. atomus *Fig. 4g*.

CR: Hohn and Hellerman (1963).

DESCRIPTION: Valves circular; central field quite broad and more or less devoid of markings with the exception of a single isolated punctum near the border; costae rather short, a variable number darker in appearance than the rest (in my specimens around 6 but reported as up to 9).

DIMENSIONS: Diameter, 5–7 μm; costae, 12–16/10 μm measured around the margin.

2. Cyclotella comta (Ehr.) Kütz. var. comta *Fig. 4f*.

CR: Hustedt (1930); Huber-Pestalozzi (1942); Lowe (1975).

DESCRIPTION: Valves circular; valves divided into a marginal costate zone and a central field which is around ⅓ to ½ of the valve diameter; central field marked by puncta that are unordered or arranged in loose radial lines; costae and striae in the marginal zone straight to slightly undulate and variable in length; some of the shorter of the markings terminated with isolated puncta; as a focus feature one can see a shadowy ring of short radial bars that represent the point of junction of internal ribs with the valve face. These ribs join the valve face with the valve mantle and are fewer in number than the costae.

DIMENSIONS: Diameter, 15–50 μm; striae (or costae), 12–15/10 μm measured around the margin; shadowy bars, around 5/10 μm.

3. **Cyclotella meneghiniana** Kütz. var. **meneghiniana** *Fig. 4j.*

CR: Hustedt (1930).

DESCRIPTION: Valve surface undulate; valve outline circular; valve divided into a marginal zone of costae and striae and a central field that may be nearly devoid of markings except a few isolated puncta or may have various combinations of large and small puncta that may be scattered or sometimes arranged in radial rows. Some authors separate exceptionally flat specimens as f. *plana* Fricke. The striae in this species are wedge-shaped with the small end toward the valve center and in some circumstances may appear to be faintly cross-lined when seen with strongly obliqued light.

DIMENSIONS: Diameter, 10–30 μm; costae (or striae), 8–9/10 μm measured around the margin.

4. **Cyclotella michiganiana** Skv. var. **michiganiana** *Fig. 4d.*

CR: Skvortzow (1937); Lowe (1975).

DESCRIPTION: Valves circular; valve markings in two zones, a marginal costate zone and a central field usually devoid of distinct markings as seen with the light microscope; central field undulate; costae apparently divided into an outer zone that is thicker than the inner zone, the transition from thick to thin quite sharp. I have seen a few specimens that varied from this description in having faint striae in the central zone but, on reexamination of the slides, was unable to locate voucher specimens to confirm this. Found thus far for certain only in McHenry County.

DIMENSIONS: Diameter, 5.0–20.4 μm; costae, 15–18/10 μm measured around the circumference of the valve.

5. Cyclotella stelligera Cl. & Grun. var. stelligera *Fig. 4i.*

CR: Hustedt (1930); Lowe (1975).

DESCRIPTION: Valves circular; valve surface divided into a fairly narrow marginal zone of costae and a central field with a star-shaped rayed mark at its center. This central star consists of a variable number of rays and a central dot. In electron micrographs taken by Lowe (1975), this star can be seen to be composed of elements that are structured like the striae. The central dot is similar in structure to the rest of the "star".

DIMENSIONS: Diameter 5–25 μm; striae (or costae), 10–12/10 μm measured around the margin of the valve.

6 Cyclotella stelligera var. tenuis Hust. *Fig. 4h.*

CR: Hustedt (1937).

DESCRIPTION: Valve surface with two zones, a marginal costate zone and a central field devoid of markings, except for a tight cluster of rounded dots at the center. I think it possible that these specimens may be only the small end of var. *stelligera*.

DIMENSIONS: Diameter, 5–7 μm; costae, 17–18/10 μm measured near the margin.

5. *Coscinodiscus* Ehrenberg 1838

The species in this large genus are almost exclusively marine. Occasional fragments of fossil cells of marine species may sometimes be found in freshwater sediments, but as these are not part of the living freshwater flora they will not be covered here. Several species do apparently occur as living components of the inland flora.

Coscinodiscus belongs to the Centrobacillariophyceae and, like the other members of this class, is characterized by radial symmetry of the valve markings. The principal valve features visible with the light microscope include marginal spines and areolae. The areolae, which appear like large puncta, may be arranged in a space-filling pattern, radial striae, or wedge-shaped groups of parallel striae. The cells are planktonic and solitary. Usually they are seen in valve view.

I have found but a single species that apparently, but not certainly, is a member of the living flora of Illinois.

1. **Coscinodiscus rothii** var. **normani** (Greg.) V.H. *Fig. 3b.*

CR: Hustedt (1930a).

DESCRIPTION: Valve surface concentrically undulate, the margin and center seldom seen in focus simultaneously; valve outline circular; valve surface divided into faint wedge-shaped sectors within which the areolae form striae parallel to each other but which are not, therefore, strictly radial; outside the wedge-shaped sectors a marginal ring of radially oriented areolae that are more densely packed than in the inner zone; marginal spines present. My specimens are smaller than those previously reported for this taxon and have a less well-developed marginal ring. They also do not exhibit distinct marginal spines, though indistinct ones may be present. Because of these differences, I must give this as a questionable identification, particularly since var. *normani,* though not exclusively marine, is not typically a member of the freshwater flora.

DIMENSIONS: Diameter, 30–110 μm (my specimens were around 25–30 μm); areolae (measured along a stria), 6–8/10 μm; areolae in the marginal ring, around 13/10 μm.

6. *Biddulphia* S. F. Gray 1821

Biddulphia belongs to the class Centrobacillariophyceae and, like the other members of this class, is characterized by valve markings which are more or less radial in symmetry. *Biddulphia* is primarily a marine genus and the one species found in Illinois is not typical. For these reasons I shall omit a detailed generic description.

1. **Biddulphia laevis** Ehr. var. **laevis** *Fig. 3c.*

CR: Hustedt (1930a).

DESCRIPTION: Valve outline broadly elliptic, the one axis only slightly shorter than the other; striae distinctly punctate and apparently radiating from a short linear region near midvalve on the longer of the axes; a large, low, elliptical protuberance present at both ends of the long axis of the valve. These protuberances can be seen in both valve and girdle views. The girdle side is highly ornamented with puncta and is very striking in appearance. The entire frustule is apparently heavily silicified and, as a consequence, looks very dark even under the oil immersion lens. *B. laevis* can be mistaken for no other species in the state.

DIMENSIONS: Diameter, 20–150 μm; puncta, around 16/10 μm as measured along a stria.

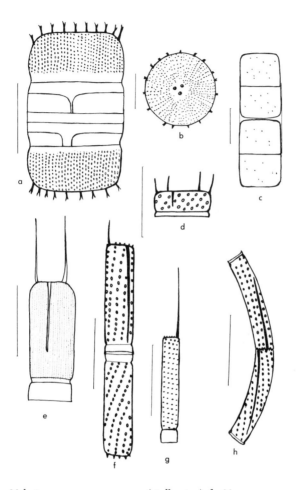

Fig. 1. *a. Melosira roeseana* v. *roeseana* (girdle view). *b. M. roeseana* v. *roeseana* (valve view). *c. M. varians* v. *varians* (two cells, girdle view). *d. M. granulata* v. *muzzanensis* (one valve, girdle view). *e. M. herzogii* v. *herzogii* (one valve, girdle view). *f. M. granulata* v. *granulata* (girdle view). *g. M. granulata* v. *angustissima* (one valve, girdle view). *h. M. granulata* v. *angustissima* f. *spiralis* (one valve each of two adjacent cells, girdle view). (Scale lines equal 10 micrometers.)

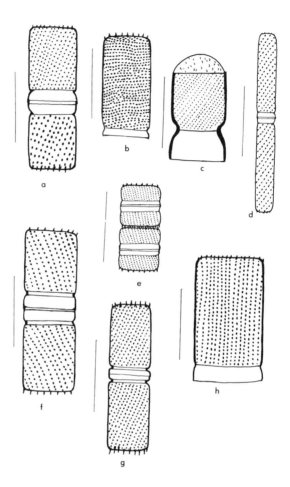

Fig. 2. a. *Melosira italica* v. *italica* (girdle view). **b.** *M. italica* v. *italica* (one valve, girdle view). **c.** *M. italica* v. *italica* (one valve of statospore, girdle view). **d.** *M. italica* v. *tenuissima* (girdle view). **e.** *M. distans* v. *alpigena* (girdle view). **f, g.** *M. ambigua* v. *ambigua* (girdle view). **h.** *M. islandica* v. *islandica* (one valve, girdle view). (Scale lines equal 10 micrometers.)

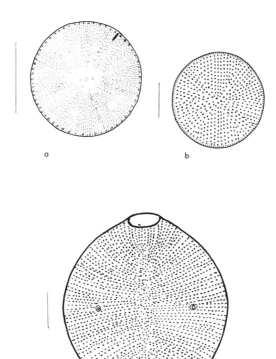

Fig. 3. a. Thalassiosira fluviatilis *v.* fluviatilis. *b.* Coscinodiscus rothii *v.* normani. *c.* Biddulphia laevis *v.* laevis. (Scale lines equal 10 micrometers.)

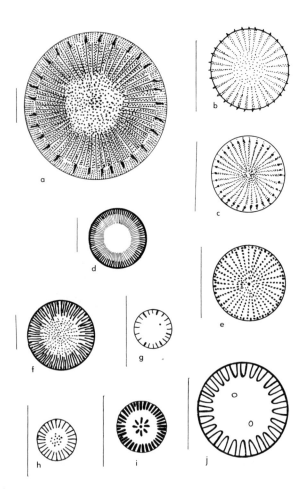

Fig. 4. a. *Stephanodiscus niagarae* v. *niagarae*. **b.** *Stephanodiscus minutus* v. *minutus*. **c.** *Stephanodiscus hantzschii* v. *hantzschii*. **d.** *Cyclotella michiganiana* v. *michiganiana*. **e.** *Stephanodiscus hantzschii* v. *hantzschii*. **f.** *Cyclotella comta* v. *comta*. **g.** *Cyclotella atomus* v. *atomus*. **h.** *Cyclotella stelligera* v. *tenuis*. **i.** *Cyclotella stelligera* v. *stelligera*. **j.** *Cyclotella meneghiniana* v. *meneghiniana*. (Scale lines equal 10 micrometers.)

7. *Tabellaria* Ehrenberg 1840

Tabellaria belongs to the order Fragilariales of class Pennatibacillariophyceae. Like other genera in the order, valve markings are bilaterally symmetric and neither valve of a cell has any sort of raphe. Valves of *Tabellaria* are symmetric to both the long and transverse axes. Each valve is associated with several internal plates or septa, each of which is open at one end by means of a large locule. The locules are not all at one end of the valve but alternate from one end to the other in successive septa. The number of septa is of taxonomic significance. It is easiest to count the septa in girdle view, though it can be done in valve view by careful focusing. Valve markings consist of fine striae.

KEY TO THE SPECIES OF Tabellaria IN ILLINOIS

1. Two septa associated with each valve _____ 1. *T. fenestrata*
1. More than two septa associated with each valve _____ 2. *T. flocculosa*

1. Tabellaria fenestrata (Lyngb.) Kütz. var. **fenestrata** *Fig. 5a–c*.

CR: Hustedt (1930).

DESCRIPTION: Valve ends capitate; valve margins strongly inflated at midvalve, becoming straight to weakly convex between the central inflation and the ends; pseudoraphe narrow; two septa associated with each valve. Please note that the septa are close together. When observing the cell or valve in girdle view, remember that the septum is thickened at one end only. With careful observation you can trace the thin continuation of the septum to the opposite end and you will find that this thin portion of the septum is distinct from the thick end of the second septum.

DIMENSIONS: Length, 30–140 μm; width at midvalve, 3–9 μm; striae, 18–20/10 μm

2. Tabellaria flocculosa (Roth) Kütz. var. **flocculosa** *Fig. 5e–g.*

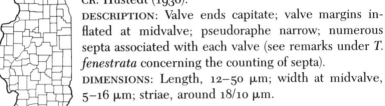

CR: Hustedt (1930).

DESCRIPTION: Valve ends capitate; valve margins inflated at midvalve; pseudoraphe narrow; numerous septa associated with each valve (see remarks under *T. fenestrata* concerning the counting of septa).

DIMENSIONS: Length, 12–50 μm; width at midvalve, 5–16 μm; striae, around 18/10 μm.

8. *Meridion* C. A. Agardh 1824

Meridion belongs to the order Fragilariales. Both valves of each cell have a narrow pseudoraphe. Valve markings include fine striae (which can be seen to be ultramicroscopically punctuate with the electron microscope) and a smaller number of coarse, thickened costae that may or may not be interrupted by the pseudoraphe. Cells are cuneate in girdle view and clavate in valve view. Internal valves are common. In nature, the cells are joined valve to valve, forming fan-shaped to helical colonies. There is one species with two varieties in Illinois.

KEY TO THE VARIETIES OF Meridion circulare IN ILLINOIS

1. Valve margin constricted just before the large end of the valve, the end therefore capitate _____ 2. *M. circulare* var. *constrictum*
1. Valve margin not constricted _____ 1. *M. circulare* var. *circulare*

1. Meridion circulare (Grev.) Ag. var. **circulare** *Fig. 6c.*

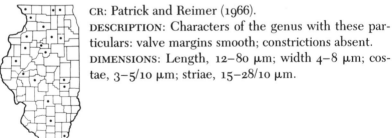

CR: Patrick and Reimer (1966).

DESCRIPTION: Characters of the genus with these particulars: valve margins smooth; constrictions absent.

DIMENSIONS: Length, 12–80 μm; width 4–8 μm; costae, 3–5/10 μm; striae, 15–28/10 μm.

2. Meridion circulare var. **constrictum** (Ralfs) V.H. *Fig. 6d.*

CR: Patrick and Reimer (1966).
DESCRIPTION: Characters of the genus with these particulars: valve ends, particularly the large end, set off by constrictions of the valve margins; margins often shallowly constricted at the costae. It is possible that this taxon and the nominate variety integrade with one another.
DIMENSIONS: See var. *circulare*.

9. *Diatoma* Bory 1824 nom. cons., non Loureiro 1790

Diatoma is a member of the order Fragilariales of class Pennatibacillariophyceae and, like other genera in the order, is characterized by the absence of the raphe on both valves of each cell. Narrow pseudoraphes are present on both valves. Symmetry is bilateral. Valves are symmetric to both the longitudinal and transverse axes. Cells are united into chains of various form by means of gelatinous strands excluded from jelly pores at the ends of the valves. Valve markings consist of fine transverse striae and thickened transverse costae. The striae can be seen to be punctate only with the aid of the electron microscope. The costae are not punctate.

KEY TO THE TAXA OF Diatoma IN ILLINOIS

1. Valves linear with nearly straight margins and subcapitate to capitate ends _____ 1. *D. tenue* var. *elongatum*
1. Valves with convex margins and more or less rostrate ends _____ _____ 2. *D. vulgare*

1. Diatoma tenue var. **elongatum** Lyngb. *Fig. 6b.*

CR: Patrick and Reimer (1966).
DESCRIPTION: Valves linear; margins straight to weakly convex; valve ends weakly capitate or subcapitate; pseudoraphe very narrow and difficult to see (I have omitted it in the drawing to emphasize this).
DIMENSIONS: Length, 40–120 μm; width, 2–4 μm; costae, 6–10/10 μm; striae, 16–18/10 μm.

2. Diatoma vulgare Bory var. vulgare *Fig. 6a.*

CR: Patrick and Reimer (1966).

DESCRIPTION: Valves linear with parallel to weakly convex margins and broadly rostrate ends; striae and costae transverse rather than diagonal; costae robust and somewhat irregular both in spacing and orientation; striae strictly transverse and evenly spaced.

DIMENSIONS: Length, 30–60 μm; width, 8–13 μm; costae, 6–8/10 μm; striae, about 16/10 μm.

10. Asterionella Hassall 1850

11. Fragilaria Lyngbye 1819

12. Synedra Ehrenberg 1832

The genera *Asterionella, Fragilaria,* and *Synedra* belong to the order Fragilariales of class Pennatibacillariophyceae and are characterized by the presence of a pseudoraphe on both valves of each cell. All three genera have valves that are marked by striae but not costae. None form septa or, as far as I have seen, internal valves.

The three genera are distinguished from each other on the basis of habit: *Asterionella* forms star-shaped colonies in nature and *Fragilaria* forms band-shaped or ribbonlike colonies. The cells of *Synedra* may be free-floating or attached but typically are not joined into colonies. Most species of the three genera are planktonic organisms. The colony-forming species maintain their position in the water column in part due to the resistance to sinking of the colony form itself, while the solitary cells of *Synedra* are often highly elongate, which has the same effect, as anyone who has watched the sinking of any long, narrow object in water can attest.

Valves of most species of *Synedra* and *Fragilaria* are symmetric to both the long and transverse axes, but one common species of *Fragilaria* (*F. pinnata* Ehr.) produces abnormal valves that are asymmetric to the transverse axis, while one species of *Synedra* (the space-parasite *S. cyclopum* Brutschi, which I did not find but expect) produces lunate cells. Valves of *Asterionella* are more or less linear but have one capitate end slightly to distinctly larger than the other.

Patrick and Reimer (1966) have pointed out that the separation of *Fragilaria* from *Synedra* may prove to be in part artificial. While

I am in agreement on this point, I will maintain the classical taxonomic separation in the present work.

Because colony form is destroyed by almost all chemical "cleaning" methods, I have produced a combined key to the three genera based on valve morphology alone. I encourage you, however, to reexamine living or preserved material or to make "burned mounts" of any sample in which these species occur. Knowing more or less certainly that you are dealing with a *Fragilaria* as opposed to a *Synedra* will often be of considerable use in deciding whether or not the identification made using the key is correct. Please note, though, that it is usually not possible to identify *Fragilaria* species in girdle view, so you will need a chemically "cleaned" preparation as well as the "burned mount."

The three genera are arranged alphabetically following the key.

KEY TO Asterionella, Fragilaria, AND Synedra CONFIRMED FOR ILLINOIS

Species numbers consist of the genus number followed by the number of the species within the genus.

1. Valves asymmetric to the transverse axis _____ 2
1. Valves symmetric to the transverse axis _____ 3
    2. Valves typically over 30 μm in length; striae over 20/10 μm _____ _____ 10–1. *Asterionella formosa* v. *formosa*.
    2. Valves typically under 30 μm in length or if over 30 μm long then striae under 15/10 μm _____ _____ 11–7. *Fragilaria pinnata* v. *pinnata* (abnormal forms)
3. Valves shorter than 50 μm _____ 4
3. Valves 50 μm long or longer (often much longer) _____ 25
    4. Valves linear to elliptic (rarely elliptic lanceolate) without constrictions or inflations of the margins; ends not protracted, rostrate, or capitate _____ 5
    4. Valves linear or linear-lanceolate with or without constrictions of the margins; ends protracted, rostrate, or capitate _____ 7
5. Valves linear or linear-elliptic with broadly rounded ends _____ _____ 11–8. *Fragilaria pinnata* v. *intercedens*.
5. Valves elliptic to elliptic-lanceolate _____ 6
    6. Striae broad and nearly reaching the midline _____ _____ 11–7. *Fragilaria pinnata* v. *pinnata*
    6. Striae thin and very short, the pseudoraphe therefore broad ____ _____ 11–5. *Fragilaria construens* v. *venter*
7. Valve margins concave near midvalve _____ 8
7. Valve margins straight or convex near midvalve _____ 10

8. Central area present and reaching the margins _____ 9
8. Central area no wider than the axial area _____
_____ 11–4. *Fragilaria construens* v. *binodis*.
9. Ends narrowly rostrate; striae around 12/10 μm _____
_____ 12–17. *Synedra ulna* v. *ramesi*
9. Ends broadly rostrate; striae 15/10 μm or more _____
_____ 11–2. *Fragilaria capucina* v. *mesolepta*
10. Valves strongly transversely inflated at midvalve, the central inflation typically one-half or more as wide as the valve is long ___ 11
10. Valves not inflated at midvalve, or if inflated then the central inflation typically no more than one-third as wide as the valve is long
_____ 12
11. Striae 5–9/10 μm ____ 11–6. *Fragilaria leptostauron* v. *leptostauron*
11. Striae 14–18/10 μm _____ 11–3. *Fragilaria construens* v. *construens*
12. Striae short, the central-axial area therefore a broad linear or lanceolate space _____ 13
12. Striae relatively long, the axial area therefore narrow _____ 14
13. Valves linear, typically considerably longer than 25 μm _____
_____ 12–11. *Synedra tabulata* v. *tabulata*
13. Valves lanceolate, less than 25 μm long _____
_____ 11–1. *Fragilaria brevistriata* v. *inflata*
14. Valve margins slightly to distinctly bulged around the central area and frequently slightly constricted at the ends of the bulge __ 19
14. Valve margins not bulged around the central area or central area absent _____ 15
15. Valves linear; ends distinctly capitate _____
_____ 12–2. *Synedra amphicephala* v. *amphicephala*
15. Valves lanceolate to linear-lanceolate; ends protracted to bulbous-subcapitate but not distinctly capitate _____ 16
16. Central area no wider than the axial area _____ 17
16. Central area distinct, often reaching the valve margins _____ 18
17. Valves under 30 μm in length; ends somewhat narrow and protracted
_____ 12–4. *Synedra parasitica* v. *parasitica*
17. Valves typically longer than 30 μm, the margins tapering gradually from midvalve to the ends _____ 12–12. *Synedra tenera* v. *tenera*
18. Striae typically over 20/10 μm; valve ends not distinctive _____
_____ 12–12. *Synedra tenera* v. *tenera*
18. Striae typically 15–18/10 μm, valve ends often slightly bulbous __
_____ 12–6. *Synedra radians* v. *radians*.
19. Striae distinctly punctate _____ 12–5. *Synedra pulchella* v. *pulchella*
19. Striae resolvable into puncta (if at all) only with strongly obliqued light
_____ 20

20. Central area reaching both valve margins; striae evenly spaced __ _____ 21

20. Central area reaching one valve margin only, or if apparently reaching both margins then the striae irregularly placed ____ 23

21. Striae more than 15/10 μm ___   12–7. *Synedra rumpens* v. *familiaris*

21. Striae fewer than 15/10 μm _____ 22

22. Striae 10–12/10 μm; length typically over 30 μm _____ _____ 12–8. *Synedra rumpens* v. *fragilarioides*

22. Striae 12–13.5/10 μm; length typically around 30–35 μm _____ _____ 12–9. *Synedra rumpens* v. *meneghiniana*

23. Valves more or less linear with rostrate ends; margin bulged on the side of the central area that reaches the margin; striae typically less than 15/10 μm _____ 24

23. Valves linear-lanceolate, the ends drawn out or subrostrate; margin bulged on the side of the central area that does not reach the margin; striae typically around 17/10 μm ____   12–10. *Synedra socia* v. *socia*

24. Valve margins regular, the striae evenly spaced _____ _____ 11–9. *Fragilaria vaucheriae* v. *vaucheriae*

24. Valve margins irregular, the striae irregularly scattered _____ _____ 11–10. *Fragilaria vaucheriae* f. *contorta*

25. Valve margins concave at midvalve __   12–17. *Synedra ulna* v. *ramesi*

25. Valve margins parallel or convex at midvalve _____ 26

26. Striae distinctly punctate (oblique light not required) _____ _____ 12–5. *Synedra pulchella* v. *pulchella*

26. Striae either not resolvable into puncta or only resolvable by means of oblique light _____ 27

27. Striae short, the pseudoraphe therefore a broad space _____ _____ 12–11. *Synedra tabulata* var. *tabulata*

27. Striae long, the pseudoraphe narrow _____ 28

28. Valve margins bulged around the central area and slightly constricted at the ends of the bulge _____ 29

28. Valve margins not bulged around the central area or central area lacking _____ 30

29. Striae fewer than 15/10 μm - 12–8. *Synedra rumpens* v. *fragilarioides*

29. Striae more than 15/10 μm ___   12–7. *Synedra rumpens* v. *familiaris*

30. Valves more or less linear, the valve margins parallel over most of the length _____ 31

30. Valves lanceolate to linear-lanceolate, margins not parallel over most of the length _____ 35

31. Valve margins inflated just before the rostrate ends _____ _____ 12–18. *Synedra ulna* v. *spathulifera*

31. Valve margins not inflated before the ends _____ 32

32. Central area not distinct from the axial area _____
_____ 12–14. *Synedra ulna* v. *amphirhynchus*
32. Central area typically distinct and devoid of striae or with short striae or "ghost striae" _____ 33
33. Valve ends rostrate _____ 12–13. *Synedra ulna* v. *ulna*
33. Valve ends more or less bulbous-subcapitate _____ 34
34. Valves typically well over 200 μm in length _____
_____ 12–16. *Synedra ulna* v. *longissima*
34. Valves typically less than 200 μm in length _____
_____ 12–15. *Synedra ulna* v. *danica*
35. Length to breadth ration 30–50:1 _____
_____ 12–3. *Synedra delicatissima* v. *delicatissima*
35. Length to breadth ratio typically less than 30:1 _____ 36
36. Striae 14/10 μm or less _____ 12–1. *Synedra acus* v. *acus*
36. Striae typically more than 15/10 μm _____ 37
37. Striae typically 20/10 μm or more; central area present or absent ____
_____ 12–12. *Synedra tenera* v. *tenera*
37. Striae typically fewer than 20/10 μm; central area almost always present _____ 12–6. *Synedra radians* v. *radians*

10. *Asterionella*

1. Asterionella formosa Hassall var. **formosa** *Fig. 5d.*

CR: Koerner (1970).
DESCRIPTION: Valves linear with more or less distinctly capitate ends, the one end (head pole) broader than the other end (foot pole), thus making the valve asymmetric to the transverse axis; pseudoraphe very narrow; striae fine and difficult to resolve; cells more or less rectangular in girdle view, the valve-to-valve distance greater at the head pole than at the foot pole; cells united by the head poles into star-shaped colonies. Koerner (1970) has demonstrated that the specimens variously called *A. formosa* var. *gracillima* (Hantz.) Grun. and *A. gracillima* Heib. var. *gracillima* are in fact part of *A. formosa* var. *formosa*.
DIMENSIONS: Length, 30–120 μm; width, head pole, 3–12 μm; width, midvalve, 1.3–6 μm; width, foot pole, 1.5–4 μm; striae, 28–32/10 μm.

11. *Fragilaria*

1. Fragilaria brevistriata var. inflata (Pant.) Hust. *Fig. 7k.*

CR: Patrick and Reimer (1966).

DESCRIPTION: Valve outline lanceolate; ends somewhat protracted; axial and central areas together form a broad lanceolate space; striae short and radiate.

DIMENSIONS: Length, 10–20 μm; width, 3.5–11.0 μm; striae, 12–16/10 μm.

2. Fragilaria capucina var. mesolepta Rabh. *Fig. 7a.*

CR: Patrick and Reimer (1966).

DESCRIPTION: Valves more or less linear in overall outline; valve margins constricted or concave at midvalve; ends vary from nondistinctive to rostrate; pseudoraphe narrow; central area rectangular and reaching both valve margins; striae parallel and faint but not extremely fine.

DIMENSIONS: Length, 25–35 μm; width at midvalve, 2–4 μm; striae 15–18/10 μm (mine reached 20/10 μm).

3. Fragilaria construens (Ehr.) Grun. var. construens *Fig. 7h.*

CR: Patrick and Reimer (1966).

DESCRIPTION: Valve margins inflated at midvalve and narrowing sharply before the rounded-subcapacitate ends; central-axial area a narrow, lanceolate space. This taxon is somewhat similar in shape to *F. leptostauron* (Ehr.) Hust., but has much finer striae.

DIMENSIONS: Length, 7–25 μm; maximum width, 5–12 μm; striae, 14–18/10 μm.

4. Fragilaria construens var. **binodis** (Ehr.) Grun. *Fig. 7f.*

CR: Hustedt (1931).

DESCRIPTION: Valve margins distinctly concave at mid-valve; valve ends narrowly rostrate, formed by a sharp constriction of the margins; central-axial area a narrow, linear space.

DIMENSIONS: Length, 7–25 μm; maximum width, 5–12 μm; striae, 14–17/10 μm.

5. Fragilaria construens var. **venter** (Ehr.) Grun. *Fig. 7j.*

CR: Patrick and Reimer (1966).

DESCRIPTION: Valves elliptic-lanceolate in outline and lacking distinctive ends; striae short, the central-axial area therefore broad.

DIMENSIONS: Length, 5–9 μm; width, 3–6 μm; striae, 14–16/10 μm.

6. Fragilaria leptostauron (Ehr.) Hust. var. **leptostauron** *Fig. 7e.*

CR: Patrick and Reimer (1966).

DESCRIPTION: Valve margins strongly inflated at mid-valve so that the maximum width in some cases almost equals the valve length; margins narrowing abruptly to form the ends; central-axial area a moderately narrow space with lanceolate shape.

DIMENSIONS: Length, 15–36 μm; width, 10–23 μm; striae, 5–9/10 μm.

7. **Fragilaria pinnata** Ehr. var. **pinnata** *Figs. 6e–g, 7i.*

CR: Patrick and Reimer (1966).

DESCRIPTION: Valve outline variable in shape but typically linear-elliptic to elliptic-lanceolate. I have found scattered populations of abnormal cells that have valves asymmetric to the transverse axis (*Fig. 6e–g*), but that are otherwise typical of the species. These may be confused with specimens of the much rarer genus *Opephora*, which has not yet been found in Illinois. I am not certain of this, but the atypical valves of *F. pinnata* may indicate, by the variable shape of their ends, an intergrading connection between the nominate variety and var. *lancettula* (Schum.) Hust., which has been separated from var. *pinnata* on the basis of slightly protracted, substrate ends.

DIMENSIONS: Length, 3–35 μm; width, 2–6 μm; striae, 7–12/10 μm.

8. **Fragilaria pinnata** var. **intercedens** (Grun.) Hust. *Fig. 7d.*

CR: Patrick and Reimer (1966).

DESCRIPTION: Valves linear with rounded ends; striae short, the central-axial area therefore a broad, linear space.

DIMENSIONS: Length, 15–34 μm; width, 4–6 μm; striae, 6–8/10 μm.

9. **Fragilaria vaucheriae** (Kütz.) Peters. var. **vaucheriae** *Fig. 7b.*

CR: Patrick and Reimer (1966).

DESCRIPTION: Valves linear with rostrate ends and somewhat convex margins; central area usually distinct and typically asymmetric; one valve margin slightly bulged around the central area. Valves of this taxon are somewhat similar in appearance to those of *Synedra socia* and the varieties of *Synedra rumpens*.

DIMENSIONS: Length, 10–40 μm; width, 2–4 μm; striae, 12–16/10 μm.

10. Fragilaria vaucheriae f. contorta Lowe *Fig. 7c.*

CR: Lowe (1972).

DESCRIPTION: Basically similar to var. *vaucheriae* with the exceptions that the valve margins may be irregular and the striae are in irregularly spaced groups. The Illinois specimens had regular margins but the scattered striae typical of f. *contorta*. This may or may not be a monstrosity. It it proves to be so, the specimens would be considered simply as abnormal examples of var. *vaucheriae*.

DIMENSIONS: Similar to var. *vaucheriae*, with the exception of the striae, which will appear coarser due to the presence of gaps in the striation.

12. *Synedra*

1. Synedra acus Kütz. var. acus *Fig. 8c.*

CR: Patrick and Reimer (1966).

DESCRIPTION: Valves lanceolate, tapering toward ends that vary from nondistinctive to subcapitate but are not needlelike; pseudoraphe narrow; central area rectangular and reaching the valve margins; striae parallel.

DIMENSIONS: Length, 90–180 μm; width, 4–6 μm; striae, 11–14/10 μm.

2. Synedra amphicephala Kütz var. amphicephala *Fig. 9e.*

CR: Patrick and Reimer (1966).

DESCRIPTION: Valves linear with weakly convex margins; ends distinctly capitate; axial area narrow; central area no wider than axial area; striae parallel.

DIMENSIONS: Length, 20–33 μm; width, 3–5 μm; striae, 13–14/10 μm.

3. **Synedra delicatissima** W. Sm. var. **delicatissima** *Fig. 8a*.

CR: Patrick and Reimer (1966).

DESCRIPTION: Valves narrowly lanceolate with very narrow, almost needlelike, ends; length to breadth ratio 30–50:1; central area rectangular, often longer than broad; striae parallel. Patrick and Reimer report that the central area is bordered by very short striae. In most of my specimens, however, these short striae were not present. Some of my specimens had striae somewhat finer than those reported by Patrick and Reimer.

DIMENSIONS: Length, 100–230 μm; width, about 3–4 μm in many cases; striae, 11–14/10 μm (mine reached 15–16/10 μm in some cases.)

4. **Synedra parasitica** (W. Sm.) Hust var. **parasitica** *Fig. 7g*.

CR: Patrick and Reimer (1966).

DESCRIPTION: Valves lanceolate with narrow, protracted ends; central-axial area a moderately broad lanceolate space; striae slender and radiate. This species is a space-parasite on other diatoms.

DIMENSIONS: Length, 10–25 μm; width, 3–5 μm; striae, 16–19/10 μm.

5. **Synedra pulchella** (Ralfs ex Kütz.) Kütz. var. **pulchella** *Fig. 8e*.

CR: Patrick and Reimer (1966).

DESCRIPTION: Valves linear-lanceolate to lanceolate, tapering gradually to slightly protracted, bulbous ends; pseudoraphe narrow; central area more or less rectangular and reaching both margins; striae parallel and seen as distinctly punctate without obliqued light.

DIMENSIONS: Length, 33–150 μm; width, 5–8 μm; striae, 12–16/10 μm, sometimes reaching 20/10 μm.

6. Synedra radians Kütz. var. **radians** *Fig. 8b.*

CR: Patrick and Reimer (1966).

DESCRIPTION: Valves linear-lanceolate, narrowing gradually toward the slightly bulbous ends; axial area narrow; central area variable, reaching the margins in some specimens, but almost absent in others (Illinois specimens were intermediate); striae more or less parallel.

DIMENSIONS: Length, 40–120 μm; width, 2.5–4 μm; striae 15–18/10 μm.

7. Synedra rumpens var. **familiaris** (Kütz.) Hust. *Fig. 9a.*

CR: Patrick and Reimer (1966).

DESCRIPTION: Valves linear-lanceolate, tapering gradually to subcapitate ends; valve margins bulged slightly around the central area; central area longer than broad and reaching both margins; axial area narrow; striae parallel.

DIMENSIONS: Length, 30–80 μm; width, 3–4 μm; striae 18–20/10 μm.

8. Synedra rumpens var. **fragilarioides** Grun. *Fig. 9c.*

CR: Patrick and Reimer (1966).

DESCRIPTION: Similar in shape to var. *familiaris* and distinguished primarily by the coarser striae.

DIMENSIONS: Length, 31–75 μm; width, 3–5 μm; striae, 10–12/10 μm.

9. Synedra rumpens var. **meneghiniana** Grun. *Fig. 9d.*

CR: Patrick and Reimer (1966).

DESCRIPTION: Similar in shape to var. *familiaris* and distinguished primarily by the coarser striae. This may or may not prove sufficient grounds for separation.

DIMENSIONS: Length, 27–38 μm; width, 3–4 μm; striae, 12–13.5/10 μm.

10. Synedra socia Wallace var. **socia** *Fig. 9b.*

CR: Patrick and Reimer (1966).

DESCRIPTION: Valves linear-lanceolate, tapering to somewhat bulbous-capitate ends; valve margins slightly constricted at the ends of the central area and slightly swollen between the constrictions; axial area narrow; central area reaching one or both margins. (The one-sided central area was the type found in Illinois.)

DIMENSIONS: Length, 15–33 μm; width, 3–4 μm; striae, around 17/10 μm.

11. Synedra tabulata (Ag.) Kütz. var. **tabulata** *Fig. 9g.*

CR: Hustedt (1932).

DESCRIPTION: Valves linear-lanceolate, tapering gradually to the somewhat bulbous ends; pseudoraphe a broad linear-lanceolate space; striae parallel and short.

DIMENSIONS: Length, 25–200 μm; width, 2–7 μm; striae, 9–14/10 μm.

12. **Synedra tenera** W. Sm. var. **tenera** *Figs. 8d, 115c.*

CR: Hustedt (1932).

DESCRIPTION: Valves narrowly linear-lanceolate, taper-ing to ends that are very narrow and almost needlelike; axial area narrow and difficult to see; striae very fine and difficult to resolve; central area typically absent, but present and reaching both margins in the Illinois specimens. In *Fig. 115c* I have shown a stylized draw-ing of a more linear example without a central area and with somewhat protracted ends. This is perhaps more typical of the species than the specimens I found.

DIMENSIONS: Length, 30–120 μm; width, 2.5–4 μm (mine were as narrow as 2 μm); striae 20–24/10 μm (mine reached 26/10 μm).

13. **Synedra ulna** (Nitz.) Ehr. var. **ulna** *Fig. 10a.*

CR: Patrick and Reimer (1966).

DESCRIPTION: Valves linear; ends moderately to broadly rostrate; axial area moderately narrow; central area present and usually reaching both margins; striae parallel, becoming slightly radiate at the ends. *Synedra ulna* and its varieties are quite variable in appearance and it may be that some or all of them are linked to-gether by intergrading forms. See also Valid Taxa Re-ported by Other Authors for additional varieties.

DIMENSIONS: Length, 50–350 μm; width, 5–9 μm; striae, 9–11/10 μm.

14. **Synedra ulna** var. **amphirhynchus** (Ehr.) Grun. *Fig. 10b.*

CR: Patrick and Reimer (1966).

DESCRIPTION: Similar in shape to var. *ulna,* but central area no wider than axial area. My specimens were much shorter than those reported by others, but I think it at least possible that var. *amphirhynchus* may be found to have a length range similar to var. *ulna.*

DIMENSIONS: Length, 180–250 μm (mine were as short as 100 μm); width, 4–7 μm; striae, 10–12/10 μm.

15. Synedra ulna var. **danica** (Kütz.) V.H. *Fig. 10d.*

CR: Patrick and Reimer (1966).

DESCRIPTION: Valves linear-lanceolate, narrowing gradually toward the somewhat bulbous-capitate ends; central area rectangular and reaching both margins; faint "ghost striae" sometimes present in the central area; axial area narrow; striae parallel. Var. *danica* is similar in appearance to var. *longissima*, but valves are shorter and, usually, narrower.

DIMENSIONS: Length, 120–200 μm; width, 4–8 μm; striae, 9–10/10 μm.

16. Synedra ulna var. **longissima** (W. Sm.) Brun *Fig. 10e.*

CR: Patrick and Reimer (1966).

DESCRIPTION: Similar in shape to var. *danica* but usually much longer. Patrick and Reimer report that the central area is usually absent, but my specimens showed a distinct rectangular central area. Valves of longer specimens are not straight, but have a somewhat irregular curvature.

DIMENSIONS: Length, 200–600 μm; width, 5–7 μm; striae, 8–12/10 μm.

17. Synedra ulna var. **ramesi** (Herib.) Hust. *Fig. 9f.*

CR: Patrick and Reimer (1966).

DESCRIPTION: Valves more or less linear, with margins concave near midvalve; ends narrowly rostrate; central area rectangular and reaching both valve margins; axial area narrow.

DIMENSIONS: Length, 45–55 μm; width, 7–8 μm at the widest point; striae, 10–12/10 μm.

18. Synedra ulna var. **spathulifera** (Grun.) V.H. *Fig. 10c.*

CR: Patrick and Reimer (1966).

DESCRIPTION: Similar to var. *ulna* with the exception that the valve margins are inflated just before narrowing to form the rostrate ends. There is considerable variation in the degree of this inflation. In some specimens it is scarcely noticeable, leading me to wonder if var. *spathulifera* intergrades with var. *ulna*.

DIMENSIONS: Similar to var. *ulna*.

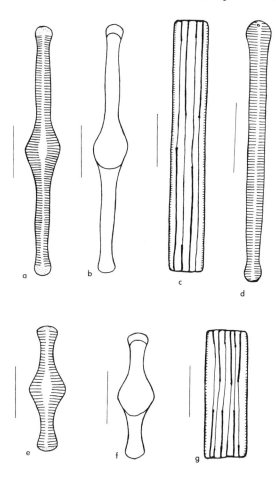

Fig. 5. a. *Tabellaria fenestrata* v. *fenestrata* (valve). **b.** *Tabellaria fenestrata* v. *fenestrata* (septum). **c.** *Tabellaria fenestrata* v. *fenestrata* (cell, girdle view). **d.** *Asterionella formosa* v. *formosa*. **e.** *Tabellaria flocculosa* v. *flocculosa* (valve). **f.** *Tabellaria flocculosa* v. *flocculosa* (septum). **g.** *Tabellaria flocculosa* v. *flocculosa* (cell, girdle view). (Scale lines equal 10 micrometers.)

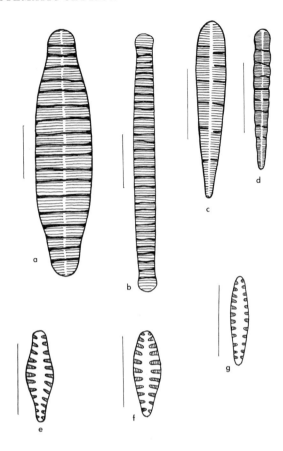

Fig. 6. a. *Diatoma vulgare* v. *vulgare*. **b.** *Diatoma tenue* v. *elongatum*. **c.** *Meridion circulare* v. *circulare*. **d.** *Meridion circulare* v. *constrictum*. **e–g.** *Fragilaria pinnata* v. *pinnata* (abnormal valve). (Scale lines equal 10 micrometers.)

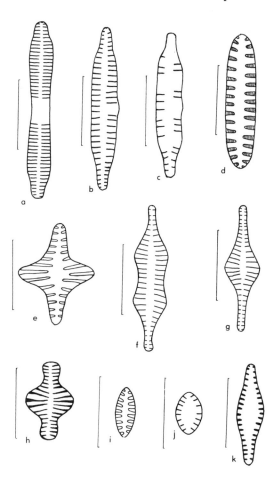

Fig. 7. a. *Fragilaria capucina* v. *mesolepta*. **b.** *Fragilaria vaucheriae* v. *vaucheriae*. **c.** *Fragilaria vaucheriae* f. *contorta*. **d.** *Fragilaria pinnata* v. *intercedens*. **e.** *Fragilaria leptostauron* v. *leptostauron*. **f.** *Fragilaria construens* v. *binodis*. **g.** *Synedra parasitica* v. *parasitica*. **h.** *Fragilaria construens* v. *construens*. **i.** *Fragilaria pinnata* v. *pinnata*. **j.** *Fragilaria construens* v. *venter*. **k.** *Fragilaria brevistriata* v. *inflata*. (Scale lines equal 10 micrometers.)

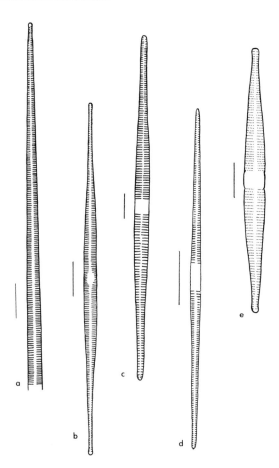

Fig. 8. a. *Synedra delicatissima* v. *delicatissima*. **b.** *S. radians* v. *radians*. **c.** *S. acus* v. *acus*. **d.** *S. tenera* v. *tenera*. **e.** *S. pulchella* v. *pulchella*. (Scale lines equal 10 micrometers.)

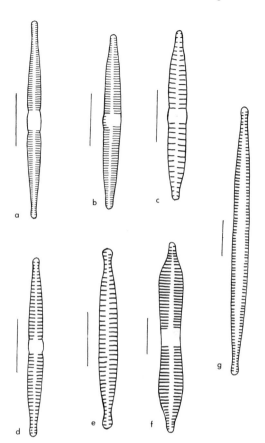

Fig. 9. a. *Synedra rumpens* v. *familiaris*. **b.** *S*. *socia* v. *socia*. **c.** *S*. *rumpens* v. *fragilarioides*. **d.** *S*. *rumpens* v. *meneghiniana*. **e.** *S*. *amphicephala* v. *amphicephala*. **f.** *S*. *ulna* v. *ramesi*. **g.** *S*. *tabulata* v. *tabulata*. (Scale lines equal 10 micrometers.)

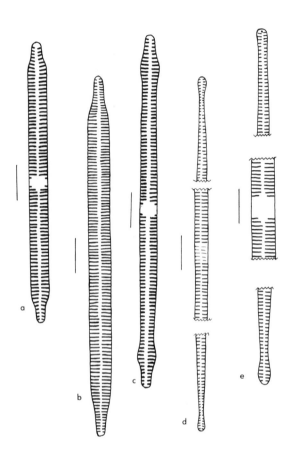

Fig. 10. a. Synedra ulna v. ulna. **b.** S. ulna v. amphirhynchus. **c.** S. ulna v. spathulifera. **d.** S. ulna v. danica. **e.** S. ulna v. longissima. (Scale lines equal 10 micrometers.)

13. *Eunotia* Ehrenberg 1837

Eunotia belongs to the order Eunotiales of class Pennatibacillario-phyceae. This genus, like others in the order, has highly reduced raphe systems on both valves of each cell. The very short raphe branches are limited to the region of the polar nodules. The valves are symmetric to the transverse axis, but not to the longitudinal axis. The valves are all weakly to strongly arc-shaped and have convex dorsal margins and weakly to strongly concave ventral margins. The polar nodules vary in size but are often quite massive. The valves have transverse striae that may or may not be visibly punctate under the light microscope. The valve margins are rarely smooth arcs, but have various degrees of curvature and may be undulate in some species. The valves may or may not have distinctive ends. Internal valves and septa are not known to occur. In girdle view the cells are rectangular. Some species may show ornamentation of the girdle. It is generally possible to make identifications only when the cells or valves are in valve view. In nature some species are found as isolated cells and others have been shown to form ribbonlike chains. These characters are not used in the taxonomy of the genus at the present time.

KEY TO THE TAXA OF Eunotia IN ILLINOIS

1. Valves with distinctive ends, rostrate or capitate, formed by a clear change of curvature near the valve extremities _____ 2
1. Valves lacking distinctive ends, the valve margins not appreciably changing in curvature as the ends are approached _____ 17
    2. Valves with distinctive swellings or humps on either margin __ 3
    2. Valves with both margins more or less smoothly curved (included here are species in which the dorsal margin is flattened at midvalve as well as those in which the margins are purely convex) ____ 11
3. Valves with dorsal or ventral swellings only _____ 4
3. Valves with both dorsal and ventral swellings _____ 9
    4. Valves with dorsal swellings only _____ 5
    4. Valves with ventral swellings only _____ 7
5. Two dorsal swellings present _____ 6
5. Three dorsal swellings present _____ 15. *E. trigibba*
    6. Ends rostrate; striae 14/10 μm or more _____
    _____ 9. *E. pectinalis* var. *minor*
    6. Ends capitate; striae 12/10 μm or fewer _____
    _____ 13. *E. praerupta* var. *bidens*

7. Valve ends rostrate _____ 11. *E. pectinalis* var. *ventralis*
7. Valve ends capitate _____ 8
   8. Ends wedge-capitate _____ 5. *E. indica*
   8. Ends rounded-capitate _____ 6. *E. maior*
9. Valves with one dorsal and one ventral swelling at midvalve _____ 10
9. Valves with three dorsal and one ventral swellings _____
   _____ 10. *E. pectinalis* var. *undulata*
   10. Valves with rostrate ends _____ 11. *E. pectinalis* var. *ventralis*
   10. Valves with rounded-capitate ends _____ 6. *E. maior*
11. Valve ends capitate _____ 12
11. Valve ends rostrate _____ 15
   12. Striae 16/10 μm or more (usually 20 or more in 10 μm); valves
     usually fairly small and slender _____ 2. *E. exigua*
   12. Striae 12/10 μm or less; valves usually large or robust and heavily
     silicified _____ 13
13. Ends formed by a sharp reflexing of the dorsal margin very close to the
   terminus; striae usually not distinctly punctate ____ 12. *E. praerupta*
13. Ends formed by a moderate reflexing of the dorsal margin beginning
   some distance from the terminus; striae usually distinctly punctate or
   at least resolvable into puncta with moderately oblique light ____ 14
   14. Ends rounded-capitate _____ 6. *E. maior*
   14. Ends wedge-capitate _____ 5. *E. indica*
15. Valve margins nearly straight at midvalve _____ 16
15. Dorsal valve margin distinctly convex at midvalve; two small, somewhat indistinct silica thickenings present on the ventral margin in addition to the polar nodules _____ 14. *E. sudetica*
   16. Dorsal margin, upon reexamination, proving to be slightly concave
     at midvalve, the margin therefore bilobed; striae 14–16/10 μm __
     _____ 9. *E. pectinalis* var. *minor*
   16. Dorsal margin straight or, in long valves, if slightly concave, then
     the central striae 7–12/10 μm _____ 8. *E. pectinalis*
17. Valves narrowing toward the ends, the valve form therefore asymmetric-lanceolate; a thin faint line sometimes extending a short distance
   from each polar nodule toward the midpoint of the valve _____
   _____ 1. *E. curvata*
17. Valves not narrowing appreciably toward the ends, the ends therefore
   broadly rounded or, if valve narrows toward the end, it gradually wid-
   ens again before the end _____ 18
   18. Striae coarse, widely spaced, fairly easily resolved into puncta;
     margins not completely parallel _____ 3. *E. fallax* var. *gracillima*

18. Striae fine (above 14/10 μm) or, if coarse, then margins very nearly parallel _____ 19
19. Margins parallel; ends broadly rounded; striae 8–16/10 μm _____
_____ 7. *E. parallela*
19. Dorsal margin not purely parallel to ventral margin; ends rounded, sometimes weakly subcapitate; a short line extending from each polar nodule toward the midpoint of the valve; striae 14–18/10 μm _____
_____ 4. *E. flexuosa*

1. Eunotia curvata (Kuetz.) Lagerst. var. **curvata** *Fig. 14d.*

CR: Patrick and Reimer (1966).
DESCRIPTION: Dorsal margin convex; ventral margin concave; margins nearly parallel at midvalve, but tapering toward the rounded to subacute ends; polar nodules small; raphe usually protracted somewhat onto the valve surface (this feature requires careful observation); striae parallel and not visibly punctate under the light microscope.
DIMENSIONS: Length, 20–150 μm; width, 3–6 μm; striae, 13–18/10, μm.

2. Eunotia exigua (Bréb.) Rabh. var. **exigua** *Fig. 12d–g.*

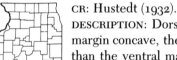

CR: Hustedt (1932).
DESCRIPTION: Dorsal margin distinctly convex; ventral margin concave, the dorsal margin slightly more arched than the ventral margin; dorsal margin reflexed at the end to form an asymmetric-capitate to hooked end; polar nodules small; striae parallel and not visibly punctate under the light microscope.
DIMENSIONS: Length, 8–67 μm; width, 2–4 μm; striae, 16–25/10 μm.

3. Eunotia fallax var. gracillima Krasske *Fig. 14c*.

CR: Hustedt (1932).

DESCRIPTION: Valves moderately lunate; margins more or less parallel; both margins weakly reflexed before the rounded ends; polar nodules moderately large; raphe branches confined to the polar nodules; striae coarse, parallel, indistinctly punctate.

DIMENSIONS: Length, 15–40 μm; width, 2–4 μm; striae, 11–13/10 μm.

4. Eunotia flexuosa Bréb. ex Kütz. var. flexuosa *Fig. 14a*.

CR: Patrick and Reimer (1966).

DESCRIPTION: Valves linear-lunate; margins nearly parallel; dorsal margin slightly reflexed to form the rounded capitate or subcapitate ends; polar nodules distinct; raphe branches typically prolonged somewhat beyond the polar nodules as short lines on the valve surface; striae not visibly punctate under the light microscope.

DIMENSIONS: Length, 90–300 μm; width, 2–5 μm; striae, 14–18/10 μm.

5. Eunotia indica Grun. var. indica *Fig. 11a–d*.

CR: Patrick and Reimer (1966).

DESCRIPTION: Ventral margin weakly concave; dorsal margin convex; valve ends wedge-capitate; polar nodules large; striae distinctly punctate.

DIMENSIONS: Length, 30–110 μm; width, 9–15 μm; striae, 8–14/10 μm.

6. Eunotia maior (W. Sm.) Rabh. var. **maior** *Fig. 11e–g.*

CR: Patrick and Reimer (1966).

DESCRIPTION: Ventral margin straight to slightly concave (in some longer specimens, the middle of the ventral margin may have a slight bulge); dorsal margin convex; ends rounded-capitate; polar nodules large; striae parallel and distinctly punctate. In longer specimens there is sometimes a thin longitudinal clear space close to the ventral margin.

DIMENSIONS: Length, 35–220 μm; width, 6–15 μm; striae, 8–14/10 μm.

7. Eunotia parallela Ehr. var. **parallela** *Fig. 14b.*

CR: Patrick and Reimer (1966).

DESCRIPTION: Valves weakly to moderately lunate; valve margins nearly parallel throughout; ends broadly rounded, not set off by constriction from the rest of the valve.

DIMENSIONS: Length, 50–150 μm; width, 5–15 μm; striae, 8–16/10 μm.

8. Eunotia pectinalis (O. F. Muell.) Rabh. var. **pectinalis** *Fig. 13a.*

CR: Patrick and Reimer (1966).

DESCRIPTION: Valves more or less linear; dorsal and ventral margins straight to very weakly concave at midvalve, the dorsal margin constricted to form the broadly rostrate ends; polar nodules distinct; striae parallel, becoming slightly curved and radiate around the polar nodules. This large *Eunotia* has a number of varieties, many of which have been found in Illinois. I would suggest examining *Fig. 13* for some of these. It is possible that some of the varieties may prove to intergrade.

DIMENSIONS: Length, 17–140 μm; width, 5–10 μm; striae, 7–12/10 μm near the center, becoming 14/10 μm at the ends.

9. Eunotia pectinalis var. **minor** (Kütz.) Rabh. *Fig. 13e, f.*

CR: Hustedt (1932); Patrick and Reimer (1966).

DESCRIPTION: Ventral margin straight to weakly concave; dorsal margin flattened to concave at midvalve; ends broadly rostrate and somewhat elongate; polar nodules fairly small; striae not visibly punctate under the light microscope.

DIMENSIONS: Length, 10–60 μm; width, 5–10 μm; striae, 14–16/10 μm.

10. Eunotia pectinalis var. **undulata** (Ralfs) Rabh. *Fig. 13c.*

CR: Patrick and Reimer (1966).

DESCRIPTION: Valves linear with broadly rostrate ends; dorsal margin with three weak swellings; ventral margin weakly concave and usually with a single central swelling; striae often fairly distinctly punctate.

DIMENSIONS: Similar to var. *pectinalis*.

11. Eunotia pectinalis var. **ventralis** (Ehr.) Hust. *Fig. 13b.*

CR: Hustedt (1932).

DESCRIPTION: Valve margins parallel throughout much of the length, but medially inflated on the ventral side (and slightly on the dorsal side); a narrow clear space extending from pole to pole near the ventral margin; valve ends rostrate; polar nodules large; striae not distinctly punctate.

DIMENSIONS: Length, 40–140 μm; width, 5–10 μm; striae, 7–12/10 μm.

12. Eunotia praerupta Ehr. var. **praerupta** *Fig. 12c.*

CR: Hustedt (1932).
DESCRIPTION: Ventral margin slightly concave; dorsal margin convex; ends broadly flat-capitate; polar nodules large; striae robust, but typically not visibly punctate under the light microscope.
DIMENSIONS: Length, 10–100 μm; width, 4–13 μm; striae, 6–12/10 μm, slightly irregularly spaced.

13. Eunotia praerupta var. **bidens** (Ehr.) Grun. *Fig. 12b.*

CR: Patrick and Reimer (1966).
DESCRIPTION: Ventral margin concave (sometimes with slight undulation near the ends); dorsal margin rounded-biundulate; ends broadly flat-capitate; polar nodules large; striae robust but not visibly punctate under the light microscope; spacing of striae somewhat irregular.
DIMENSIONS: Similar to var. *praerupta*.

14. Eunotia sudetica O. Muell. var. **sudetica** *Fig. 13d.*

CR: Hustedt (1932).
DESCRIPTION: Ventral margin straight to weakly concave; dorsal margin convex; ends rostrate to subrostrate. In addition to the two subterminal polar nodules, there are two additional small thickenings on the ventral margin.
DIMENSIONS: Length, 15–50 μm; width, 6–8 μm; striae, 8–11/10 μm near center, becoming finer at the ends. Puncta not visible with the light microscope.

15. Eunotia trigibba Hust. var. **trigibba** *Fig. 12a.*

CR: A.S.A. P*l.* 286, Fig. s. 16–18.

DESCRIPTION: Ventral margin concave; dorsal margin triundulate; ends capitate; polar nodules distinct; striae distinctly punctate.

DIMENSIONS (my specimen): Length, 28 μm; width, 7.5 μm; striae, 14–16/10 μm.

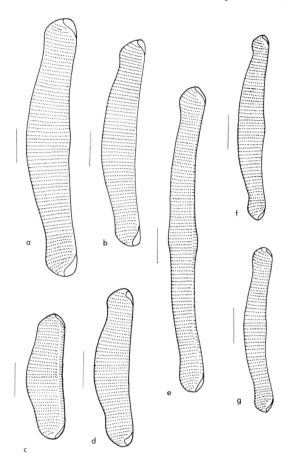

Fig. 11. *a–d. Eunotia indica* v. *indica. e–g. E. maior* v. *maior.* (Scale lines equal ten micrometers.)

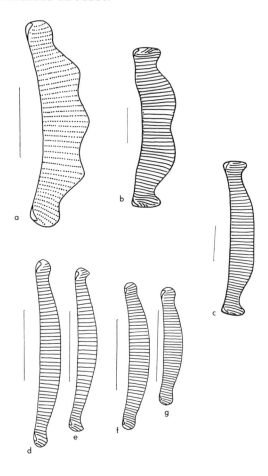

Fig. 12. a. *Eunotia trigibba* v. *trigibba*. **b.** *E. praerupta* v. *bidens*. **c.** *E. praerupta* v. *praerupta*. **d–g.** *E. exigua* v. *exigua*. (Scale lines equal 10 micrometers.)

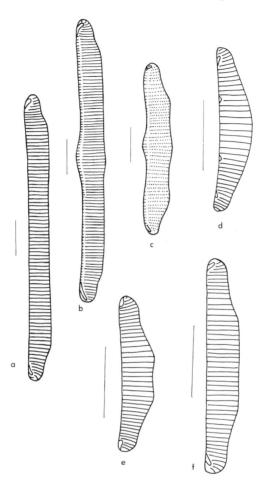

Fig. 13. a. *Eunotia pectinalis* v. *pectinalis*. **b.** *E. pectinalis* v. *ventralis*. **c.** *E. pectinalis* v. *undulata*. **d.** *E. sudetica* v. *sudetica*. **e, f.** *E. pectinalis* v. *minor*. (Scale lines equal 10 micrometers.)

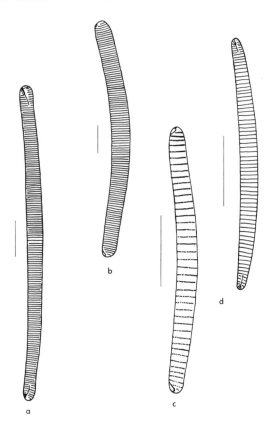

Fig. 14. a. *Eunotia flexuosa* v. *flexuosa*. **b.** *E. parallela* v. *parallela*. **c.** *E. fallax* v. *gracillima*. **d.** *E. curvata* v. *curvata*. (Scale lines equal 10 micrometers.)

14. Achnanthes Bory 1822

Achnanthes belongs to the order Achnanthales of class Pennatibac-illariophyceae. One valve of each cell has a fully developed true raphe and the other valve a clear axial space or pseudoraphe. Complete cells can be seen in either valve or girdle view. In girdle view the cells or valves most often appear bent around the transverse axis of the cell or valve. In valve view, this bending sometimes makes it difficult to focus simultaneously on the ends of the valve and the center. By contrast, valves of the closely related genus *Cocconeis* are arched around the long axis and in valve view it will sometimes be difficult to maintain the midline and the margins in simultaneous focus.

Most species of *Achnanthes* have linear or lanceolate valves with distinctive ends. The few *Achnanthes* species with elliptical cells found in this area also have distinct horseshoe-shaped marks or depressions at midvalve on one side of the pseudoraphe. *Cocconeis* species in Illinois are all linear-elliptic to elliptic in outline and none has a horseshoe-shaped mark on the pseudoraphe valve.

Technically it is necessary to have complete frustules to make identifications of *Achnanthes* species, but practically it is usually possible to do so if both the raphe valves and pseudoraphe valves are present in the same sample. Isolated raphe valves are difficult to separate from other naviculoid diatoms, but it is sometimes (though not always!) possible to identify *Achnanthes* from the pseudoraphe valve.

KEY TO THE TAXA OF Achnanthes IN ILLINOIS

1. Both valves with distinctly and coarsely punctate striae; pseudoraphe lies close to one margin of the valve _____ 2
1. Neither valve with distinctly punctate striae, or if striae can be resolved into puncta, then the pseudoraphe lying along the longitudinal midline of the valve _____ 3
   2. Valve margins typically concave at midvalve _____ 1. *A. coarctata*
   2. Valve margins highly convex at midvalve _____ 7. *A. inflata*
3. Valves lanceolate to elliptic-lanceolate, the pseudoraphe valve bearing a large, horseshoe-shaped mark on one side of the central area _____ 4
3. Valves variously shaped, pseudoraphe valve never having a horseshoe-shaped mark on one side of the central area _____ 5
   4. Valves lanceolate or elliptic-lanceolate, ends neither rostrate nor drawn-out (that is, the margins do not change from convex to concave as ends are approached) _____ 8. *A. lanceolata*

    4. Valves with rostrate or drawn-out ends _____
_____ 9. *A. lanceolata* var. *dubia*

5. Central area on raphe valve reaching both valve margins _____ 6

5. Central area enclosed on raphe valve, not reaching either valve margin
_____ 7

    6. Pseudoraphe valve with a narrow axial area not widened to form a
central area; valve ends slightly and broadly subrostrate _____
_____ 6. *A. hungarica*

    6. Pseudoraphe valve with a narrow axial area and a central area
formed by shortening the central pair of striae; ends distinctly and
narrowly rostrate to rostrate-capitate _____
_____ 2, 3, 4. *A. exigua* and varieties

7. Valves linear with nearly straight margins and broadly rounded ends not
set off from the rest of the valve; pseudoraphe very narrow _____
_____ 10. *A. linearis*

7. Valves lanceolate or, if somewhat linear, then ends distinctive _____ 8

    8. Valves narrowly linear or linear-lanceolate; ends vary from rostrate
to capitate; pseudoraphe very narrow; striae 30/10 μm or more ___
_____ 11. *A. minutissima*

    8. Valves lanceolate; pseudoraphe moderately narrow to broadly lan-
ceolate; ends somewhat protracted or subrostrate, but not capitate;
striae much fewer than 30/10 μm _____ 9

9. Valves broadly lanceolate; pseudoraphe moderately narrow; striae fewer
than 20/10 μm _____ 5. *A. hauckiana*

9. Valves moderately lanceolate to linear-lanceolate; ends fairly broad,
weakly to distinctly subrostrate; pseudoraphe broad and lanceolate;
striae 20–25/10 μm _____ 12. *A. subrostrata*

    1. Achnanthes coarctata (Bréb. in W. Sm.) Grun. var. **coarctata**
Fig. 16a, b.

CR: Patrick and Reimer (1966).

DESCRIPTION: Valves more or less linear with margins concave at midvalve; raphe valve with a narrow axial area and a central area that reaches both margins; pseudoraphe valve with a narrow pseudoraphe lying near one margin; striae distinctly punctate on both valves.

DIMENSIONS: Length, 20–48 μm; width, 7–12 μm; striae on raphe valve, 13–15/10 μm (mine reached 18/10 μm); striae on pseudoraphe valve, 10–14/10 μm; puncta on both valves, 14–18/10 μm.

2. Achnanthes exigua Grun. var. **exigua** *Fig. 15m, n.*

CR: Patrick and Reimer (1966).

DESCRIPTION: Valve margins more or less straight at midvalve, narrowing abruptly to form narrowly rostrate ends; raphe valve with a narrow axial area and a central area that reaches both margins; pseudoraphe valve with a narrow axial area (pseudoraphe) and a central area formed by the shortening of a few striae; axial areas on both valves lie along the longitudinal midline; striae weakly radiate; puncta usually not visible.

DIMENSIONS: Length, 7–17 μm; width, 4.5–6 μm; striae on raphe valve, 24–25/10 μm; striae on pseudoraphe valve, 20–22/10 μm.

3. Achnanthes exigua var. **constricta** (Grun.) Hust. *Fig. 15k, l.*

CR: Patrick and Reimer (1966).

DESCRIPTION: Similar to var. *exigua* with the exception that the valve margins are weakly to moderately concave at midvalve. This taxon may or may not be connected with var. *exigua* by intergrading forms.

DIMENSIONS: As for var. *exigua*.

4. Achnanthes exigua var. **heterovalva** Krasske *Fig. 15o, p.*

CR: Patrick and Reimer (1966).

DESCRIPTION: Valves shaped like those of var. *exigua*, but margins sometimes slightly convex at midvalve. This taxon differs from var. *exigua* in the density of striae on the raphe valve.

DIMENSIONS: length and width, as for var. *exigua*; striae on raphe valve, 30/10 μm at midvalve becoming 34/10 μm at the ends; striae on pseudoraphe valve, 22/10 μm.

5. Achnanthes hauckiana Grun. var. **hauckiana** *Fig. 15i, j.*

CR: Patrick and Reimer (1966).

DESCRIPTION: Valves lanceolate with narrow and somewhat protracted ends; raphe valve with a moderately narrow axial area opening into a broad, rounded central area; pseudoraphe valve with a moderately narrow axial area (pseudoraphe), but no distinct central area.

DIMENSIONS: Length, 9–31 μm; width, 5–9 μm; striae (both valves), 10–12/10 μm at midvalve, becoming 12–15/10 μm at the ends.

6. Achnanthes hungarica (Grun.) Grun. var. **hungarica** *Fig. 16c, d.*

CR: Patrick and Reimer (1966).

DESCRIPTION: Valve outline linear-elliptic to elliptic-lanceolate; valve ends blunt, sometimes subrostrate; raphe valve with a narrow axial area and a central area that reaches both valve margins; pseudoraphe valve with a narrow axial area (pseudoraphe), but no distinct central area; striae slightly radiate on the pseudoraphe valve and distinctly radiate on the raphe valve.

DIMENSIONS: Length, 14–45 μm; width, 6–8 μm; striae (both valves), 19–23/10 μm.

7. Achnanthes inflata (Kütz.) Grun. var. **inflata** *Fig. 16e, f.*

CR: Patrick and Reimer (1966).

DESCRIPTION: Valve outline broadly lanceolate as a result of the pronounced lateral inflation of the margins at midvalve; raphe valve with a moderately narrow axial area and a central area that reaches both margins; pseudoraphe valve lacking central area and with the axial area displaced to one margin; both valves with moderately radiate striae that are coarsely punctate.

DIMENSIONS: Length, 30–65 μm; width, 10–18 μm; striae on raphe valve, 10–13/10 μm; striae on pseudoraphe valve, 9–11/10 μm; puncta, about 10/10 μm on both valves.

8. Achnanthes lanceolata (Bréb.) Grun. var. **lanceolata** *Fig. 15a–d.*

CR: Patrick and Reimer (1966).

DESCRIPTION: Valve outline lanceolate to elliptical; ends not distinctive; raphe valve with a narrow axial area and a transversely expanded central area; pseudoraphe valve with a narrow axial area and a central area that is marked by a horseshoe-shaped clear area on one side.

DIMENSIONS: Length, 12–31 μm; width, 4.5–8 μm; striae (both valves), 11–14/10 μm; puncta, not resolvable with the light microscope.

9. Achnanthes lanceolata var. **dubia** Grun. *Fig. 15g, h.*

CR: Patrick and Reimer (1966).

DESCRIPTION: Differs from var. *lanceolata* primarily in the shape of the ends, which are distinctly rostrate to subrostrate in var. *dubia*. The valves I have seen are also usually somewhat more linear in appearance. The horseshoe-shaped mark on the pseudoraphe valve is like that of var. *lanceolata*.

DIMENSIONS: Length, 8–16 μm; width, 3.6–5 μm; striae (both valves) 10–14/10 μm.

10. Achnanthes linearis (W. Sm.) Grun. var. **linearis** *Fig. 15e, f.*

CR: Patrick and Reimer (1966).

DESCRIPTION: Valves linear with weakly convex margins; ends not set off from the rest of the valve by constriction or expansion; axial area of both valves narrow; raphe valve with a central area produced by the shortening of a single stria on each side of the valve; pseudoraphe valve lacking a distinct central area. Care must be taken in separating specimens of this taxon from short specimens of the much more common *A. minutissima* Kütz., which always have distinctive ends, even if this distinction is subtle.

DIMENSIONS: Length, 4–20 μm; width, 2.4–3 μm; striae (both valves), 23–29/10 μm.

11. Achnanthes minutissima Kütz. var. **minutissima** *Fig. 15q, r.*

CR: Patrick and Reimer (1966).

DESCRIPTION: Valves linear to linear-lanceolate with distinctive ends varying from rostrate through subcapitate to distinctly capitate; axial area narrow on both valves; raphe valve with a central area formed by shortening of a few striae on each side of the valve; pseudoraphe valve without a distinct central area. *A. minutissima* is quite variable in form, but apparently always has ends that are at least slightly set off from the rest of the valve even in short cells. I am almost convinced that the range of variation of *A. minutissima* merges with that of *A. microcephala* Kütz., a taxon which has been reported for Illinois by other authors.

DIMENSIONS: Length, 5–40 μm; width, 2–4 μm; striae (both valves), 30–38/10 μm.

12. Achnanthes subrostrata Hust. var. **subrostrata** *Fig. 16g, h.*

CR: Hustedt (1942).

DESCRIPTION: Valves elliptic-lanceolate with subrostrate ends; raphe valve with a narrow axial area widening suddenly into a moderately large, rounded, central area; pseudoraphe valve with central and axial areas combined into a broad linear space.

DIMENSIONS: Length, about 20–22.5 μm; width, 4–5 μm; striae (both valves), 20–24/10 μm.

15. Cocconeis Ehrenberg 1837

Cells of *Cocconeis* generally grow on organic and inorganic substrates as individual cells closely appressed to the substrate.

Cocconeis belongs to the order Achnanthales and, like other genera in that order, has only one valve, the raphe valve, with a fully developed raphe. The other valve entirely lacks a raphe and is called the pseudoraphe valve. *Cocconeis* cells are broadly elliptic in outline and lack any sort of distinctive ends when seen in valve view. Due to the shape, *Cocconeis* cells are almost never seen in girdle view. Unlike *Achnanthes*, in which the valvar plane is bent around its transverse axis, cells of *Cocconeis* are either nearly plane or, if arched, then the bending is around the long axis of the cell. Practically, the elliptical shape, coupled with a lack of horseshoe-shaped clear areas on the pseudoraphe valve, is sufficient to distinguish *Cocconeis* from all *Achnanthes* found in Illinois.

The valve markings on *Cocconeis* are bilaterally symmetric to the long axis and are of different character on the two valves.

KEY TO THE TAXA OF Cocconeis IN ILLINOIS

1. Valve surface nearly flat; center and margins therefore may be in focus simultaneously _____ 2, 3, 4, 5, *C. placentula* and varieties
1. Valve surface highly arched; center and margin rarely in focus simultaneously _____ 1. *C. pediculus*

1. **Cocconeis pediculus** Ehr. var. **pediculus** *Fig. 17e,f.*

CR: Patrick and Reimer (1966).

DESCRIPTION: Valve outline elliptical; valve plane highly arched; raphe valve with a narrow axial area and an elongate-elliptic central area; pseudoraphe valve with central and axial areas combined in a moderately broad biundulate space; striae clearly dot-punctate on the raphe valve; striae of pseudoraphe valve composed of dashlike puncta that are aligned to form undulating longitudinal striae as well.

DIMENSIONS: Length, 11–56 μm; width, 6–37 μm; striae (raphe valve), 16–20/10 μm; striae (pseudoraphe valve), 15–18/10 μm.

2. Cocconeis placentula Ehr. (and varieties) *Fig. 17a–d*.

CR: Patrick and Reimer (1966).

DESCRIPTION: There are three named varieties of this taxon, all of which are distinguished from each other only on the basis of differences in the pseudoraphe valve. My experience has been that these varieties intergrade with one another to an extent that suggests that they may either be growth forms of one genetic species or at best only minor genetic variants of one another. Unless the pseudoraphe valve is present, identification of the cells beyond species is not possible.

For all varieties, the raphe valve may be described as follows: valve outline elliptic without transapical constrictions of any kind; axial area narrow; central area small and elliptic; striae punctate and curved (which gives one the impression that they are radiate throughout); striation interrupted by two hyaline bands (one marginal, the other submarginal) that are separated from one another by a narrow striated band. These features, as well as the rather flat valves, are sufficient to distinguish the present taxon from *C. pediculus*.

DIMENSIONS: Length, 10–70 μm; width, 8–40 μm; striae (raphe valve), 19–23/10 μm; striae (pseudoraphe valve), 24–26/10 μm.

These characters are common to all three varieties. The varieties differ in the number of longitudinal striae on the pseudoraphe valve. These differences and the distribution maps are given under each variety name.

3. Cocconeis placentula Ehr. var. placentula *Fig. 17a, b*.

CR: Patrick and Reimer (1966).

DESCRIPTION: Raphe valve as described for the species; striae on pseudoraphe valve formed of rounded instead of dashlike puncta; puncta numerous (usually 15 or more rows on each side of the axial area); striae curved in the same way as they are on the raphe valve; pseudoraphe (axial area) narrow.

DIMENSIONS: See species description.

4. Cocconeis placentula var. **euglypta** (Ehr.) Cl. *Fig. 17a, c.*

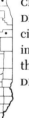

CR: Patrick and Reimer (1966).
DESCRIPTION: Raphe valve as described for the species. Pseudoraphe valve marked by rows of dashlike puncta arranged into striae. There are 2–5 longitudinal rows of these puncta on each side of the pseudoraphe.
DIMENSIONS: See species description.

5. Cocconeis placentula var. **lineata** (Ehr.) V.H. *Fig. 17a, d.*

CR: Patrick and Reimer (1966).
DESCRIPTION: Raphe valve as described for the species. Pseudoraphe marked by dashlike puncta arranged into striae. There are around 10 longitudinal rows of these puncta on each side of the pseudoraphe.
DIMENSIONS: See species description.

16. Rhoicosphenia Grunow 1860

Rhoicosphenia belongs to the order Achnanthales of class Pennatibacillariophyceae. One valve of each cell has a fully developed two-branch raphe. The other valve has short, scarcely noticeable, raphe rudiments at the extreme ends of the valve. *Rhoicosphenia* thus has a raphe valve and a rudimentary raphe valve instead of a raphe valve and pseudoraphe valve as is the case in other genera in the order.

Cells of *Rhoicosphenia* can be seen in both valve and girdle views. The valves have a clavate outline. Pseudosepta are present at both the head and foot poles. In girdle view cells and valves have the form of a curved wedge, and this curvature can be seen as a focus feature in valve view as well. The curvature is useful in separating *Rhoicosphenia* from *Gomphonema* in either view. There is but a single species.

1. Rhoicosphenia curvata (Kütz.) Grun, var. **curvata** *Fig.*
17g–i.

CR: Patrick and Reimer (1966).

DESCRIPTION: See generic description above.

DIMENSIONS: Length, 12–75 μm; maximum width, 4–8 μm; striae on raphe valve, 9–15/10 μm at midvalve, becoming 16–20/10 μm at the ends; striae on rudimentary raphe valve, 11–13/10 μm at midvalve, becoming 16–18/10 μm at the ends.

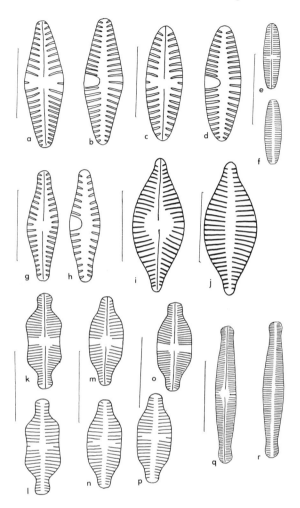

Fig. 15. *a. Achnanthes lanceolata* v. *lanceolata* (raphe valve). *b. A. lanceolata* v. *lanceolata* (pseudoraphe valve). *c. A. lanceolata* v. *lanceolata* (raphe valve). *d. A. lanceolata* v. *lanceolata* (pseudoraphe valve). *e. A. linearis* v. *linearis* (raphe valve). *f. A. linearis* v. *linearis* (pseudoraphe valve). *g. A. lanceolata* v. *dubia* (raphe valve). *h. A. lanceolata* v. *dubia* (pseudoraphe valve). *i. A. hauckiana* v. *hauckiana* (raphe valve). *j. A. hauckiana* v. *hauckiana* (pseudoraphe valve). *k. A. exigua* v. *constricta* (raphe valve). *l. A. exigua* v. *constricta* (pseudoraphe valve). *m. A. exigua* v. *exigua* (raphe valve). *n. A. exigua* v. *exigua* (pseudoraphe valve). *o. A. exigua* v. *heterovalva* (raphe valve). *p. A. exigua* v. *heterovalva* (pseudoraphe valve). *q. A. minutissima* v. *minutissima* (raphe valve). *r. A. minutissima* v. *minutissima* (pseudoraphe valve). (Scale lines equal 10 micrometers.)

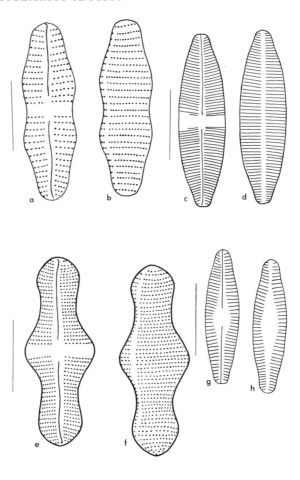

Fig. 16. a. *Achnanthes coarctata* v. *coarctata* (raphe valve). **b.** *A. coarctata* v. *coarctata* (pseudoraphe valve). **c.** *A. hungarica* v. *hungarica* (raphe valve). **d.** *A. hungarica* v. *hungarica* (pseudoraphe valve). **e.** *A. inflata* v. *inflata* (raphe valve). **f.** *A. inflata* v. *inflata* (pseudoraphe valve). **g.** *A. subrostrata* v. *subrostrata* (raphe valve). **h.** *A. subrostrata* v. *subrostrata* (pseudoraphe valve). (Scale lines equal 10 micrometers.)

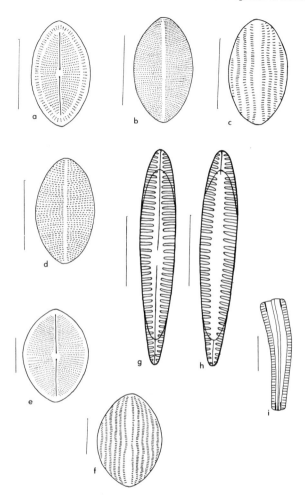

Fig. 17. a. *Cocconeis placentula* (raphe valve of all varieties). **b.** *Cocconeis placentula* v. *placentula* ((pseudoraphe valve). **c.** *Cocconeis placentula* v. *euglypta* (Pseudoraphe valve). **d.** *Cocconeis placentula* v. *lineata* (pseudoraphe valve). **e.** *Cocconeis pediculus* v. *pediculus* (raphe valve). **f.** *Cocconeis pediculus* v. *pediculus* (pseudoraphe valve). **g.** *Rhoicosphenia curvata* v. *curvata* ((raphe valve). **h.** *Rhoicosphenia curvata* v. *curvata* (rudimentary raphe valve). **i.** *Rhoicosphenia curvata* v. *curvata* (cell in girdle view). (Scale lines equal 10 micrometers.)

17. *Mastogloia* Thwaites ex. W. Sm. 1856

Mastogloia belongs to the order Naviculales of class Pennatibacillariophyceae and, like all members of that order, is characterized by the bilateral symmetry of valve markings and by the presence of a fully developed true raphe on both valves of each cell.

Cells of *Mastogloia* are symmetric to the long and transverse axis in valve view and to the long and pervalvar axes in girdle view. Striae are distinctly punctate. The distinguishing character for the genus is the presence of a complex septum associated with each valve. The septum has a locule running along the long axis. The ends of this central locule are more or less spoon-shaped. There is a row of small cell-like chambers bordering each side of the central locule.

KEY TO THE SPECIES OF Mastogloia CONFIRMED FOR ILLINOIS

1. Valve ends more or less cuneate and slightly protracted; ends of central locule of septum broadly rounded _____ 1. *M. elliptica* var. *danseii*
1. Valve ends more or less rostrate; ends of central locule of septum acute _____ 2. *M. Smithii* v. *lacustris*

1. **Mastogloia elliptica** var. **danseii** (Thwaites) Cl. *Fig. 18g, h.*

CR: Patrick and Reimer (1966); Stoermer (1967).

DESCRIPTION: Valves elliptic-lanceolate with subrostrate or cuneate ends; striae alternating longer and shorter at the central area. In my specimens this alternation was regular, but it may be otherwise. The ends of the central locule of the septum are broadly rounded. Stoermer has found unusual populations of this taxon in which the striae were doubly punctate on one valve and singly punctate on the other, leading to a suspicion that there may be intergrades between this taxon and *M. grevillei* W. Sm., which has doubly punctate striae. I have not found *M. grevillei* in Illinois and have found no populations of *M. elliptica* var. *danseii* with doubly punctate striae, but you should be on the lookout for them, especially in highly alkaline habitats.

DIMENSIONS: Length, 20–51 μm; width, 11–15 μm; striae, 16–18/10 μm; chambers along central locule of septum, 8–10/10 μm.

2. Mastogloia smithii var. **lacustris** Grun. *Fig. 18e, f.*

CR: Patrick and Reimer (1966).

DESCRIPTION: Valves elliptic-lanceolate with rostrate ends; central area rounded; striae consist of single rows of puncta; ends of central locule of septum acute.

DIMENSIONS: Length, 20–45 μm; width, 8–11 μm; striae, 15–16/10 μm; chambers along central locule of septum, 6–8/10 μm.

18. *Amphipleura* Kützing 1844

Amphipleura belongs to the order Naviculales of class Pennatibacillariophyceae. Like all members of the order, valve markings are bilaterally symmetric and both valves bear a true raphe. Valves are symmetric to the long and transverse axes.

The raphe structure is the major generic character. The raphe branches are short, extending only a short distance from the ends. Each branch is embedded in a silica rib. The two branches are connected by the narrow, highly elongate central nodule. Striae are in two systems, transverse and longitudinal, as a result of the positioning of the puncta. In the one species found in Illinois, however, the striae are so fine that they are not normally visible without strongly obliqued light. In fact, this species, *A. pellucida* (Kütz.) Kütz., is famous as a test object for microscope lenses. It is said that a good lens will resolve the transverse striae, a great lens the longitudinal striae, and the legendary Zeiss n.a. 1.6 apochromat the puncta. Rather ordinary electron microscopes will do all three.

1. Amphipleura pellucida (Kütz.) Kütz. var. **pellucida** *Fig. 28c.*

CR: Patrick and Reimer (1966).

DESCRIPTION: As for the genus.

DIMENSIONS: Length, 74–140 μm; width, 7–9 μm; transverse striae, 37–40/10 μm; longitudinal striae, in excess of 40/10 μm and typically not visible.

19. *Frustulia* Rabenhorst 1853 nom. cons., non Agardh 1824

Frustulia belongs to the order Naviculales of class Pennatibacilla-riophyceae. Valves are symmetric to both the long and transverse axes. Valve markings consist of puncta so arranged as to produce two perpendicular systems of striae, the one transverse, the other longitudinal. While some of the species I found in Illinois have striae that are so fine that the valves appear nonstriate, oblique light always easily reveals at least one system. *Frustulia* species are thus not the difficult challenge for the lens that *Amphipleura pellucida* is.

In addition to the systems of striae, *Frustulia* is characterized by raphe branches that usually approach each other at midvalve and that are embedded in silica ribs. Even in the one variety that has short raphe branches, these branches are long by comparison with those of *A. pellucida*.

KEY TO Frustulia CONFIRMED FOR ILLINOIS

1. Proximal raphe ends straight _____ 2
1. Proximal raphe ends deflected to one side of the valve _____
   _____ 6. *F. weinholdii*
   2. Valve ends capitate _____ 3
   2. Valve ends nondistinctive or rostrate _____ 4
3. Valve margins undulate _____ 4. *F. rhomboides* v. *saxonica* f. *undulata*
3. Valve margins smooth _____ 3. *F. rhomboides* v. *saxonica* f. *capitata*
   4. Valves linear; transverse striae faintly radiate at midvalve; longitu-
      dinal striae curving around the central area _____ 5. *F. vulgaris*
   4. Valves lanceolate to elliptic lanceolate _____ 5
5. Valves lanceolate with rather broad subacute ends; raphe branches fairly
   short, the central nodule therefore elongate; striae around 20/10 μm at
   midvalve _____ 1. *F. rhomboides* v. *amphipleuroides*.
5. Valves elliptic-lanceolate with slightly protracted ends; raphe branches
   approaching each other at midvalve; striae resolvable typically only with
   oblique light _____ 2. *Frustulia rhomboides* v. *saxonica*

1. Frustulia rhomboides var. **amphipleuroides** (Grun.) Cl. *Fig. 28a*.

CR: Patrick and Reimer (1966).

DESCRIPTION: Valves lanceolate, the ends, however, not pointed but rounded; both longitudinal and transverse striae clearly visible; central area no wider than the axial area and in some cases slightly constricted at midvalve. Var. *amphipleuroides* is separated from all other varieties by the length of the central nodule (measured as the distance separating the raphe branches), which in var. *amphipleuroides* is around 8 μm, but in other varieties is under 5 μm.

DIMENSIONS: Length, 70–160 μm; width, 15–30 μm; transverse striae, 22–24/10 μm; longitudinal striae, 18–24/10 μm.

2. Frustulia rhomboides var. **saxonica** (Rabh.) De T. *Fig. 27a*.

CR: Hustedt (1930).

DESCRIPTION: Valves elliptic-lanceolate with slightly protracted, but not capitate, ends; axial area moderately narrow; central area no wider than axial area and sometimes constricted at midvalve; both transverse and longitudinal striae present but very fine. Under axial illumination, the striated zone appears simply as a gray area, the striae becoming visible under oblique light. I have shown the striae in the drawing as they might appear under oblique light.

DIMENSIONS: Length, 40–70 μm; width, 12–20 μm; transverse striae, 34–36/10 μm; longitudinal striae, approaching 40/10 μm.

3. Frustulia rhomboides var. **saxonica** f. **capitata** A. Mayer *Fig. 27d.*

CR: Hustedt (1930).
DESCRIPTION: Differs from var. *saxonica* in shape: valves linear with distinctly capitate ends; valves often somewhat narrower for a given length than in var. *saxonica*. I suspect that the present taxon may intergrade with f. *undulata* Hust.
DIMENSIONS: Similar to var. *saxonica*.

4. Frustulia rhomboides var. **saxonica** f. **undulata** Hust. *Fig. 27b.*

CR: Hustedt (1930).
DESCRIPTION: Differs from var. *saxonica* in shape: valves elliptic-lanceolate with distinctly but shallowly undulate margins and rostrate to subcapitate ends.
DIMENSIONS: Similar to var. *saxonica*.

5. Frustulia vulgaris (Thwaites) De T. var. **vulgaris** *Fig. 28b.*

CR: Patrick and Reimer (1966).
DESCRIPTION: Valve outline linear with slightly convex margins and ends that vary from nondistinctive to broadly rostrate; central area elongate-elliptic; proximal raphe ends straight.
DIMENSIONS: Length, 50–70 μm; width, 10–13 μm; transverse striae, 24/10 μm at center becoming 34/10 μm at ends; longitudinal striae, 26–35/10 μm.

6. Frustulia weinholdii Hust. var. **weinholdii** *Fig. 27c.*

CR: Patrick and Reimer (1966).

DESCRIPTION: Valve outline linear with slightly convex margins narrowing to nondistinctive or broadly rostrate ends; transverse and longitudinal striae visible, but often faint; proximal raphe ends long and bent at right angles to the long axis of the valve, both proximal ends bent to the same side of the valve.

DIMENSIONS: Length, 40–60 μm; width, 7–10 μm; striae, 30–34/10 μm near midvalve becoming 40/10 μm near the ends.

20. *Gyrosigma* Hassall nom. cons.

Gyrosigma belongs to the order Naviculales of class Pennatibacillariophyceae. As with other genera in this order, both valves of each cell have a fully developed two-branched raphe that is not part of a keel-type or canal-type structure, but is a simple slit in the valve surface. The transverse axis is straight, but the longitudinal axis is more or less sigmoid. With the axes construed in this way, the valves are symmetric to both axes in shape as well as markings.

There are two systems of striae, longitudinal and transverse, intersecting more or less at right angles. In a few cases only one system is easily resolvable.

KEY TO THE SPECIES OF Gyrosigma CONFIRMED FOR ILLINOIS

1. Valve ends long and beaklike; valves lightly silicified, the striae therefore difficult to see _ 3. *G. macrum*
1. Valve ends broad or narrow, but not elongate and beaklike _ _ _ _ _ _ _ 2
    2. Both longitudinal and transverse striae around 17–19/10 μm _ _ _ 3
    2. The two striae systems having different densities _ _ _ _ _ _ _ _ _ _ _ 4
3. Central area more or less diagonally oriented _ _ _ _ _ _ _ 6. *G. sciotense*
3. Central area with its long axis along the long axis of the valve _ _ _ _ _ _ _
   _ 1. *G. acuminatum*
    4. Both striae systems easily resolved _ _ _ _ _ _ _ _ _ _ _ _ _ _ _ _ _ _ _ 5
    4. Longitudinal striae typically not resolvable; transverse striae faint, 26–30/10 μm; raphe distinctly sigmoid, but valve only moderately sigmoid _ 4. *G. obscurum*
5. Large to very large cells, 150 μm long or more; striae of both systems under 17/10 μm _ 2. *G. attenuatum*
5. Cells under 100 μm long; striae of both systems more than 20/10 μm _
   _ 5. *G. scalproides*

1. **Gyrosigma acuminatum** (Kuetz.) Rabh. var. **acuminatum**
   Fig. 30c.

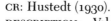

CR: Hustedt (1930).

DESCRIPTION: Valves distinctly sigmoid-lanceolate; valve ends neither long attenuate nor distinctive; raphe lying along valve midline; raphe straight, not undulate, as it enters the central area; proximal raphe ends hooked in opposite directions.

DIMENSIONS: Length, 100–200 μm; width, 15–20 μm; striae (both systems), about 18/10 μm.

2. **Gyrosigma attenuatum** (Kütz.) Rabh. var. **attenuatum** *Fig. 29b.*

CR: Patrick and Reimer (1966).

DESCRIPTION: Valves moderately sigmoid-lanceolate, large and very robust in construction; valve surface arched so that margins usually not completely in focus with points near the long axis of the valve; valves tapering evenly toward the ends without sudden constriction; raphe sigmoid but not undulate; proximal raphe fissures curved in opposite directions; central area rather large.

DIMENSIONS: Length, 150–250 μm; width, 23–27 μm; transverse striae, 14–16/10 μm; longitudinal striae, 10–12/10 μm.

3. **Gyrosigma macrum** (W. Sm.) Griff. & Henfr. var. **macrum**
   Fig. 30a.

CR: Patrick and Reimer (1966).

DESCRIPTION: Long axis fairly straight within the middle parts of the valve, but curving sharply in the long, very narrow ends; middle part of valve linear-lanceolate, narrowing fairly suddenly to form the elongate, beaklike ends, axial area fairly narrow; central area little wider than the axial area; striae barely visible except under strongly obliqued light. This taxon is unique in shape among species to be found here. The present specimens are significantly shorter than specimens previously reported for the United States.

DIMENSIONS: Length, 140–270 μm; width, 10–13 μm; transverse striae, 26–28/10 μm; longitudinal striae, 30–33/10 μm.

4. Gyrosigma obscurum (W. Sm.) Griff. & Henfr. var. **obscurum** *Fig. 29c, d.*

CR: Patrick and Reimer (1966).

DESCRIPTION: Valves weakly sigmoid, linear-lanceolate, with acute ends; axial and central areas narrow; raphe highly sigmoid; proximal raphe ends curved in opposite directions; longitudinal striae not visible; transverse striae fairly distinct (moderately oblique light may be required). The transverse striae appear somewhat undulate due to curvature of the valve surface.

DIMENSIONS: Length, 85–150 μm; width, 10–15 μm; transverse striae, 26–30/10 μm.

5. Gyrosigma scalproides (Rabh.) Cl. var. **scalproides** *Fig. 30b.*

CR: Hustedt (1930).

DESCRIPTION: Valve outline slightly sigmoid and tapered gradually to rather blunt ends; raphe lying on longitudinal midline of valve; proximal raphe fissures straight ("T-shaped"); distal raphe fissures abruptly hooked at the ends, meeting the margins in a subterminal position; both striae systems distinct.

DIMENSIONS: Length, 25–70 μm; width, 5.5–10 μm; transverse striae, 22–24/10 μm; longitudinal striae, 28–31/10 μm.

6. Gyrosigma sciotense (Sulliv. & Wormley) Cl. var. **sciotense**
Fig. 29e.

CR: Patrick and Reimer (1966).

DESCRIPTION: Valves moderately sigmoid elliptic-lanceolate with rounded ends; raphe lying more or less on the longitudinal midline of the valve; raphe branches slightly undulate before the central area, proximal raphe fissures curved in opposite directions; axial area narrow; central area somewhat diagonal; both striae systems distinct.

DIMENSIONS: Length, 100–160 μm; width, 15–18 μm; transverse striae, 16–17/10 μm; longitudinal striae, 17–19/10 μm.

21. Pleurosigma William Smith 1852, nom. cons.

Pleurosigma belongs to order Naviculales and, like other genera in the order, is characterized by the presence of a true raphe on both valves of each cell. The transverse axis of the valve is straight, but the longitudinal axis is sigmoid (S-shaped). The raphe is also sigmoid and lies more or less on the sigmoid longitudinal midline of the valve. As with *Gyrosigma*, there are two systems of striae. Unlike with *Gyrosigma* (in which the systems are perpendicular to each other), the striae systems of *Pleurosigma* intersect diagonally.

Most species of this genus occur only in salt or brackish water, but a few can be found in freshwater streams.

I found only the following species:

1. Pleurosigma delicatulum W. Sm. var. **delicatulum** *Fig. 29a.*

CR: Patrick and Reimer (1966).

DESCRIPTION: Valves moderately sigmoid and rather narrowly lanceolate; valve ends acute; both striae systems distinct; axial area narrow; central area small.

DIMENSIONS: Length, 130–280 μm; width, 13–19 μm; striae (both systems), 20–23/10 μm.

22. *Stauroneis* Ehrenberg 1843

Stauroneis belongs to order Naviculales of class Pennatibacillario-phyceae. Like other genera in the order, both valves bear a two-branched raphe which does not form part of a keel-type or canal-type raphe structure. The valve outline is symmetrical to both the longitudinal and transverse axes and the valve markings are bilaterally symmetric. Striae are punctate (though this can in some cases be seen only with strongly obliqued light and in others only with the electron microscope.)

The distinguishing character of this genus is the transversely widened central nodule or stauros, which is typically also accompanied by a transversely widened central area. The stauros is best seen under low-power lenses with greater depth of focus than the high-oil lens. Such observation will help distinguish between species of *Stauroneis* and those of other genera that have the widened central area but not the widened central nodule.

KEY TO Stauroneis CONFIRMED IN ILLINOIS

1. Valves with pseudosepta visible near the ends of the valves _____ 2
1. Valves lacking pseudosepta _____ 6
   2. Striae distinctly and usually rather coarsely punctate _____ 3
   2. Striae typically not resolvable into puncta with the light microscope _____ 5
     3. Striae around 16/10 μm near midvalve __8. *S. obtusa* var. *catarinensis*
     3. Striae around 20/10 μm at midvalve _____ 4
       4. Ends rounded, not subrostrate or subcapitate _____
       _____ 4. *S. borrichii* var. *borrichii*
       4. Ends subrostrate to subcapitate _ 5. *S. borrichii* var. *subcapitata*
     5. Valve margins triundulate, the central undulation wider than the others _____ 11. *S. smithii* var. *smithii*
     5. Valves lanceolate, the margins weakly concave at midvalve _____
     _____ 12. *S. smithii* var. *incisa*
      6. Valve ends distinctly capitate _____ 7
      6. Valve ends not capitate _____ 13
     7. Valves lanceolate; striae 26/10 μm or more; length usually over 40 μm
     _____ 2. *S. anceps* f. *gracilis*
     7. Valves linear to linear-elliptic _____ 8
      8. Valve margins straight at midvalve _____ 9
      8. Valve margins convex at midvalve _____ 11

9. Striae distinctly radiate at midvalve, central area therefore butterfly-shaped _____ 10
9. Striae nearly parallel at midvalve and very faint, around 30/10 μm ___
_____ 6. *S. kriegeri* var. *kriegeri*
10. Tiny valves under 5 μm in width _____
_____ 13. *S. thermicola* var. *thermicola*
10. Valves usually wider than 8 μm _____ 3. *S. anceps* f. *linearis*
11. Striae radiate at midvalve, central area therefore butterfly-shaped; valves usually wider than 8 μm _____ 1. *S. anceps* var. *anceps*
11. Striae nearly parallel at midvalve or if slightly radiate then valves narrower than 7 μm _____ 12
12. Striae around 30/10 μm and parallel at midvalve _____
_____ 6. *S. kriegeri* var. *kriegeri*
12. Striae around 24/10 μm and radiate at midvalve _____
_____ 13. *S. thermicola* var. *thermicola*
13. Valve margins slightly undulate just before the ends; striae irregularly punctate near midvalve _____ 7. *S. nobilis* var. *baconiana*
13. Valve margins not undulate near the ends; striae regularly punctate near midvalve _____ 14
14. Large to very large valves with striae usually less than 17/10 μm _
_____ 9. *S. phoenicenteron* var. *phoenicenteron*
14. Large valves with striae 17–20/10 μm _____
_____ 10. *S. phoenicenteron* f. *gracilis*

1. Stauroneis anceps Ehr. var. **anceps** *Fig. 25c*.

CR: Patrick and Reimer (1966).

DESCRIPTION: Valve outline elliptic-lanceolate; ends rostrate to rostrate-capitate; valve margins moderately to strongly convex; axial area narrow; central area a butterfly-shaped fascia; raphe lateral.

DIMENSIONS: Length, 24–75 μm; width, 9–15 μm; striae, 20–30/10 μm; puncta, fine.

2. Stauroneis anceps f. gracilis Rabh. *Fig. 25b.*

CR: Patrick and Reimer (1966).

DESCRIPTION: Valves lanceolate with somewhat protracted, subcapitate ends, striae radiate and finely punctate; central area a rectangular fascia; raphe narrowly bandlike to filiform; striae usually finer than those of var. *anceps* and puncta less distinct.

DIMENSIONS: Length, 40–55 μm; width, 8–10 μm; striae, 26–30/10 μm.

3. Stauroneis anceps f. linearis (Ehr.) Hust. *Fig. 25f.*

CR: Patrick and Reimer (1966).

DESCRIPTION: Similar to var. *anceps* in every particular except shape; valves with nearly straight margins in the center part of the valve instead of the convex margins of var. *anceps*. I found only a very few specimens that correspond to this description. I think that it is possible that this represents nothing more than the extreme of the normal shape variation of the species, but I have chosen to retain the name until a convincing series of intergrade forms can be demonstrated.

DIMENSIONS: As for var. *anceps*.

4. Stauroneis borrichii (Peters.) Lund var. **borrichii** *Fig. 26f.*

CR: Hustedt (1959).

DESCRIPTION: Valves linear-elliptic, the shape modified by the slight concavity of both valve margins at the central area; a pseudoseptum evident at both ends of the valve; axial area narrow; central area a butterfly-shaped fascia. Striae radiate and indistinctly punctate throughout. This taxon is similar to *S. obtusa* var. *catarinensis* but has finer striae near midvalve.

DIMENSIONS: Length, 10–23 μm; width, 3–4.5 μm; striae, 19–25/10 μm.

5. **Stauroneis borrichii** var. **subcapitata** (Peters.) Lund *Fig. 26g*.

CR: Lund (1946).

DESCRIPTION: Differs from the nominate variety mainly in the shape of the ends, which are distinctly rostrate to rostrate capitate. In addition my specimens commonly lacked the concavity of margins at the central area that is typical of v. *borrichii*.

DIMENSIONS: See var. *borrichii*.

6. **Stauroneis kriegeri** Patr. var. **kriegeri** *Fig. 25e*.

CR: Patrick and Reimer (1966).

DESCRIPTION: Valve margins weakly convex; ends narrow and distinctly capitate; axial area narrow; central area rectangular and reaching both valve margins; striae nearly parallel at midvalve, becoming radiate at the ends. I could not resolve the striae into puncta, though Patrick and Reimer report the striae as finely punctate.

DIMENSIONS: Length, 20–23 μm; width, 4–6 μm; striae, 26–28/10 μm.

7. **Stauroneis nobilis** var. **baconiana** (Stodd.) Reim. *Fig. 25a*.

CR: Patrick and Reimer (1966).

DESCRIPTION: Valves lanceolate; valve margins slightly undulate immediately before the ends; axial area narrow; central area a butterfly-shaped fascia; striae radiate and punctate throughout; striae closest to the central area are irregularly punctate; raphe bandlike.

DIMENSIONS: Length, 96–115 μm; width, 13–22 μm; striae, 16/10 μm near center, becoming 20/10 μm at the ends.

8. Stauroneis obtusa var. **catarinensis** Krasske *Fig. 26a, b.*

CR: Hustedt (1959).

DESCRIPTION: Valves linear-elliptic to elliptic-lanceolate with broadly rounded ends; valve margins of shorter valves somewhat concave at the central area; a short pseudoseptum visible at both ends of the valve; axial area narrow; central area and stauros broadly butterfly-shaped; striae radiate and punctate throughout. This taxon differs from *S. borrichii* in having coarser striae near midvalve.

DIMENSIONS: Length, 21–36 μm; width, 4.4–5 μm; striae, 16/10 μm center becoming 24/10 μm at ends.

9. Stauroneis phoenicenteron (Nitz.) Ehr. var. **phoenicenteron** *Fig. 24a.*

CR: Patrick and Reimer (1966).

DESCRIPTION: Valves lanceolate; central area a butterfly-shaped fascia; striae radiate and clearly punctate throughout; raphe broad and bandlike; pseudosepta always absent; cells usually large, often very large, and almost always heavily silicified.

DIMENSIONS: Length, 70–380 μm; width, 16–53 μm; striae, 12–17/10 μm; puncta, similar in number to the striae.

10. Stauroneis phoenicenteron f. **gracilis** (Ehr.) Hust. *Fig. 24b.*

CR: Patrick and Reimer (1966).

DESCRIPTION: Valves lanceolate to elliptic-lanceolate in outline; ends rounded and very slightly drawn out; striae radiate and punctate throughout; central area a butterfly-shaped fascia; raphe bandlike. This taxon is usually smaller than var. *phoenicenteron* and is also less strongly silicified and has finer striae.

DIMENSIONS: Length, 80–160 μm; width, 16–21 μm; striae, 17–20/10 μm.

11. Stauroneis smithii Grun. var. **smithii** *Fig. 26d*.

CR: Patrick and Reimer (1966).

DESCRIPTION: Overall valve outline lanceolate, modified by constrictions into a triundulate-lanceolate form with slightly protracted, rostrate ends; a pseudoseptum evident at both ends of the valve; axial area moderately narrow, flaring somewhat just before the central area; central area a weakly butterfly-shaped fascia. The striated parts of the valve are unusual in appearance, the spaces between the striae different in contrast from the axial and central areas. The striae are quite faint though not exceptionally fine. Oblique light is usually required for observation of striae.

DIMENSIONS: Length, 14–40 μm; width, 4–9 μm; striae, 26–30/10 μm.

12. Stauroneis smithii var. **incisa** Pant. *Fig. 26e*.

CR: Patrick and Reimer (1966).

DESCRIPTION: Valves narrowly lanceolate; valve margins slightly concave at midvalve; ends narrowly subrostrate. Pseudosepta present at both ends of the valve. The appearance of the striated areas is similar to var. *smithii*.

DIMENSIONS: Length, 18–25 μm; width, 4–5.5 μm; striae, 32–33/10 μm.

13. Stauroneis thermicola (Peters.) Lund var. **thermicola** *Fig. 25d*.

CR: Hustedt (1959).

DESCRIPTION: Valves narrow with weakly convex margins and distinctly capitate to rostrate-capitate ends; axial area narrow; central area a butterfly-shaped fascia; striae radiate throughout and not visibly punctate under the light microscope. This tiny *Stauroneis* is easy to confuse with some small *Naviculae*. When in doubt, it is best to observe the specimen with a high-power dry objective. The *Navicula* species show a dotlike central nodule while, if the specimen is a *Stauroneis*, you will see that the central nodule is broadened into a stauros. It is easier to see the contrast between thick and thin areas of the valve using a dry objective.

DIMENSIONS: Length, 8–17 μm; width, 3–4 μm; striae, about 24/ 10 μm.

23. *Capartogramma* Kufferath 1956

Capartogramma belongs to the order Naviculales of class Pennatibacillariophyceae. It is closely related to *Stauroneis* (some workers still recognize it as part of *Stauroneis*). The primary difference is in the nature of the stauros, which is X-shaped in *Capartogramma* but a solid butterfly- or rectangular-shaped structure in *Stauroneis*.

1. **Capartogramma crucicula** (Grun. ex Cl.) Ross var. **crucicula**
   Fig. 26c.

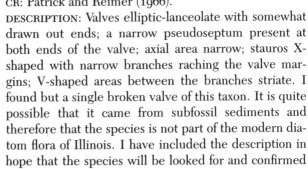

CR: Patrick and Reimer (1966).

DESCRIPTION: Valves elliptic-lanceolate with somewhat drawn out ends; a narrow pseudoseptum present at both ends of the valve; axial area narrow; stauros X-shaped with narrow branches raching the valve margins; V-shaped areas between the branches striate. I found but a single broken valve of this taxon. It is quite possible that it came from subfossil sediments and therefore that the species is not part of the modern diatom flora of Illinois. I have included the description in hope that the species will be looked for and confirmed in the living condition.

DIMENSIONS: Length, 20–36 μm; width, 7.5–9 μm; striae, 24/10 μm.

24. *Anomoeoneis* Pfitzer 1871

Anomoeoneis belongs to the order Naviculales of Class Pennatibacillariophyceae. Like the other genera in the order, both valves of each cell bear a two-branched raphe that is not part of a keel-type or canal-type raphe structure. The valve outline is symmetric to both the longitudinal and transverse axes. Valve markings consist of striae that are distinctly punctate, the puncta so arranged that the spaces between them form longitudinal clear spaces. In some species at least, these longitudinal spaces are said to be thickened costae. *Anomoeoneis* has a distinctive overall appearance that is unfortunately difficult to describe. You will soon learn to recognize the few species found in Illinois.

KEY TO Anomoeoneis CONFIRMED FOR ILLINOIS

1. Striae around 30/10 μm; longitudinal clear spaces difficult to resolve in some cases _ 3. *A. vitrea* var. *vitrea*
1. Striae fewer than 20/10 μm _ 2
   2. Puncta becoming more widely and irregularly spaced from the valve margins toward the axial area _ _ _ 2. *A. sphaerophora* var. *sculpta*
   2. Puncta more or less regularly spaced along the striae _ _ _ _ _ _ _ _ _ _ _
_ 1. *A. sphaerophora* var. *sphaerophora*

1. **Anomoeoneis sphaerophora** (Ehr.) Pfitz. var. **sphaerophora**
   Fig. 23b.

CR: Patrick and Reimer (1966).

DESCRIPTION: Valves elliptic-lanceolate with somewhat protracted, fairly narrow capitate or subcapitate ends; central area asymmetric, one side widening gradually and reaching the margin, the other side not widening appreciably and not reaching the margin; transverse striae radiate and punctate, the puncta so placed as to form, in addition, undulate longitudinal striae.

DIMENSIONS: Length, 30–80 μm; width, 13–22 μm; transverse striae, 15–17/10 μm.

2. **Anomoeoneis sphaerophora** var. **sculpta** O. Muell. *Fig. 23a.*

CR: Patrick and Reimer (1966).

DESCRIPTION: Valves broadly lanceolate with some-what elongate, rather narrowly rostrate ends; trans-verse striae radiate, the puncta closely spaced near the valve margins but becoming progressively more widely and irregularly spaced toward the axial area. The puncta are so few finally that there is a fairly broad space that is almost devoid of puncta. Along both sides of the axial area, close to it, and interrupted only by the central area, are single longitudinal rows of closely spaced puncta.

DIMENSIONS: Length, 65–200 μm; width, 25–36 μm; transverse striae, 11–16/10 μm.

3. **Anomoeoneis vitrea** (Grun.) Ross var. **vitrea** *Fig. 23c.*

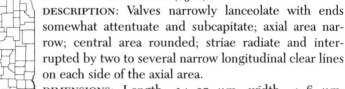

CR: Patrick and Reimer (1966).

DESCRIPTION: Valves narrowly lanceolate with ends somewhat attentuate and subcapitate; axial area narrow; central area rounded; striae radiate and interrupted by two to several narrow longitudinal clear lines on each side of the axial area.

DIMENSIONS: Length, 14–35 μm; width, 4–6 μm; striae, 30–35/10 μm.

25. *Neidium* Pfitzer 1871

Neidium belongs to the order Naviculales of class Pennatibacillariophyceae. Indeed, until comparatively recently, the members of the genus were named as species of *Navicula*.

Cells of *Neidium* are symmetric to all axes. Cells must be seen in valve view for identification to the species level, although it is often possible to infer the identity of cells in girdle view based on the identity of specimens in the same sample that are seen in valve view.

Valve markings consist of striae that usually, though not always, are distinctly punctate when seen with the light microscope. The striae are crossed by at least one longitudinal band on each side of the valve. This primary longitudinal band is typically located near the valve margin. The band appears different in thickness and structure from adjacent areas of the valve. The proximal raphe fissures are of importance in taxonomy. Most species have proximal raphe fissures directed to opposite sides of the valve. Two species found so far in Illinois have "straight" fissures. In most cases it is fairly easy to decide which type is present, but it is easiest to learn the two "straight" fissured species by sight. Both are very distinctive in appearance and are also comparatively rare. The distal raphe fissures, when seen, usually are V-shaped. One last feature that you may notice is the *défaut régulier,* a "notch" in the striation near both ends of the valve in some species. The *défaut* is formed by the shortening of a single stria. To date this character has not been used in defining species.

Neidium affine (and its varieties) is the most commonly encountered species in Illinois. The large-celled species *N. iridis* (which is commonly reported in the Midwest) was not encountered in the present study, its place being taken apparently by the less well known *N. decens*. Reimer (personal communication) notes that the primary longitudinal band of *N. iridis* is massive and very distinct, while that of the fairly similar *N. decens* is weakly expressed. For further information on *Neidium*, see Patrick and Reimer (1966).

KEY TO Neidium CONFIRMED FOR ILLINOIS

1. Proximal raphe ends straight _____ 2
1. Proximal raphe ends curved in opposite directions _____ 3
    2. Valve margins concave at midvalve; striae not easily resolved into puncta _____ 7. *N. binode* var. *binode*
    2. Valve margins convex at midvalve; striae distinctly punctate ____ _____ 10. *N. dubium* var. *dubium*
3. Valves with narrowly to broadly rostrate ends or capitate ends ____ 4
3. Valve ends either not distinctive or at least not distinctly rostrate or capitate _____ 14
    4. Valve ends capitate or subcapitate _____ 5
    4. Valve ends rostrate or substrate _____ 7
5. Valves with weakly to strongly triundulate margins; length usually over 50 μm and width over 14 μm _____ 6. *N. affine* var. *undulatum*
5. Valve margins convex or straight and if straight, then length under 50 μm and width under 14 μm _____ 6
    6. Valves with more or less straight margins at midvalve; width generally under 9 μm _____ 4. *N. affine* var. *longiceps*
    6. Valves with more or less convex margins at midvalve; width generally 9–12 μm _____ 2. *N. affine* var. *amphirhynchus*
7. Valve ends moderately narrowly rostrate _____ 8
7. Valve ends broadly substrate to broadly rostrate _____ 11
    8. Valve margins straight to very weakly convex, the outline therefore appearing linear _____ 4. *N. affine* var. *longiceps*
    8. Valve margins slightly to moderately convex at midvalve, the valves therefore linear-elliptic with ends distinctly narrower than the maximum valve width _____ 9
9. Valve width 9–12 μm _____ 2. *N. affine* var. *amphirhynchus*
9. Valve width 14 μm or more _____ 10
    10. Striae 14–18/10 μm _____ 9. *N. decens* var. *decens*
    10. Striae 20–22/10 μm _____ 3. *N. affine* var. *humerus*
11. Valves linear, the ends set off by very slight constrictions _____ _____ 11. *N. hankensis* var. *hankensis*

11. Valves linear to elliptic-lanceolate, the ends broadly rostrate or cuneate
   subrostrate _____ 12
   12. Ends cuneate, sometimes slightly protracted _____
      _____ 5. *N. affine* var. *tenuirostris*
   12. Ends broadly rostrate _____ 13
13. Striae 26–30/10 μm _____ 8. *N. bisulcatum* var. *nipponicum*
13. Striae 22–24/10 μm _____ 1. *N. affine* var. *affine*
   14. Valves lanceolate; striae diagonal throughout, even at the ends __
      _____ 12. *N. hercynicum* var. *hercynicum*
   14. Valves linear to linear-lanceolate; striae diagonal, becoming con-
      vergent at one or both ends (Look carefully, especially with small
      cells!), or parallel, but not diagonal throughout _____ 15
15. Valves linear with broadly rounded ends _____
   _____ 11. *N. hankensis* var. *hankensis*
15. Valves with cuneate ends; width 14 μm or more _____
   _____ 5. *N. affine* var. *tenuirostris*
15. Valves with slightly protracted ends; width under 6 μm _____
   _____ 8. *N. bisulcatum* var. *nipponicum*

1. **Neidium affine** (Ehr.) Pfitz. var. **affine** *Fig. 19c, d.*

CR: Patrick and Reimer (1966).

DESCRIPTION: Valves linear with weakly convex margins and broadly rostrate ends; axial area moderately
narrow; central area transversely elliptic to almost rectangular; proximal raphe ends curved in opposite directions; striae parallel to oblique (in smaller specimens
strongly oblique) throughout most of the valve, becoming convergent in part at one or both ends; primary longitudinal band at or near margins. This species is highly
variable in outline and many varieties have been described. It is possible that some or all of these may be
part of the normal variation of the var. *affine,* but this has not been
proved. The smallest specimens I have seen are considerably
smaller than previously reported. These pose some problem in
identification since the proximal raphe fissures are weakly expressed and the striae are strongly oblique, which may lead one into
misidentifying the specimens as *N. hercynicum* f. *subrostratum*
Wallace. But note that in this latter taxon, the striae are oblique
throughout.

DIMENSIONS: Length, 24–65 μm; width, 7.5–13 μm; striae, 22–24/
10 μm (mine reached 26/10 μm in the shortest specimens); puncta,
around 24/10 μm.

2. Neidium affine var. amphirhynchus (Ehr.) Cl. *Fig. 20b, e.*

CR: Patrick and Reimer (1966).

DESCRIPTION: Valve margins typically convex, sometimes nearly straight at midvalve; valve ends typically somewhat protracted rostrate, more rarely the ends subcapitate to capitate (*Fig. 20e*) and the cells approaching var. longiceps in shape; central area rounded and somewhat expanded transversely; proximal raphe fissures curved in opposite directions; striae weakly radiate in most of the valve, becoming convergent at the ends; puncta distinct; primary band submarginal.

DIMENSIONS: Length, 30–45 μm; width, 9–12 μm; striae, 22–24/10 μm.

3. Neidium affine var. humerus Reim. *Fig. 20a.*

CR: Patrick and Reimer (1966).

DESCRIPTION: Valves linear-elliptic with weakly convex margins; margins narrowing rather abruptly to produce moderately narrow rostrate-cuneate ends; axial area narrow; central area rounded to rounded-rhombic; proximal raphe fissures curved in opposite directions; ghost-striae present or absent in central area. Keep in mind that there may be intergrade forms between the present taxon and some other varieties of this species.

DIMENSIONS: Length, 41–78 μm; width, 14–22 μm; striae and puncta, around 20–22/10 μm.

4. Neidium affine var. longiceps (Greg.) Cl. *Fig. 20d.*

CR: Patrick and Reimer (1966).

DESCRIPTION: Similar to var. *amphirhynchus*, but with margins nearly always straight and ends that may be subcapitate as well as rostrate; cells usually narrower for a given length than with var. *amphirhynchus*.

DIMENSIONS: Length, 25–38 μm; width, 6–9 μm; striae, 24–26/10 μm; puncta, 24–26/10 μm.

5. Neidium affine var. **tenuirostris** A. Mayer *Fig. 20c.*

CR: Stoermer (1963).

DESCRIPTION: Valve margins weakly convex; ends cuneate to somewhat protracted-cuneate; axial area narrow; central area roughly transverse-elliptic; proximal raphe fissures curved in opposite directions; striae weakly radiate, becoming weakly convergent at the ends; primary band marginal to submarginal.

DIMENSIONS: Length, 36–55 μm; width, 14–15 μm; striae, 20–22/10 μm.

6. Neidium affine var. **undulatum** (Grun.) Cl. *Fig. 19a, b.*

CR: Patrick and Reimer (1966).

DESCRIPTION: Valves linear with rostrate to capitate ends; valve margins quite variable, in most specimens distinctly triundulate but the degree of undulation varying from this to a point where one would be inclined to call the margins straight. This latter condition is typical of the Illinois specimens. Sometimes it is necessary to focus carefully to detect the undulations. The only other large *Neidium* similar in shape to the linear specimens of the present taxon is *N. productum* (W. Sm.) Cl., which is typically 20 μm wide or more.

DIMENSIONS: Length, 55–80 μm; width, 14–17 μm; striae, 20–22/10 μm; puncta, around 20/10 μm.

7. Neidium binode (Ehr.) Hust. var. **binode** *Fig. 22e.*

CR: Patrick and Reimer (1966).

DESCRIPTION: Valve with narrowly rostrate ends and with margins transapically constricted at midvalve; axial area narrow; central area rounded; proximal raphe fissures straight; striae radiate and not visibly punctate.

DIMENSIONS: Length, 15–30 μm; width, 6–9 μm; striae, 22–24/10 μm near center, becoming 26–28/10 μm at the ends.

8. Neidium bisulcatum var. nipponicum Skv. *Fig. 22b.*

CR: Patrick and Reimer (1966).

DESCRIPTION: Valves linear with broadly subrostrate to rostrate ends (Patrick and Reimer note that cuneate ends are more typical); striae weakly radiate near center, becoming convergent at the ends; axial area narrow; central area rounded; proximal raphe fissures curved in opposite directions.

DIMENSIONS: Length, 18–26 μm; (mine was 30 μm); width, 5–6 μm; striae, 26–30/10 μm; puncta, 24–28/10 μm.

9. Neidium decens (Pant.) Stoermer var. **decens** *Fig. 21a, b.*

CR: Stoermer (1963).

DESCRIPTION: Valves robust; outline broadly elliptic-lanceolate; margins weakly convex; ends broadly to narrowly rostrate. This taxon is similar in appearance to some varieties of *N. iridis* (Ehr.) Cl., but differs from all members of that species in having the primary longitudinal bands weakly expressed, that is, though they are visible they do not stand out strongly. It is my feeling that there is considerable shape-polymorphism in *Neidium* (compare, for example, *N. affine*) and, though the original description reports the valve ends as broadly rostrate, I have seen and illustrated also a narrowly rostrate form. I would prefer not to name it as a forma or variety but to consider it as part of the natural variation of var. *decens*.

DIMENSIONS: Length, 56–98 μm; width, 17–22 μm; striae, 14–18/10 μm; puncta, around 16/10 μm.

10. Neidium dubium (Ehr.) Cl. var. **dubium** *Fig. 22c.*

CR: Patrick and Reimer (1966).

DESCRIPTION: Valves with weakly convex margins at midvalve; valves narrowing suddenly to form short rostrate ends; axial area moderately narrow; central area irregularly rounded; proximal raphe fissures straight; striae nearly parallel and finely punctate; primary band submarginal.

DIMENSIONS: Length, 30–50 μm; width, 10–16 μm; striae and puncta, around 18–20/10 μm.

11. Neidium hankensis Skv. var. **hankensis** *Fig. 22a.*

CR: Skvortzow (1928); Stoermer (1963).

DESCRIPTION: Valves linear with broadly rounded to subcuneate ends; central area transverse-elliptic; axial area moderate in width; proximal raphe fissures curved in opposite directions; striae weakly radiate, becoming convergent at the ends. I cannot see the justification for separation of the variety *elongatum* Skv., as has been done by some. The original descriptions for both varieties are hard to interpret. Stoermer maintains that var. *elongatum* is coarser than var. *hankensis*, but does not cite any numerical data to prove it.

DIMENSIONS: Length, 32–60 μm; width, around 10 μm; striae, 19–22/10 μm; puncta, up to 24/10 μm.

12. Neidium hercynicum A. Mayer var. **hercynicum** *Fig. 22d.*

CR: Patrick and Reimer (1966).

DESCRIPTION: Valves elliptic-lanceolate in outline; axial area narrow; central area almost rhombic; proximal raphe ends curved in opposite directions; striae oblique throughout; puncta visible, though fine; longitudinal band marginal.

DIMENSIONS: Length, 25–50 μm; width, 5–8 μm; striae, 26–28/10 μm.

26. *Diploneis* Ehrenberg 1844

Diploneis belongs to the order Naviculales of class Pennatibacillariophyceae. As with the other genera in the order, both valves of each cell bear a two-branched raphe. The raphe branches are enclosed in longitudinal silica ribs. The valve surface is depressed along both sides of the axis to form longitudinal furrows (sometimes called canals). The striae are formed of alveoli and alternate with thickened costae. All species in Illinois have elliptic to linear-elliptic valve outlines.

There are species of *Navicula* that have a similar appearance to *Diploneis*, but it is usually fairly simple to separate them on the basis of species characters.

KEY TO SPECIES OF Diploneis CONFIRMED FOR ILLINOIS

1. Longitudinal furrows narrow, curving outward at midvalve around the central area; alveoli easily resolved _ 1. *D. oblongella* var. *oblongella*
1. Longitudinal furrows typically not curving outward around the central area; alveoli resolvable only with difficulty _ _ _ _ _ _ _ _ _ _ _ _ _ _ _ _ _ _ _ 2
   2. Striae more or less parallel at midvalve, valve margins more or less straight at midvalve _ _ _ _ _ _ _ _ _ _ _ _ _ _ 2. *D. oculata* var. *oculata*
   2. Striae slightly radiate throughout; valve margins more or less convex at midvalve _ _ _ _ _ _ _ _ _ _ _ _ _ 3. *D. peterseni* Hust. var. *peterseni*

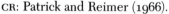

1. **Diploneis oblongella** (Näge. ex Kütz.) Ross var. **oblongella** *Fig. 18a, b.*

CR: Patrick and Reimer (1966).

DESCRIPTION: Valves elliptic to broadly elliptic in outline; larger valves appear to be heavily silicified and have distinctly arched valve faces, while smaller valves often appear flatter and less heavily silicified; striae radiate throughout; striae in larger valves may appear rough, but are always composed of a single row of alveoli, never a double row; costae of varying degrees of thickness alternate with striae; central area usually wider than the adjacent axial area. Hustedt (1930), among others, recognizes the separation of *D. oblong-ella* from *D. ovalis* (Hilse) Cl. but Patrick and Reimer do not.

DIMENSIONS: Length: 10–100 μm; width, 6–35 μm; striae, 10–19/10 μm; alveoli, 13–20/10 μm.

2. **Diploneis oculata** (Breb.) Cl. var. **oculata** *Fig. 18c.*

CR: Patrick and Reimer (1966).

DESCRIPTION: valves linear-elliptic with broadly rounded ends; valves lightly silicified and oblique light may be needed to resolve the striae; striae more or less parallel at midvalve; alveoli difficult or impossible to resolve with the light microscope; longitudinal furrows straight, not curving outward around the central area.

DIMENSIONS: Length, 10–20 μm; width, 6–8 μm (mine were as narrow as 5 μm); striae, 20–28/10 μm.

3. Diploneis peterseni Hust. var. **peterseni** *Fig. 18d.*

CR: Hustedt (1937a).

DESCRIPTION: valves elliptic to elliptic-lanceolate with convex margins at midvalve; striae slightly radiate throughout; alveoli difficult or impossible to resolve with the light microscope; longitudinal furrows more or less narrowly lanceolate, but not curving outward to form a distinct central area.

DIMENSIONS: Length, 15–19 μm (mine were as short as 12 μm); width, 5–6 μm; striae, around 24/10 μm at midvalve, becoming 28–30/10 μm at the ends.

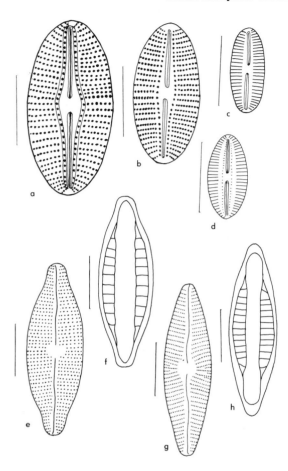

Fig. 18. a, b. *Diploneis oblongella* v. *oblongella.* **c.** *Diploneis oculata* v. *oculata.* **d.** *Diploneis peterseni* v. *peterseni.* **e.** *Mastogloia smithii* v. *lacustris* (valve). **f.** *Mastogloia smithii* v. *lacustris* (septum). **g.** *Mastogloia elliptica* v. *danseii* (valve). **h.** *Mastogloia elliptica* v. *danseii* (septum). (Scale lines equal 10 micrometers.)

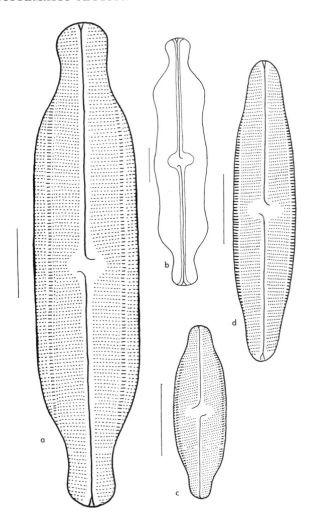

Fig. 19. a. *Neidium affine* v. *undulatum*. **b.** *N. affine* v. *undulatum* (striae not shown). **c, d.** *N. affine* v. *affine*. (Scale lines equal 10 micrometers.)

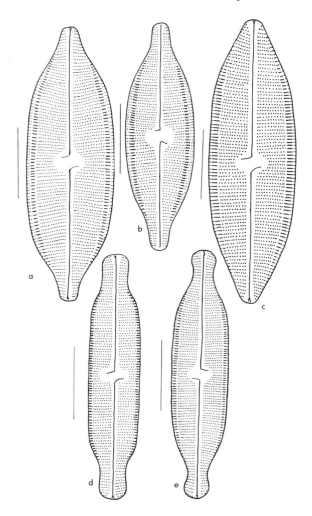

Fig. 20. a. *Neidium affine* v. *humerus*. **b.** *N. affine* v. *amphirhynchus*. **c.** *N. affine* v. *tenuirostris*. **d.** *N. affine* v. *longiceps*. **e.** *N. Affine* v. *amphirhynchus*. (Scale lines equal 10 micrometers.)

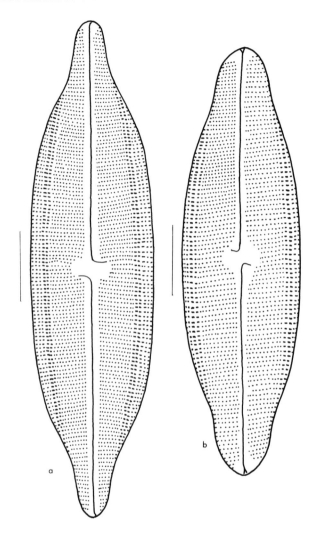

Fig. 21. a, b. *Neidium decens* v. *decens*. (Scale lines equal 10 micrometers.)

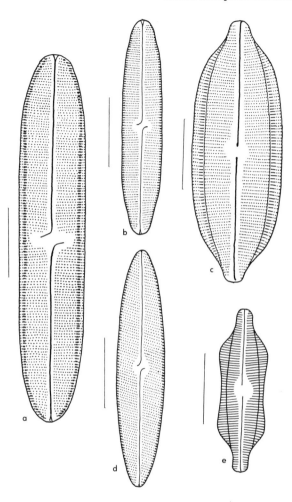

Fig. 22. a. *Neidium hankensis* v. *hankensis*. **b.** *N. bisulcatum* v. *nipponicum*. **c.** *N. dubium* v. *dubium*. **d.** *N. hercynicum* v. *hercynicum*. **e.** *N. binode* v. *binode*. (Scale lines equal 10 micrometers.)

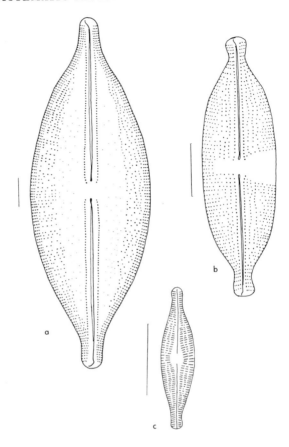

Fig. 23. a. Anomoeoneis sphaerophora v. sculpta. **b.** A. sphaerophora v. sphaerophora. **c.** A. vitrea v. vitrea. (Scale lines equal 10 micrometers.)

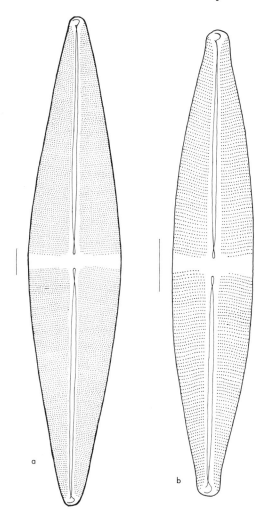

Fig. 24. a. *Stauroneis phoenicenteron* v. *phoenicenteron.* **b.** *S. phoenicenteron* f. *gracilis.* (Scale lines equal 10 micrometers.)

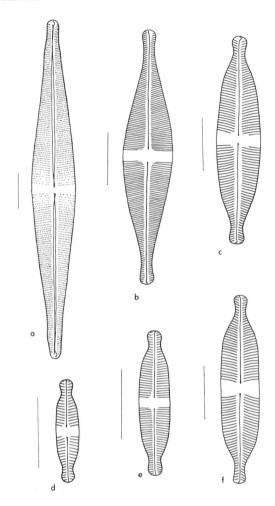

Fig. 25. a. *Stauroneis nobilis* v. *baconiana*. **b.** S. *anceps* f. *gracilis*. **c.** S. *anceps* v. *anceps*. **d.** S. *thermicola* v. *thermicola*. **e.** S. *kriegeri* v. *kriegeri*. **f.** S. *anceps* f. *linearis*. (Scale lines equal 10 micrometers.)

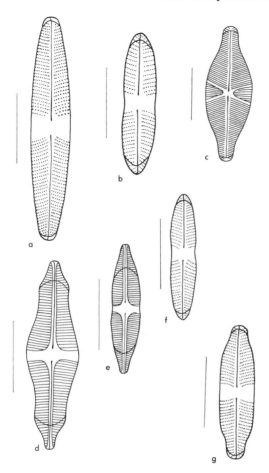

Fig. 26. a, b. *Stauroneis obtusa* v. *catarinensis*. **c.** *Capartogramma crucicula* v. *crucicula*. **d.** *Stauroneis smithii* v. *smithii*. **e.** *Stauroneis smithii* v. *incisa*. **f.** *Stauroneis borrichii* v. *borrichii*. **g.** *Stauroneis borrichii* v. *subcapitata*. (Scale lines equal 10 micrometers.)

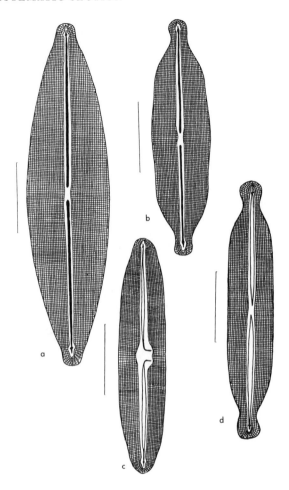

Fig. 27. a. *Frustulia rhomboides* v. saxonica. **b.** *F. rhomboides* v. *saxonica* f. *undulata*. **c.** *F. weinholdii* v. *weinholdii*. **d.** *F. rhomboides* v. *saxonica* f. *capitata*. (Scale lines equal 10 micrometers.)

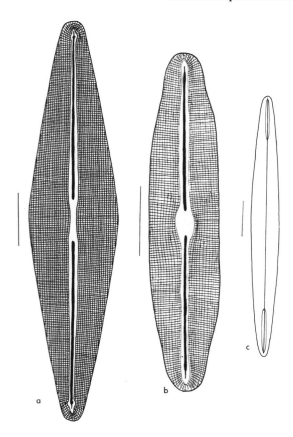

Fig. 28. a. *Frustulia rhomboides* v. *amphipleuroides*. **b.** *Frustulia vulgaris* v. *vulgaris*. **c.** *Amphipleura pellucida* v. *pellucida*. (Scale lines equal 10 micrometers.)

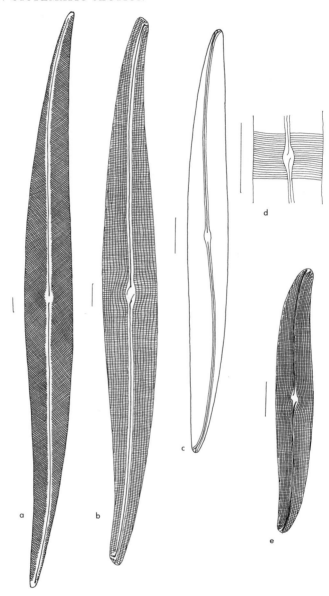

Fig. 29. *a. Pleurosigma delicatulum* v. *delicatulum*. *b. Gyrosigma attenuatum* v. *attenuatum*. *c. Gyrosigma obscurum* v. *obscurum*. *d. Gyrosigma obscurum* v. *obscurum* (detail of central area). *e. Gyrosigma sciotense* v. *sciotense*. (Scale lines equal 10 micrometers.)

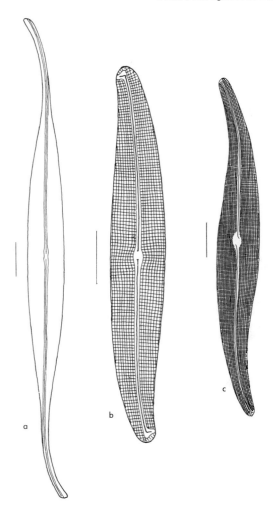

Fig. 30. a. *Gyrosigma macrum* v. *macrum*. **b.** *G. scalproides* v. *scalproides*. **c.** *G. acuminatum* v. *acuminatum*. (Scale lines equal 10 micrometers.)

27. *Navicula* Bory 1824

The genus *Navicula* is the least homogeneous of all the genera of freshwater diatoms. It is almost certainly a completely artificial assemblage of diatoms that have the basic characters of the family Naviculaceae; that is, symmetry of the valves to all axes and a fully developed true raphe on both valves of each cell. This is one of the earliest described genera of diatoms that is still in modern use and dates from a time when little could be seen of the structure of many diatoms beyond such coarse features as shape. Over the years a number of genera have been carved out of the assemblage called *Navicula* based on distinctive characters that have been revealed by better microscopes and preparations. It is not to be doubted that the wider availability of the electron microscopes, both transmission and scanning, will ultimately result in the setting up of still more genera. Friedrich Hustedt was in the process of completing work on this genus for his massive work on European diatoms at the time of his death and had arranged the genus into a number of sections that could in many cases be established as genera in their own right. Hustedt admitted though that some of the sections were set up for convenience and were not intended as natural units. There does not at present appear to be any urgency on the part of diatom taxonomists to carry Hustedt's work on this genus to the logical conclusion of separation into natural genera. This is not surprising, since to do so would properly require the examination of thousands of often very rare species with the electron microscope—perhaps a lifetime's work for more than just one researcher. Currently the resources of those who are in a position to carry out such an effort are directed toward other groups of equal or greater importance, and work on *Navicula* continues to be piecemeal and slow. Thus it remains the case that *Navicula* is defined as the residue of species left when the more distinctive naviculoid genera have been removed. There are few if any positive characters that can be offered as applicable to the genus as a whole beyond symmetry and the presence of a true raphe on both valves, neither of which characters is sufficient to eliminate most other genera from consideration. Instead of saying what *Navicula* is not, I would suggest that you read the descriptions of other naviculoid genera and learn what they are, leaving *Navicula* as what it is, a large and motley residue of often very distinctive species awaiting the definition of distinctive genera.

KEY TO THE TAXA OF Navicula CONFIRMED FOR ILLINOIS

The taxonomic sections of *Navicula* are not particularly useful in making keys to the genus. I have therefore divided the species I found into purely artificial groupings based on the characters observable with good light microscopes. I emphasize this last point because it is frequently necessary to measure striae densities that exceed 30/10 μm. The terms I have used are defined and illustrated in the Glossary, but I call your attention especially to the terms "lineate" and "linelike," which are not synonymous (the first meaning "marked by crosslines" and the second "looking like a line"). I apologize for the use of relative terms such as "coarse" or "fine" that may mean different things to different people. In these cases it may be useful to examine illustrations of species so described. I have tried in any case to be as consistent as possible in the use of relative terms. The significance of "oblique light" or "obliquing" is explained in the Glossary. The word "striae" used without a qualifying term means "transverse striae." A species name not followed by a variety name is the nominate variety.

Let me reemphasize that the "sections" are entirely artificial and have little or no taxonomic significance. In fact, I have entered several species in more than one "section" key if I thought it useful to do so.

KEY TO ARTIFICIAL SECTIONS OF Navicula

1. Transverse striae parallel near midvalve _____ Section I
1. Transverse striae radiate near midvalve or striae not visible _____ 2
   2. Striae visible over most of the valve length (oblique light may be required) _____ 3
   2. Striae not resolvable even with oblique light, or if resolvable, then only at midvalve _____ Section II
3. Striae distinctly punctate _____ Section III
3. Striae linelike, lineate, or indistinctly punctate _____ 4
   4. Striae indistinctly punctate (moderately oblique light may be required) _____ Section IV
   4. Striae lineate (each stria marked with numerous crosslines), or striae linelike (puncta typically not visible even with oblique light) ___ 5
5. Striae linelike _____ Section V
5. Striae lineate _____ Section VI

SECTION I. **Valves with parallel striae near midvalve**

1. Both transverse and longitudinal striae easily resolved _____ 2
1. Transverse striae only can be resolved _____ 3
   2. Valves lanceolate without distinctive ends ____ 22. *N. cuspidata*
   2. Valves elliptic-lanceolate with rostrate to rostrate-capitate ends __
   _____ 23. *N. cuspidata* var. *ambigua*
3. Valve outline linear to linear-elliptic or valve margins distinctly concave at midvalve; valve ends not distinctive _____ 4
3. Valve outline more or less lanceolate; valve ends distinctive or nondistinctive _____ 9
   4. Valve margins concave at midvalve; ends broadly rounded; small valves with striae difficult to resolve even with obliqued light ___
   _____ 20. *N. contenta* var. *biceps*
   4. Valve outline linear to linear-elliptic _____ 5
5. Axial area broad (around 1/3 of width); striae not distinctly punctate _
   _____ 2. *N. americana*
5. Axial area narrow or if fairly broad then striae distinctly punctate __ 6
   6. Striae distinctly punctate or formed of dashlike units _____ 7
   6. Striae linelike, the puncta usually not resolvable _____
   _____ 81. *N subhamulata*
7. Striae punctate; valves linear elliptic __ 53. *N. monmouthiana-stodderi*
7. Striae composed of dashlike units _____ 8
   8. Axial area narrow _____ 43. *N. insociabilis*
   8. Axial area moderately wide, at midvalve almost 1/3 of valve breadth
   _____ 87. *N. tenera*
9. Valves with protracted or distinctive ends _____ 11
9. Valves lanceolate without distinctive ends _____ 10
   10. Striae around 23/10 μm throughout _____ 96. *N. vaucheriae*
   10. Striae 26–28/10 μm at midvalve, becoming 36–40/10 μm at the ends _____ 11. *N. biconica*
11. Valves with somewhat protracted-subrostrate ends; striae 30/10 μm at midvalve, approaching 36/10 μm at the ends _____ 52. *N. miniscula*
11. Valves with distinctly protracted ends or with rostrate to rostrate-capitate ends; striae fewer than 30/10 μm at midvalve _____ 12
   12. Valves lanceolate with protracted ends _____ 13
   12. Valves lanceolate to elliptic-lanceolate with rostrate to subcapitate ends _____ 14
13. Striae 20–22/10 μm _____ 35. *N. halophila* f. *tenuirostris*
13. Striae 24–28/10 μm at midvalve and somewhat finer toward the ends
   _____ 9. *N. belliatula*

14. Striae somewhat more widely spaced at midvalve than at the ends; ends rostrate _____ 1. *N. accomoda*

14. Striae more or less evenly spaced along the valve; ends typically subcapitate _____ 32. *N. gregaria*

SECTION II. **Striae not visible with obliqued light or visible only around the central area**

1. Striae not visible anywhere on the valve even under strongly obliqued light _____ 2

1. Striae visible at least around the central area under strongly obliqued light _____ 5

2. Valve margins triundulate _____ 92. *N. tridentula*

2. Valve margins straight or convex _____ 3

3. Valve ends rostrate to rostrate-capitate _____ 4

3. Valves lanceolate to elliptic-lanceolate without distinctive ends _____ _____ 41. *N. indifferens*

4. Valve margins convex at midvalve _____ 34. *N. gysingensis*

4. Valve margins nearly straight at midvalve _____ 6. *N. arvensis*

5. Striae faintly visible around the central area, but nowhere else, even under strongly obliqued light _____ 84. *N. subtilissima*

5. Striae visible almost all the way to the ends under very strongly obliqued light _____ 80. *N. subarvensis*

SECTION III. **Striae distinctly punctate (resolvable into puncta with little or no obliquing of the light)**

1. Pseudosepta visible near both ends of the valve _____ 2

1. Pseudosepta not present _____ 3

2. Valve ends rostrate; valve margins undulate before the ends _____ _____ 44. *N. integra*

2. Valve ends subcapitate; margins not undulate before the ends ___ _____ 74. *N. sanctaecrucis*

3. Striae composed of an outer doubly-punctate zone and an inner zone of coarse, dashlike units _____ 95. *N. tuscula*

3. Striae not showing two types of punctation _____ 4

4. Puncta arranged in two diagonally intersecting striae systems ___ _____ 62. *N. placenta*

4. Striae transverse only or, if two systems present, then these are transverse and longitudinal _____ 5

5. Central area asymmetric, reaching both margins but longer on one margin than the other; small lanceolate cells _____ 42. *N. ingenua*

5. Central area not reaching the margins and not asymmetric _____ 6
   6. Valves with a broad axial space marked only by a single row of puncta along one or both sides of the raphe _____ 7
   6. Axial space narrow or broad, but not as above _____ 9
7. Longitudinal row of puncta on one side of the raphe only _____ 8
7. Longitudinal row of puncta along both sides of the raphe _____ _____ 7. *N. auriculata*
   8. Puncta rounded; valves lanceolate _____ 18. *N. circumtexta*
   8. Puncta dashlike; valves elliptic to linear-elliptic ___ 87. *N. tenera*
9. Proximal raphe ends sharply deflected to one side of the valve ___ 10
9. Proximal raphe ends straight or curved but not sharply deflected ____ _____ 11.
   10. Valve margins weakly convex or straight _____ 89. *N. terrestris*
   10. Valve margins undulate _____ _____ 90. *N. terrestris* var. *relicta* f. *triundulata*
11. Valves elliptic; ends rounded to subrostrate; striation interrupted by an H-shaped clear area, one "upright" on each side of the valve, the central area forming the "crossbar" _____ 68. *N. pygmaea*
11. Cells lacking such an H-shaped clear area _____ 12
   12. Striae coarsely punctate; an isolated punctum or stigma present on one side of the central area_____ 13
   12. Striae coarsely or finely punctate, but stigmata or isolated puncta lacking _____ 20
13. Valve ends rostrate or capitate _____ 14
13. Valves without distinctive ends _____ 18
   14. Valve margins undulate _____ 15
   14. Valve margins straight to convex at midvalve _____ 16
15. Striae 12/10 μm near midvalve _____ 16. *N. charlatii*
15. Striae 17–24/10 μm near midvalve _____ 57. *N. mutica* var. *nivalis*
   16. Stigma present in the central area but not close to the end of one stria _____ 17
   16. Isolated punctum present at the end of one central stria _____ _____ 91. *N. texana*
17. Valve ends distinctly capitate; valve margins sharply widened at midvalve and weakly undulate between the central bulge and the ends __ _____ 39. *N. heufleriana*
17. Valve ends rostrate, subrostrate, or capitate, but outline not as described above _____ 56. *N. mutica*
   18. Striae 28–30/10 μm _____ 58. *N. muticoides*
   18. Striae 23/10 μm or less _____ 19

19. Width 8–15 μm _____ 88. *N. terminata*
19. Width 4.5–7.5 μm _____ 56. *N. mutica*
   20. Valves broadly elliptic, almost as wide as long; striae alternating longer and shorter at midvalve and coarsely punctate _____
   _____ 75. *N. scutelloides*
   20. Valves elliptic to narrowly elliptic but not almost circular ____ 21
21. Valves with rostrate or capitate ends _____ 22
21. Valves without distinctive ends _____ 25
   22. Central-axial area a moderately narrow space; striae weakly radiate and finely punctate _____ 22. *N. cuspidata* (abnormal valves)
   22. Central area distinctly wider than the axial area or central-axial area a broad lanceolate space _____ 23
23. Central-axial area a broad lanceolate space _____ 19. *N. confervacea*
23. Central area transversely widened and distinct from the axial area ___
   _____ 24
   24. Valves 35 μm long or longer; striae coarsely punctate _____
   _____ 3. *N. amphibola*
   24. Valves under 25 μm long; striae finely punctate _ 33. *N. grimmei*
25. Puncta dashlike _____ 26
25. Puncta dotlike _____ 27
   26. Axial area narrow _____ 43. *N. insociabilis*
   26. Axial and central areas forming a fairly broad lanceolate space ___
   _____ 83. *N. subsulcata*
27. Valves linear-elliptic or elliptic _____ 28
27. Valves lanceolate; puncta fine, striae coarse; large cells _____
   _____ 22. *N. cuspidata* (abnormal valves)
   28. Axial area narrow, the central area distinctly wider than the axial area _____ 5. *N. annexa*
   28. Axial and central areas forming a single broad space _____ 29
29. Valves 40 μm long or longer _____ 53. *N. monmouthiana-stodderi*
29. Valves 30 μm long or less _____ 19. *N. confervacea*

SECTION IV. **Striae indistinctly punctate**

1. Striation interrupted by an H-shaped clear area, one "upright" on each side of the valve, the central area forming the "crossbar" _____
   _____ 68. *N. pygmaea*
1. H-shaped clear areas absent _____ 2
   2. Proximal raphe ends sharply deflected to one side of valve ____ 3
   2. Proximal raphe ends not sharply deflected _____ 4
3. Valve margins more or less straight _____ 89. *N. terrestris*

3. Valve margins undulate __ 90. *N. terrestris* var. *relicta* f. *triundulata*
4. Valves without distinctive ends _____ 5
4. Valve ends attenuate, rostrate or capitate _____ 9
5. Valves lanceolate or elliptic-lanceolate; ends subacute _____ 6
5. Valves linear-elliptic with broadly rounded ends _____ 5. *N. annexa*
6. Central-axial area a broad lanceolate space __ 19. *N. confervacea*
6. Axial area narrow _____ 7
7. Striae clearly radiate at the ends _____ 85. *N. symmetrica*
7. Striae parallel or convergent at the ends _____ 8
8. Striae around 10/10 μm at midvalve and somewhat finer at the ends
   _____ 47. *N. lanceolata*
8. Striae around 16/10 mm at midvalve _____
   _____ 70. *N. tenelloides*
9. Terminal striae broken into dashlike units _____ 40. *N. ignota*
9. Terminal striae not different from the rest _____ 10
10. Striae alternating longer and shorter at the central area _____ 11
10. Striae not alternating longer and shorter _____ 13
11. One central stria tipped by an isolated punctum ___ 24. *N. decussis*
11. Isolated puncta absent _____ 12
12. Valve ends narrow and subrostrate; raphe lying in an axial band of
    different thickness from the adjacent parts of the valve _____
    _____ 36. *N. hambergii*
12. Valve ends broadly subcapitate; raphe not lying in a distinct axial
    band _____ 49. *N. laterostrata*
13. Valve margins undulate; valve ends rostrate _____
    _____ 27. *N. elginensis* var. *neglecta*
13. Valve margins straight or convex _____ 14
14. Valves more or less linear, the margins straight to weakly convex at
    midvalve _____ 15
14. Valves lanceolate to elliptic-lanceolate _____ 18
15. Valve ends narrowly rostrate to rostrate-capitate __ 26. *N. elginensis*
15. Valve ends subrostrate to broadly rostrate _____ 16
16. Striae around 12/10 μm at midvalve, becoming 24/10 μm at the
    ends _____ 63. *N. protracta*
16. Striae 12–16/10 μm at midvalve, but not noticeably finer at the
    ends _____ 17
17. Striae 12–14/10 μm; width typically over 5 μm _____ 4. *N. angusta*
17. Striae 16–17/10 μm; width typically under 5 μm _____ 60. *N. notha*
18. Valves narrowly lanceolate; striae faint and around 20/10 μm ____
    _____ 32. *N. gregaria*

18. Valves elliptic to lanceolate; striae distinct and usually not over 16/ 10 μm _____ 19
19. Valves broadly elliptic-lanceolate with short subrostrate to rostrate ends _____ 31. N. gastrum
19. Valves lanceolate with protracted ends _____ 20
   20. Valves broadly lanceolate and usually over 7 μm wide _____ _____ 47. N. lanceolata
   20. Valves lanceolate and usually no wider than 6 μm _____ _____ 21. N. cryptocephala

SECTION V. **Striae visible, at least at the central area, line-like (not evidently punctate or lineate), and radiate near midvalve**

1. Valves marked by an H-shaped clear space formed by the central area and a longitudinal clear space in the striation on both sides of the valve _____ 68. N. pygmaea
1. H-shaped clear area lacking _____ 2
   2. Striae visible only at the central area even under very strongly obliqued light _____ 84. N. subtilissima
   2. Striae visible at least most of the way to the ends (strongly obliqued light may be required) _____ 3
3. Terminal striae broken into dashlike units _____ 40. N. ignota
3. Terminal striae no different from other striae _____ 4
   4. Proximal raphe ends sharply deflected (almost at right angles) to one side of the valve _____ 5
   4. Proximal raphe ends not sharply deflected _____ 6
5. Valve margins almost straight _____ 89. N. terrestris
5. Valve margins undulate __ 90. N. terrestris var. relicta f. triundulata
   6. Terminal regions of the valve devoid of striae and refracting light differently from adjacent areas _____ 7
   6. Valves striate almost to the ends, the terminal regions not refracting light differently _____ 10
7. Valve margins straight at midvalve _ 66. N. pupula var. rectangularis
7. Valve margins convex at midvalve _____ 8
   8. Valves elliptic to elliptic-lanceolate without distinctive ends _____ _____ 65. N. pupula var. elliptica
   8. Valves linear to lanceolate with more or less distinctive ends __ 9
9. Valves lanceolate with narrowly rostrate-capitate ends _____ _____ 67. N. pupula f. rostrata

9. Valves more or less linear with broadly subcapitate ends _____
_____ 64. *N. pupula*

 10. Valves with attenuate, subrostrate, rostrate, subcapitate, or capitate ends _____ 11

 10. Valves without distinctive ends _____ 25

11. Striae 30/10 μm or more at a point halfway from midvalve to the ends
_____ 12

11. Striae typically no more than 25/10 μm at a point halfway from midvalve to the ends _____ 13

 12. Striae 30/10 μm at midvalve, becoming nearly 40/10 μm at the ends _____ 80. *N. subarvensis*

 12. Striae 24/10 μm at midvalve, becoming 30–36/10 μm near midvalve _____ 13. *N. brockmannii*

13. Valve margins triundulate; ends rostrate-capitate _____
_____ 27. *N. elginensis* var. *neglecta*

13. Valve margins straight or convex, but not triundulate _____ 14

 14. One central stria tipped by an isolated punctum __ 24. *N. decussis*

 14. Isolated puncta lacking _____ 15

15. Valve margins straight or nearly straight at midvalve; ends subrostrate and not much narrower than the adjacent parts of the valve _____ 16

15. Valves not linear or, if margins nearly straight at midvalve, then ends distinctly narrower than adjacent parts of the valve _____ 17

 16. Striae 12–14/10 μm _____ 4. *N. angusta*

 16. Striae 16–17/10 μm _____ 60. *N. notha*

17. Valve ends capitate_____ 18

17. Valve ends attenuate to rostrate _____ 23

 18. Ends not much narrower than the maximum valve width ____ 19

 18. Ends distinctly narrower than the maximum valve width ____ 20

19. Valves 3 μm wide or less _____ 37. *N. hassiaca*

19. Valves over 3 μm wide _____ 25. *N. disjuncta*

 20. Valve margins nearly straight at midvalve _____ 21

 20. Valve margins convex at midvalve _____ 22

21. Striae moderately radiate over most of the valve __ 26. *N. elginensis*

21. Striae sharply radiate at midvalve, becoming strongly convergent toward the ends _____ 10. *N. bicephala*

 22. Ends distinctly and rather broadly capitate; one central stria on each side of the valve reaching almost to the central nodule _____
_____ 49. *N. laterostrata*

 22. Ends attenuate-subcapitate; central striae no longer than adjacent striae _____ 21. *N. cryptocephala*

23. Valve ends broadly rostrate; striae 12–14/10 μm in the central area,

becoming 24/10 μm at the ends _____ 63. *N. protracta*
23. Valve ends attenuate to narrowly rostrate; striae more or less uniformly
    spaced _____ 24
    24. One central stria on each side of the valve reaching almost to the
        central nodule _____ 36. *N. hambergii*
    24. Central striae typically shorter than those adjacent to them, the
        central area therefore rounded _____ 21. *N. cryptocephala*
25. Valves linear to lanceolate; ends subacute_____ 26
25. Valve elliptic to linear-elliptic; ends rounded _____ 36
    26. Valves linear with slightly convex margins and subcacute ends;
        striae distinctly radiate all the way to the ends and fewer than 18/
        10 μm _____ 85. *N. symmetrica*
    26. Valves lanceolate (in shorter valves elliptic-lanceolate) _____ 27
27. Axial area narrow and central area indistinct _____ 31
27. Axial area narrow and central area transversely widened or axial and
    central areas together forming a broad lanceolate space _____ 28
    28. Central-axial area a broad lanceolate space _____ 29
    28. Axial area narrow; central area distinct _____ 30
29. Striae very strongly radiate throughout _____ 79. *N. subadnata*
29. Striae weakly radiate throughout _____ 45. *N. krasskei*
    30. Central area butterfly-shaped, almost reaching the margins _____
        _____ 48. *N. lapidosa*
    30. Central area large, rhombic-lanceolate, and almost as broad as long
        _____ 8. *N. bacilloides*
31. Central area formed by irregular shortening of a few striae; length typ-
    ically over 14 μm _____ 70. *N. tenelloides*
31. Central area no wider than axial area and central striae all of approxi-
    mately equal length _____ 32
    32. Striae around 30–32/10 μm, becoming somewhat finer at the ends
        _____ 29. *N. fluens*
    32. Striae coarser than 26/10 μm near midvalve or if finer then no more
        than 30/10 μm at the ends _____ 33
33. Striae around 28/10 μm throughout _____ 97. *N. ventosa*
33. Striae 26/10 μm or less _____ 34
    34. Striae gradually becoming finer from midvalve toward the ends __
        _____ 35
    34. Striae around 20/10 μm and not noticeably finer near the ends __
        _____ 30. *N. frugalis*
35. Striae weakly radiate, in some cases nearly parallel _____
    _____ 96. *N. vaucheriae*
35. Striae moderately radiate _____ 82. *N. subminiscula*

36. Striae crossed on both sides of the valve by a thin longitudinal line that curves outward around the central area __ 54. *N. monoculata*

36. Striae not crossed by such lines _____ 37

37. Striae over 25/10 μm _____ 38

37. Striae under 25/10 μm _____ 41

38. Valve margins straight; central area more or less butterfly-shaped _____ 86. *N. tantula*

38. Valve margins convex at midvalve _____ 39

39. Central area no wider than axial area _____ 55. *N. muralis*

39. Central area distinct from axial area _____ 40

40. Striae 26–30/10 μm _____ 51. *N. minima*

40. Striae 34–38/10 μm _____ 59. *N. nigrii*

41. Valves over 22–45 μm in length _____ 46. *N. laevissima*

41. Valves under 20 μm long _____ 42

42. Axial area narrow; central area more or less distinct _____ 44

42. Axial area moderately broad; central area indistinct _____ 43

43. Striae 12/10 μm at midvalve, becoming 24/10 μm at the ends _____ _____ 28. *N. excelsa*

43. Striae 20–23/10 μm throughout _____ 12. *N. brevissima*

44. Striae moderately strongly radiate at midvalve; valves narrowly linear _____ 78. *N. seminulum* var. *radiosa*

44. Striae weakly to moderately radiate at midvalve; valves linear-elliptic to elliptic _____ 45

45. Central area more or less regular in form and typically rectangular __ _____ 77. *N. seminulum*

45. Central area more or less irregular in form and in some cases indistinct _____ 76. *N. seminuloides*

SECTION VI. Valves with lineate striae

1. Valve ends marked by thickened bars _____ 2

1. Valve ends not thicker than other parts of the valve _____ 3

2. Valve ends subacute or blunt but not distinctly capitate _____ _____ 15. *N. capitata* var. *hungarica*

2. Valve ends distinctly capitate _____ 14. *N. capitata*

3. Striae parallel or convergent at the ends _____ 8

3. Striae distinctly radiate at the ends _____ 4

4. Valves lanceolate without distinctive ends ___ 85. *N. symmetrica*

4. Valve ends rostrate to rostrate-capitate _____ 5

5. Valve margins undulate _____ 27. *N. elginensis* var. *neglecta*

5. Valve margins straight or convex at midvalve, but not undulate ___ 6

6. Central striae alternating longer and shorter, the centermost stria on each side almost reaching the central nodule _____ _____ 36. *N. hambergii*

6. Central striae of various lengths but, if alternating longer and shorter, then the centermost stria not reaching almost to the central nodule _____ 7

7. Valve margins convex at midvalve; central striae irregularly shortened, sometimes alternating longer and shorter _____ 31. *N. gastrum*

7. Valve margins nearly straight at midvalve _____ 26. *N. elginensis*

8. Valve margins undulate; ends rostrate; terminal striae broken into dashlike units _____ 40. *N. ignota*

8. Valve margins not undulate; terminal striae not broken into dashlike units _____ 9

9. Valve margins nearly straight at midvalve _____ 10

9. Valve margins convex at midvalve _____ 11

10. Striae 8–10/10 μm; width 10 μm or more _____ _____ 100. *N. viridula* var. *linearis*

10. Striae 12–14/10 μm; width 5–7 μm _____ 4. *N. angusta*

11. Valves linear to linear-lanceolate without distinctive ends _____ 12

11. Valves lanceolate or elliptic-lanceolate, frequently with distinctive ends _____ 16

12. Valves typically longer than 70 μm; striae massive and coarse (6–9/10 μm) but with crosslines that are difficult to resolve without strongly obliqued light _____ 61. *N. oblonga*

12. Valves typically shorter than 60 μm; striae typically over 10/10 μm _____ 13

13. Striae 8–10/10 μm at midvalve, becoming 17/10 μm near the ends __ _____ 17. *N. cincta*

13. Striae more or less evenly spaced over much of the valve length _ 14

14. Valve ends rounded-truncate _____ 38. *N. heufleri*

14. Valve ends subacute to acute _____ 15

15. Central area broad, often appearing almost square; striae weakly radiate over much of the length; valves linear-lanceolate _____ _____ 93. *N. tripunctata*

15. Central area linear-elliptic or rounded and not particularly large; striae radiate to distinctly radiate over much of the length; valves more lanceolate or elliptic-lanceolate _ 94. *N. tripunctata* var. *schizonemoides*

16. Valve ends distinctly rostrate to rostrate-capitate _____ 17

16. Valves without distinctive ends or with ends that are substrostrate or attenuate _____ 21

17. Striae 13/10 μm or fewer _____ 101. *N. viridula* var. *rostellata*
17. Striae usually 14/10 μm or more _____ 18
   18. Central striae alternating longer and shorter; an isolated punctum
      at the end of one central stria _____ 24. *N. decussis*
   18. Central striae variously arranged but none with an isolated
      punctum _____ 19
19. Width generally under 7 μm; striae faintly lineate, the crosslines visible only under oblique light _____ 21. *N. cryptocephala*
19. Width generally 8–12 μm; striae usually distinctly lineate _____ 20
   20. Centermost stria on each side typically almost reaching the central
      nodule, the central area therefore indistinct; ends narrowly rostrate-subcapitate _____ 73. *N. salinarum* var. *intermedia*
   20. Centermost striae not noticeably longer than adjacent striae, the
      central area rounded; ends rostrate _____ 72. *N. salinarum*
21. Striae 12/10 μm or fewer at midvalve _____ 22
21. Striae 14–20/10 μm _____ 26
   22. Valves broadly lanceolate with attenuate ends _ 47. *N. lanceolata*
   22. Valves lanceolate to elliptic-lanceolate without distinctive ends or,
      if ends are slightly set off, then they are not attenuate _____ 23
23. Central area rounded or irregular; valves lanceolate _____ 24
23. Central area transversely elliptic or butterfly-shaped _____ 25
   24. Valves lanceolate and typically over 40 μm in length _____
      _____ 69. *N. radiosa*
   24. Valves elliptic-lanceolate and typically under 40 μm in length ___
      _____ 50. *N. menisculus*
25. Central area butterfly-shaped, not reaching the margins; ends truncate
      _____ 99. *N. viridula* var. *avenacea*
25. Central area more or less transversely rectangular, not reaching the
      margins; ends very slightly set off–substrate _____ 98. *N. viridula*
   26. Valves almost linear with slightly protracted ends __ 60. *N. notha*
   26. Valves distinctly lanceolate _____ 27
27. Width 4 μm or less _____ 70. *N. tenelloides*
27. Width 5–7 μm _____ 71. *N. radiosa* var. *tenella*

1. Navicula accomoda Hust. var. **accomoda** *Fig. 33g.*

CR: Patrick and Reimer (1966).

DESCRIPTION: Valves broadly elliptic-lanceolate with narrow, short-rostrate ends; axial area narrow; central area little wider than axial area; striae nearly parallel at midvalve, sometimes slightly radiate just before the ends; puncta not visible.

DIMENSIONS: Length, 19–35 μm; width, 7–10 μm; striae, 20–25/10 μm at midvalve and around 32/10 μm at the ends.

2. Navicula americana Ehr. var. **americana** *Fig. 34a.*

CR: Hustedt (1961).

DESCRIPTION: Valves linear with broadly rounded ends; axial area broad (around 1/3 of valve width); striae parallel over most of the valve, becoming slightly radiate at the ends; puncta typically not visible.

DIMENSIONS: Length 30–140 μm; width, 10–30 μm; striae, 13–18/10 μm.

3. Navicula amphibola Cl. var. **amphibola** *Fig. 40a.*

CR: Patrick and Reimer (1966).

DESCRIPTION: Valves elliptic to linear-elliptic; ends narrowly rostrate to rostrate-capitate; axial area moderately narrow; central area transversely expanded and butterfly-shaped but not reaching the valve margins; striae radiate and distinctly punctate.

DIMENSIONS: Length, 37–70 μm; width, 19–28 μm; striae, 7–10/10 μm; puncta, 12–16/10 μm.

4. Navicula angusta Grun. var. **angusta** *Fig. 50a.*

CR: Patrick and Reimer (1966).

DESCRIPTION: Valves linear with broadly subrostrate to rostrate ends; axial area narrow; central area irregularly rounded; striae radiate over most of the valve, becoming convergent at the ends; striae appear lineate under oblique light.

DIMENSIONS: Length, 43–65 μm; width, 5–7 μm; striae, 12–14/10 μm.

5. Navicula annexa Hust. var. **annexa** *Fig. 36h.*

CR: Hustedt (1966).

DESCRIPTION: Valves linear-elliptic to elliptic-lanceolate with slightly convex margins and blunt or rounded ends; axial area narrow; central area transversely expanded but not reaching the margins; striae radiate and distinctly punctate.

DIMENSIONS: Length, 16–17.5 μm; width, around 4 μm; striae, around 20/10 μm.

6. Navicula arvensis Hust. var. **arvensis** *Fig. 35b, c.*

CR: Patrick and Reimer (1966).

DESCRIPTION: Valves with straight to slightly convex margins and moderately narrow rostrate ends; raphe lying in a narrow axial band that is different in thickness from the adjacent parts of the valve.

DIMENSIONS: Length, 5–13 μm; width, 2–5 μm; striae not visible even with strongly obliqued light.

7. **Navicula auriculata** Hust. var. **auriculata** *Fig. 43d.*

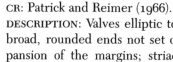

CR: Patrick and Reimer (1966).

DESCRIPTION: Valves elliptic to elliptic-lanceolate with broad, rounded ends not set off by constriction or expansion of the margins; striae punctate, radiate and short; a single longitudinal row of puncta present along both sides of each raphe branch, the space between these rows and the striae devoid of markings. This taxon is similar in appearance to *N. tenera* Hust., which differs in having longitudinal rows of puncta only along one side of each raphe branch. Observation of this distinction in the case of complete frustules requires careful focusing.

DIMENSIONS: Length, 13–16 μm; width, 5–6 μm; striae, 16–20/10 μm.

8. **Navicula bacilloides** Hust. var. **bacilloides** *Fig. 38a.*

CR: Hustedt (1961).

DESCRIPTION: Valves broadly lanceolate to elliptic-lanceolate, with either rounded ends (my specimens) or with ends that are very slightly protracted (Hustedt's specimens); striae radiate throughout and not visibly punctate; raphe lying in a narrow thickened band that continues through the rounded central area.

DIMENSIONS: Length, 26–30 μm; width, 10–12 μm; striae, 22–24/10 μm.

9. **Navicula belliatula** Arch. var. **belliatula** *Fig. 33b.*

CR: Archibald (1971).

DESCRIPTION: Valves broadly lanceolate with narrow, protracted ends; axial area narrow; central area with more or less concave margins and slightly narrower than the adjacent axial area; striae nearly parallel throughout and not visibly punctate.

DIMENSIONS: Length, 36–40 μm; width, 6.5–8 μm; striae, 24–28/10 μm near midvalve and somewhat finer at the ends.

10. Navicula bicephala Hust. var. **bicephala** *Fig. 36c*.

CR: Patrick and Reimer (1966).

DESCRIPTION: Valves with straight to slightly convex margins that are narrowed and constricted abruptly to form rostrate-capitate ends; axial area narrow; central area indistinct, formed by irregular shortening of a few striae; striae fairly strongly radiate near midvalve, becoming convergent toward the ends; puncta not visible.

DIMENSIONS: Length, 20–26 μm; width, 3–4 μm; striae, 16–18/10 μm near midvalve, becoming 20–22/10 μm at the ends.

11. Navicula biconica Patr. var. **biconica** *Fig. 38h*.

CR: Patrick and Reimer (1966).

DESCRIPTION: Valves lanceolate to elliptic-lanceolate without distinctive ends; striae parallel and more widely spaced at midvalve than at the ends; puncta visible, if at all, only with strongly obliqued light.

DIMENSIONS: Length, 12–14 μm; width, 4–5 μm; striae, 26–28/10 μm at midvalve, approaching 40/10 μm at the ends.

12. Navicula brevissima Hust. var. **brevissima** *Fig. 39f*.

CR: Hustedt (1962).

DESCRIPTION: Valves elliptic; central and axial areas forming a linear space; raphe branches somewhat arched; striae radiate throughout; puncta not visible.

DIMENSIONS: Length, 8–14 μm; width, 3.5–4.5 μm; striae, 20–23/10 μm.

13. Navicula brockmannii Hust. var. **brockmannii** *Fig. 36j.*

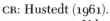

CR: Hustedt (1961).

DESCRIPTION: Valves linear with rostrate-capitate ends; axial area narrow; central area transverse; striae more or less radiate throughout and more widely spaced at midvalve than elsewhere; puncta not visible. Hustedt's figures show the ends as more broadly rostrate or sub-capitate than is the case with my specimens, which had ends more similar in shape to those of *N. bryophila* var. *lapponica* Hust. The terminal striae of *N. bryophila* var. *lapponica* are, however, distinctly convergent. It is possible that my specimens should not be assigned to *N. brockmannii* but, given the already considerable reported shape variation for this species, I am content to leave them here for the present.

DIMENSIONS: Length, 17–24 μm; width, 4.5–6 μm; striae, 24/10 μm at midvalve, 30–36/10 μm elsewhere.

14. Navicula capitata Ehr. var. **capitata** *Fig. 44e.*

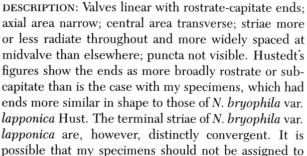

CR: Patrick and Reimer (1966).

DESCRIPTION: Valve margins convex; ends rostrate-capitate; each valve end with a thickened polar cap; axial area narrow at the ends but widening gradually to merge with the indistinct central area; striae radiate, becoming parallel or convergent at the ends; crosslines on the striae are easily seen.

DIMENSIONS: Length, 12–47 μm; width, 5–10 μm; striae, 8–10/10 μm.

15. Navicula capitata var. **hungarica** (Grun.) Ross *Fig. 44d.*

CR: Patrick and Reimer (1966).

DESCRIPTION: Valves elliptic-lanceolate; ends not distinctive. The valve structure is otherwise similar to var. *capitata*.

DIMENSIONS: Length, 10–36 μm; width, 4–10 μm; striae, 8–11/10 μm.

16. Navicula charlatii M. Peragallo var. **charlatii** *Fig. 42a.*

CR: Hustedt (1966).

DESCRIPTION: Valves elliptic-lanceolate with triundulate margins and narrow rostrate-capitate ends; striae radiate and distinctly punctate; central area transversely widened but not reaching the margins; a single large stigma present in the central area; proximal raphe ends curved slightly away from the side of the valve with the stigma.

DIMENSIONS: Length, 32–45 μm; width, 11–16 μm; striae, 12/10 μm at midvalve, becoming around 15/10 μm at the ends; puncta, 12–16/10 μm.

17. Navicula cincta (Ehr.) Ralfs var. **cincta** *Fig. 50h.*

CR: Patrick and Reimer (1966).

DESCRIPTION: Valves elliptic-lanceolate with rounded ends; axial area narrow; central area fairly small and irregular in shape; striae radiate, becoming convergent at the ends and more widely spaced at midvalve than elsewhere; lineae faint.

DIMENSIONS: Length, 10–42 μm; width, 4–8 μm; striae, 8–10/10 μm at midvalve, becoming 17/10 μm at the ends.

18. Navicula circumtexta Meist. ex Hust. var. **circumtexta** *Fig. 43a.*

CR: Patrick and Reimer (1966).

DESCRIPTION: Valves lanceolate with slightly protracted ends; axial and central areas forming a broad lanceolate space; each raphe branch bordered on one side by a longitudinal row of puncta; striae on both sides of the valve crossed by a thin longitudinal line that lies near the central-axial space.

DIMENSIONS: Length, 26–27 μm; width, 6–7 μm; striae, 16–18/10 μm.

19. Navicula confervacea (Kütz.) Grun. var. **confervacea** *Fig.* 43e, f.

CR: Patrick and Reimer (1966).

DESCRIPTION: Valves elliptic with rounded-subacute or slightly protracted ends; axial and central areas forming a broad lanceolate space; striae radiate and distinctly or indistinctly punctate.

DIMENSIONS: Length, 12–20 μm; width, 5–8 μm; striae, 20–24/10 μm.

20. Navicula contenta var. **biceps** (Arnott ms. Grun. in V.H.) Cl. *Fig. 36f.*

CR: Hustedt (1962).

DESCRIPTION: Valve margins more or less concave (the specimen I have illustrated is about average, though some have a greater concavity and others have margins that are nearly straight); ends rounded; striae visible with difficulty under oblique light. The striae can sometimes, as I have shown, be resolved all the way to the valve margins, but in most cases they appear to terminate before reaching the margins. This distinction does not appear to be of taxonomic significance.

DIMENSIONS: Length, 6–30 μm, but rarely over 15 μm; width at widest point, 2–6 μm; striae, 25/10 μm to (typically) 36/10 μm.

21. Navicula cryptocephala Kütz var. cryptocephala *Figs. 33f(?); 49a–f.*

CR: Patrick and Reimer (1966).

DESCRIPTION: Valves lanceolate with more or less attenuate-rostrate or rostrate-capitate ends; axial area narrow; central area rounded; striae radiate, becoming parallel or convergent at the ends. Valves may vary in degree of silicification. The striae may be distinctly or indistinctly lineate and thin or thick. I have illustrated one example (*Fig. 33f*) that has the form of *N. cryptocephala* but has a structure that is much too fine. I have included this in the hope that someone may look for and find intergrading examples between it and the more typical specimens shown in *Fig. 49.*

DIMENSIONS: Length, 14–40 μm; width, 4–7 μm; striae, 14–16/10 μm, perhaps 18/10 μm.

22. Navicula cuspidata (Kütz.) Kütz. var. cuspidata *Figs. 31a; 32a, b.*

CR: Hustedt (1930); Patrick and Reimer (1966).

DESCRIPTION: Valves broadly lanceolate with somewhat blunt-subacute ends that may or may not be slightly protracted; axial area moderately narrow; central area no wider than the axial area; transverse striae parallel; longitudinal striae perpendicular to the transverse ones. There is another, rarer, arrangement of striae in which the transverse striae are radiate and the longitudinal striae not evident. This has been shown in *Fig. 32b.* Hustedt has reported the radiate forms as var. *heribaudii* Peragallo. I have seen complete cells, however, with one valve of "normal" type and the other of "*heribaudii*" type and therefore do not accept the variety. Some specimens of var. *cuspidata* also have internal silica "craticular plates" that may sometimes be seen separately in "cleaned" preparations. I have illustrated one such plate as *Fig. 32a.*

DIMENSIONS: Length, 30–170 μm; width, 15–37 μm; transverse striae, 14–24/10 μm; longitudinal striae, 22–26/10 μm.

23. Navicula cuspidata var. **ambigua** (Ehr.) Cl. *Fig. 31b.*

CR: Hustedt (1930).

DESCRIPTION: Valves broadly elliptic-lanceolate with rostrate to rostrate-capitate ends; other details as for var. cuspidata. This may be, as Patrick and Reimer (1966) indicate, only a minor shape variant of var. *cuspidata*. I am retaining it here only because I did not find a convincing series of intergrading forms in Illinois.

DIMENSIONS: As for var. *cuspidata* with the exception that longitudinal striae may reach 29/10 μm.

24. Navicula decussis Oestr. var. **decussis** *Fig. 51e.*

CR: Patrick and Reimer (1966).

DESCRIPTION: Valves broadly elliptic-lanceolate with rostrate-capitate ends; axial area narrow; central area irregular in shape; striae radiate, becoming convergent at the ends; striae in central area may alternate longer and shorter; resolution of crosslines on striae requires oblique light; a single isolated punctum present in the central area.

DIMENSIONS: Length, 16–27 μm; width, 6–8 μm; striae, 16–20/10 μm.

25. Navicula disjuncta Hust. var. **disjuncta** *Fig. 36d, e.*

CR: Hustedt (1961).

DESCRIPTION: Valves linear with weakly convex margins that are fairly sharply reflexed to form the relatively broad, capitate ends; axial area narrow; central area transverse, formed by the regular shortening of the central striae; striae radiate throughout. My specimens were similar in size to *N. disjuncta* f. *anglica* Hust. but had the striae arrangement of the nominate variety.

DIMENSIONS: Length, 21–28 μm (mine were as short as 13 μm); width, 4–5 μm (mine were as narrow as 3.3 μm; striae, 22–28/10 μm.

26. Navicula elginensis (Greg.) Ralfs var. **elginensis** *Fig. 51f.*

CR: Patrick and Reimer (1966).

DESCRIPTION: Valve margins straight to slightly convex at midvalve; valve ends narrowly rostrate to rostrate-capitate; axial area narrow; central area transversely elliptic to irregular; striae radiate throughout; lineae visible only under oblique light.

DIMENSIONS: Length, 20–40 μm; width, 8–14 μm; striae, 9–11/10 μm at midvalve, becoming 14/10 μm at the ends.

27. Navicula elginensis var. **neglecta** (Krasske) Patr. *Fig. 51g.*

CR: Patrick and Reimer (1966).

DESCRIPTION: Similar to var. *elginensis* with the exception that the valve margins are distinctly triundulate.

DIMENSIONS: Length, 22–23 μm; maximum width, 7–9 μm; striae, 10–14/10 μm.

28. Navicula excelsa Krasske var. **excelsa** *Fig. 39e.*

CR: Hustedt (1962).

DESCRIPTION: Valves elliptic with a broad, lanceolate central-axial area; striae radiate throughout; raphe branches appear as rather thick lines. The valves of this species are lightly silicified and may be overlooked. Oblique light may be required to see the striae even though they are fairly widely spaced.

DIMENSIONS: Length, 11–16 μm; 5–7 μm; striae, 12/10 μm at midvalve, becoming around 24/10 μm at the ends.

29. Navicula fluens Hust. var. **fluens** *Fig. 38g.*

CR: Hustedt (1962).

DESCRIPTION: Valves elliptic-lanceolate to lanceolate with subacute ends; axial area narrow and central area no wider than axial area; striae radiate throughout and somewhat more widely spaced at midvalve than at the ends. This species is similar in form to *N. vaucheriae* Peters., but has finer striae. To date found only in Pulaski Co.

DIMENSIONS: Length, 10–16 μm; width, around 5 μm; striae 30–32/10 μm.

30. Navicula frugalis Hust. var. **frugalis** *Fig. 37q.*

CR: Hustedt (1962).

DESCRIPTION: Valves lanceolate to elliptic-lanceolate; axial area narrow; central area no wider than axial area; striae radiate throughout; puncta not visible.

DIMENSIONS: Length, 7.0–12.5 μm; width, 3.5–4.0 μm; striae, around 20/10 μm.

31. Navicula gastrum (Ehr.) Kütz. var. **gastrum** *Fig. 51a–d.*

CR: Patrick and Reimer (1966).

DESCRIPTION: Valves broadly elliptic-lanceolate with rostrate to subrostrate ends; axial area narrow; central area rounded to irregular in shape; central striae regularly or irregularly shortened, sometimes alternating longer and shorter; striae radiate throughout; lineae visible with oblique light.

DIMENSIONS: Length, 18–60 μm; width, 7.5–20 μm; striae, 8–10/10 μm at midvalve, becoming 12/10 μm at the ends.

32. Navicula gregaria Donk. var. **gregaria** *Fig. 33c–e.*

CR: Patrick and Reimer (1966.)

DIMENSIONS: Valves elliptic-lanceolate with protracted rostrate ends; axial and central areas narrow; striae more or less parallel, sometimes very slightly radiate at midvalve and very slightly convergent at the ends; puncta not visible.

DIMENSIONS: Length, 15–35 μm; width, 5–9 μm; striae, 16–22/10 μm.

33. Navicula grimmei Krasske var. **grimmei** *Fig. 43i.*

CR: Hustedt (1966).

DESCRIPTION: Valve margins convex; ends rostrate-capitate and distinctly narrower than the maximum valve width; axial area narrow; central area transverse but not reaching the margins; striae radiate and distinctly punctate.

DIMENSIONS: Length, 15–21 μm; width, 5–6 μm; striae, 20/10 μm at midvalve, becoming 25/10 μm at the ends.

34. Navicula gysingensis Foged var. **gysingensis** *Fig. 35a.*

CR: Patrick and Reimer (1966).

DESCRIPTION: Valves elliptic-lanceolate with fairly narrow attenuate-subcapitate ends; raphe lying in a narrow axial band of different thickness from the adjacent parts of the valve.

DIMENSIONS: Length, 12–17 μm; width, 4–5 μm; striae not visible with the light microscope.

35. Navicula halophila f. tenuirostris Hust. *Fig. 33a.*

CR: Patrick and Reimer (1966.)

DESCRIPTION: Valves lanceolate with protracted rostrate to rostrate-capitate ends; axial area narrow; central area typically no wider than the axial area; striae more or less parallel throughout and not visibly punctate. This taxon is similar in appearance to *N. belliatula* Arch. but has coarser striae.

DIMENSIONS: Length, 38–50 μm; width, 8–10 μm; striae, 20–22/10 μm near midvalve and somewhat finer at the ends.

36. Navicula hambergii Hust. var. hambergii *Fig. 52e.*

CR: Patrick and Reimer (1966).

DESCRIPTION: Valves broadly elliptic-lanceolate with somewhat protracted rostrate ends; axial area narrow; one central stria on each side of the valve almost reaching the central nodule, the central area therefore irregular in shape; raphe branches lying in a narrow axial band of different thickness from the adjacent parts of the valve; striae radiate; crosslines on striae visible only with oblique light.

DIMENSIONS: Length, 15–25 μm; width, 5.5–7.5 μm; striae, 14–16/10 μm.

37. Navicula hassiaca Krasske var. hassiaca *Fig. 36g.*

CR: Hustedt (1962).

DESCRIPTION: Valve margins convex at midvalve; ends capitate; axial area narrow; central area reaching both valve margins.

DIMENSIONS: Length, 9–12 μm; maximum width, 2.5–3 μm (mine were as narrow as 2 μm); striae 16–18/10 μm (mine reached 20/10 μm).

38. Navicula heufleri Grun. var. heufleri *Fig. 50g*.

CR: Patrick and Reimer (1966).

DESCRIPTION: Valves elliptic-lanceolate with rather blunt, somewhat protracted ends; axial area narrow; central area more or less elliptical. This species is similar to *N. cincta* (Ehr.) Ralfs, but has finer striae at midvalve.

DIMENSIONS: Length, 18–32 μm; width, 4–6 μm; striae 10–14/10 μm throughout.

39. Navicula heufleriana (Grun.) Cl. var. heufleriana *Fig. 41h*.

CR: Hustedt (1966).

DESCRIPTION: Valves lanceolate with distinctly capitate ends and sharply inflated midvalve margins; margins sometimes very weakly undulate between the central inflation and the ends; axial area narrow; central area reaching the margins; striae radiate; puncta coarse and almost dashlike; a single stigma present in one side of the central area.

DIMENSIONS: Length, 20–35 μm; width, 8–12 μm; striae, 16–20/10 μm.

40. Navicula ignota Krasske var. ignota *Fig. 52d*.

CR: Lund (1946).

DESCRIPTION: Valves linear with triundulate margins and comparatively broad, rostrate-capitate ends; axial area narrow; central area formed by the shortening of a few striae; striae radiate; lineae visible under oblique light. The terminal striae are broken into dashlike units and the raphe branches appear to terminate short of the ends of the valve at about the point where the nature of the striae changes.

DIMENSIONS: Length, 13–19 μm; width, 4–5 μm; striae, 15–18/10 μm.

41. Navicula indifferens Hust. var. **indifferens** *Fig. 35g.*

CR: Hustedt (1961).
DESCRIPTION: Valves elliptic-lanceolate without distinctive ends; raphe lying in a narrow, thickened axial band.
DIMENSIONS: Length, 6–11.5 μm; width, 2.8–4 μm; striae not visible with the light microscope.

42. Navicula ingenua Hust var. **ingenua** *Fig. 43h.*

CR: Hustedt (1962).
DESCRIPTION: Valves lanceolate without distinctive ends; central area reaching both valve margins and asymmetric, longer on one margin than the other; striae radiate and distinctly punctate.
DIMENSIONS: Length, 7.5–9.0 μm; width, 2–4 μm; striae, 24/10 μm (mine were 18/10 μm).

43. Navicula insociabilis Krasske var. **insociabilis** *Fig. 43g.*

CR: Hustedt in A.S.A., Taf. ⁴⁰⁰/₁₀₃₋₁₀₅; Hustedt (1962).
DESCRIPTION: Valves linear with broadly rounded ends; striae parallel to weakly radiate, each stria apparently broken into two dashlike units, the breaks so arranged as to produce a thin longitudinal line. There is considerable difference between the illustrations in Hustedt in A.S.A. and Hustedt (1962). Hustedt did not clarify the matter in the latter work. My specimens match the appearance of those in A.S.A. in having a narrow central-axial area. The illustrations in Hustedt (1962) show a moderately wide lanceolate central-axial area.
DIMENSIONS (from Hustedt, 1962): Length, 10–22 μm (mine were as short as 7 μm); width, 4.5–7 μm (mine were as narrow as 3 μm); striae, 20–25/10 μm (mine were somewhat finer).

44. Navicula integra (W. Sm.) Ralfs var. **integra** *Fig. 44a*.

CR: Patrick and Reimer (1966).

DESCRIPTION: Valves elliptic-lanceolate, the margins undulate just before the rostrate ends; pseudosepta visible at both valve ends; axial area narrow; central area rounded; striae radiate and punctate throughout.

DIMENSIONS: Length, 25–45 μm; width, 8–10 μm; striae, 12/10 μm at midvalve, becoming more closely spaced toward the ends where they reach 24/10 μm.

45. Navicula krasskei Hust. var. **krasskei** *Fig. 38d, e*.

CR: Hustedt (1962).

DESCRIPTION: Valves lanceolate with subacute to slightly protracted ends; central-axial area a rather broad, lanceolate space; striae radiate throughout and very difficult to resolve even with oblique light.

DIMENSIONS: Length, 6–15 μm; width, 3.5–5 μm; striae, 36–40/10 μm.

46. Navicula laevissima Kütz. var. **laevissima** *Fig. 34h*.

CR: Patrick and Reimer (1966).

DESCRIPTION: Valves more or less linear with broadly rounded ends. Typical examples have margins slightly convex at midvalve and slightly concave before the ends. Axial area narrow; central area irregular in shape; striae radiate throughout, the striation continuing to the ends. This species is quite similar in appearance to *N. pupula* Kütz., but lacks the striae-free areas at the ends of the valves.

DIMENSIONS: Length, 22–45 μm; width, 6.5–10 μm; striae, 12–15/10 μm near midvalve, becoming 20–22/10 μm at the ends.

47. Navicula lanceolata (Ag.) Kütz. var. **lanceolata** *Fig. 48a–d.*

CR: Patrick and Reimer (1966).

DESCRIPTION: Valves lanceolate with ends that are slightly protracted-subrostrate in short valves and attenuate-rostrate in longer ones; axial area narrow; central area transversely elliptic to irregular; striae radiate, becoming parallel or convergent at the ends; lineae on striae usually distinct. This taxon has been described in various ways. I recommend reading Patrick and Reimer (1966) and Germain (1964).

DIMENSIONS: Length, 23–50 μm; width, 6.5–12 μm; striae, around 10/10 μm at midvalve, becoming 14/10 μm toward the ends.

48. Navicula lapidosa Krasske var. **lapidosa** *Fig. 38c.*

CR: Hustedt (1962).

DESCRIPTION: Valves lanceolate to elliptic-lanceolate with rounded ends; axial area narrow; central area butterfly-shaped but not reaching the margins; striae radiate throughout; puncta not visible.

DIMENSIONS: Length, 12–24 μm; width, 6–9 μm; striae, 24–28/10 μm.

49. Navicula laterostrata Hust. var. **laterostrata** *Fig. 36b.*

CR: Hustedt (1962).

DESCRIPTION: Valve margins convex at midvalve, narrowing before the beginning of the broadly capitate ends; axial area narrow; central area irregular in shape, not reaching the valve margins; striae frequently alternating longer and shorter in the central area. The striae are radiate except at the ends where they are distinctly convergent. Hustedt reports that the striae are distinctly punctate, but in my specimens the puncta could be resolved only with strongly obliqued light.

DIMENSIONS: Length, 20–30 μm; width, 5–10 μm; striae, 15–21/10 μm over much of the valve, becoming somewhat finer at the ends.

50. Navicula menisculus Schumann var. **menisculus** *Fig. 45c.*

CR: Hustedt (1930); Schoeman (1973).

DESCRIPTION: Valves lanceolate without distinctive ends; axial area narrow; central area rounded; striae radiate over much of the valve and radiate, parallel, or convergent at the ends; lineae faintly visible on the striae. Schoeman has reported that var. *menisculus* completely intergrades with var. *upsaliensis* (Grun.) Grun. This has been my experience also. Valves of *N. menisculus* are similar in appearance to short specimens of *N. lanceolata* (Ag.) Kütz., but lack the protracted-subrostrate ends that are evident even in short valves of *N. lanceolata*.

DIMENSIONS: Length, 23–50 μm; width, 8–12 μm; striae, 9–12/10 μm at midvalve and somewhat finer at the ends.

51. Navicula minima Grun. var. **minima** *Fig. 37d–f.*

CR: Hustedt (1962).

DESCRIPTION: Valves elliptic to linear-elliptic with broadly rounded ends; axial area narrow; central area widened but not reaching the valve margins; striae radiate throughout; puncta not visible with the light microscope.

DIMENSIONS: Length, 6–17 μm; width, 2.5–4.5 μm; striae, 26–30/10 μm.

52. Navicula miniscula Grun. var. **miniscula** *Fig. 36i.*

CR: Patrick and Reimer (1966).

DESCRIPTION: Valves elliptic-lanceolate with rostrate ends; axial area narrow; central area no wider than axial area; striae parallel near midvalve and radiate elsewhere. The striae are very fine and can typically be resolved only with oblique light; puncta not visible with the light microscope.

DIMENSIONS: Length, 10–15 μm; width, 3–5 μm; striae, 30–34/10 μm.

53. **Navicula monmouthiana-stodderi** Yerm. var. **monmouth-iana-stodderi** *Fig. 43j.*

CR: Patrick and Reimer (1966).

DESCRIPTION: Valves linear with weakly convex margins and subacute ends; central-axial area a fairly broad linear-lanceolate space, sometimes expanded slightly at midvalve; striae punctate and radiate throughout; proximal raphe ends slightly curved to one side of the valve.

DIMENSIONS: Length, 28–70 μm; width, 5–10 μm; striae, 16/10 μm at midvalve, becoming 21/10 μm at the ends.

54. **Navicula monoculata** Hust. var. **monoculata** *Fig. 39c.*

CR: Hustedt (1962).

DESCRIPTION: Valves elliptic to linear-elliptic with broadly rounded ends; central area little wider than the narrow axial area; striae weakly to moderately radiate throughout; puncta not visible; striae crossed on each side of the axial area by a longitudinal line that lies close to the axial area near the poles but curves outward at midvalve. This outward curvature frequently tricks the eye into seeing a widened central area where there is none.

DIMENSIONS: Length, 9–18 μm; width, 3.5–5.0 μm; striae, 23–28/10 μm.

55. **Navicula muralis** Grun. var. **muralis** *Fig. 38j.*

CR: Patrick and Reimer (1966); Van Heurck type slide A-VH 144.

DESCRIPTION: Valves elliptic with broadly rounded ends; axial area narrow; central area no wider than the axial area; striae radiate, fine and frequently only visible with oblique light.

DIMENSIONS: Length, 5–12 μm; width, 3–5 μm; striae, 30–36/10 μm.

56. Navicula mutica Kütz var. mutica *Fig. 41a–e, 41f (?)*.

CR: Hustedt (1966).

DESCRIPTION: Valve shape quite variable, but most commonly lanceolate with broadly rounded to truncate ends. The elliptic to elliptic-lanceolate specimens are somewhat more rare, while valves with rostrate or rostrate-capitate ends are quite uncommon. The specimen with capitate ends shown in *Fig. 41f* may be too extreme to include in the variation series of *N. mutica*. In fact specimens with this shape have been described as a species, *N. neoventricosa* Hust. I have shown the distribution of the capitate cells in a separate map, labeled "A." All of the variations of *N. mutica* have a narrow axial area, a more or less butterfly-shaped central area, a stigma in one side of the central area and radiate, coarsely punctate, striae. The *"mutica"* group might prove interesting for a study in diatom speciation.

DIMENSIONS: Length, 6–25 μm; width, 4.5–7.5 μm; striae, 15–23/10 μm.

57. Navicula mutica var. nivalis (Ehr.) Hust. *Fig. 41g*.

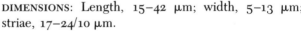

CR: Hustedt (1930, 1966).

DESCRIPTION: Valves with a general appearance similar to that of var. *mutica*, but with rostrate-capitate to capitate ends and triundulate margins. The undulations in some specimens are slight and these specimens tend toward the appearance of the capitate forms of var. *mutica*. I can, therefore, accept var. *nivalis* as a variety of *N. mutica,* as Hustedt (1930) has done, but not as a species (*N. nivalis* Ehr.) as presented in Hustedt (1966.)

DIMENSIONS: Length, 15–42 μm; width, 5–13 μm; striae, 17–24/10 μm.

58. Navicula muticoides Hust. var. **muticoides** *Fig. 42b–d.*

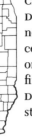

CR: Hustedt (1966).

DESCRIPTION: Valves elliptic to elliptic-lanceolate with nondistinctive to subrostrate ends; axial area narrow; central area transversely widened; stigma present on one side of the central area; striae radiate and rather finely punctate compared with those of *N. mutica* Kutz.

DIMENSIONS: Length, 10–23 μm; width, 6–9 μm; striae, 28–30/10 μm.

59. Navicula nigrii De Notaris var. **nigrii** *Fig. 37i, j.*

CR: Granetti (1968) according to Begres (1971).

DESCRIPTION: Valves elliptic to elliptic-lanceolate with rounded to broadly subacute ends; axial area narrow; central area irregular, formed by the shortening of a few central striae; striae radiate throughout; puncta not resolvable with the light microscope.

DIMENSIONS: Length, 2.5–12.0 μm; width, 2.5–3.5 μm; striae, 34–38/10 μm.

60. Navicula notha Wallace var. **notha** *Fig. 50e.*

CR: Patrick and Reimer (1966).

DESCRIPTION: Valves linear-lanceolate with margins that are weakly convex to almost straight at midvalve; ends somewhat attenuate-rostrate; axial area narrow; central area irregular in shape; striae radiate near midvalve, becoming parallel to convergent at the ends. To date I have found *N. notha* only in Williamson County.

DIMENSIONS: Length, 19–32 μm; width, 4–5 μm; striae, 16–17/10 μm.

61. Navicula oblonga (Kütz.) Kütz. var. **oblonga** *Fig. 52a.*

CR: Patrick and Reimer (1966).

DESCRIPTION: Valves linear with straight to slightly convex margins and broadly rounded ends; axial area slightly wider than the broad, bandlike raphe; striae weakly to moderately radiate, becoming convergent toward the ends. The striae are broad and appear somewhat undulate due to the curvature of the valve surface. The lineae on the striae are faint but always visible. This species is often mistaken for a *Pinnularia*, but no true *Pinnularia* has lineate striae.

DIMENSIONS: Length, 70–220 μm; width, 13–24 μm; striae, 6–9/10 μm.

62. Navicula placenta Ehr. var. **placenta** *Fig. 40c.*

CR: Hustedt (1962).

DESCRIPTION: Valves elliptic with narrowly rostrate-capitate ends; axial area narrow; central area rounded; puncta distinct and so arranged as to produce two diagonally intersecting systems of striae.

DIMENSIONS: Length, 28–60 μm; width, 14–20 μm; striae (both systems), 14–20/10 μm.

63. Navicula protracta Grun. var. **protracta** *Fig. 36a.*

CR: Hustedt (1962).

DESCRIPTION: Valve linear with broadly rostrate ends; margins straight to slightly convex at midvalve; axial area narrow; central area irregular in shape and moderately small; striae radiate throughout, the striae around the central area often apparently more widely spaced than elsewhere; puncta visible (if at all) only with strongly obliqued light.

DIMENSIONS: Length, 17–55 μm; width, 5–10 μm; striae, 12–14/10 μm at midvalve, around 24/10 μm elsewhere.

64. Navicula pupula Kütz. var. **pupula** *Fig. 34c.*

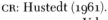

CR: Hustedt (1961).

DESCRIPTION: Valves more or less linear, the margins slightly convex at midvalve and slightly concave before the broadly rounded to subcapitate ends; axial area narrow; central area transverse but not reaching the margins; striae radiate, the striation not reaching the extreme ends of the valve (Compare with *N. laevissima* above!); terminal regions formed by the transversely expanded polar nodule. *N. pupula* varies considerably in shape and the named varieties may well prove to intergrade completely.

DIMENSIONS: Length, 13–66 μm; width, 5–16 μm; striae, 22–26/10 μm.

65. Navicula pupula var. **elliptica** Hust. *Fig. 34d, g.*

CR: Hustedt (1961).

DESCRIPTION: Similar in form to var. *pupula* with the exception that the valves are elliptic to linear-elliptic in outline with ends typically rounded and rarely very slightly protracted.

DIMENSIONS: Similar to var. *pupula*.

66. Navicula pupula var. **rectangularis** (Greg.) Grun. *Fig. 34b.*

CR: Patrick and Reimer (1966).

DESCRIPTION: Similar to var. *pupula* with the exception that the margins are straight and the ends nondistinctive. This variety may be only one extreme of the normal shape variation of var. *pupula*.

DIMENSIONS: Similar to var. *pupula*.

67. Navicula pupula f. rostrata (Hust.) Hust. *Fig. 34e, f.*

CR: Hustedt (1961).

DESCRIPTION: Valves relatively narrow with convex margins at midvalve and ends that are narrowly rostrate to rostrate-capitate. Due to the narrowness of the ends, the terminal expansion of the polar nodules is less noticeable than in var. *pupula*, so care must be taken in observation.

DIMENSIONS: Length, generally under 20 μm; width, as narrow as 4.2 μm; striae, similar to var. *pupula*.

68. Navicula pygmaea Kütz. var. **pygmaea** *Fig. 39a, b.*

CR: Hustedt (1964).

DESCRIPTION: Valves elliptic to elliptic-lanceolate with rounded or sometimes slightly protracted-subrostrate ends; true axial area narrow; striae radiate throughout; puncta visible with oblique light. The most noticeable character of the *Naviculae lyratae*, of which *N. pygmaea* is a part, is the H-shaped area in the striation. This mark is formed by two longitudinal clear spaces, one on each side of the valve, connected by the transverse central area.

DIMENSIONS: Length, 20–45 μm; width, 8–14 μm; striae, 24/10 μm near midvalve, becoming 30/10 μm at the ends.

69. Navicula radiosa Kütz var. **radiosa** *Fig. 45a.*

CR: Patrick and Reimer (1966).

DESCRIPTION: Valves lanceolate with acute to subacute ends; axial area narrow; central area rounded to longitudinally elliptic; raphe narrowly bandlike; striae radiate, becoming parallel or convergent at the ends; lineae fairly distinct.

DIMENSIONS: Length, 40–120 μm; width, 10–19 μm; striae, 10–12/10 μm.

70. **Navicula tenelloides** Hust. var. **tenelloides** *Fig. 50c, d, f(?)*.

CR: Patrick and Reimer (1966).

DESCRIPTION: Valve narrowly lanceolate; axial area narrow; central area small and sometimes indistinct; striae radiate, becoming convergent at the ends. The striae are rather thin. Lineae typically cannot be resolved with the light microscope. I have seen some specimens, however, in which linea are faintly visible. I believe that in these cases the "lineae" are either artifacts of observation or result from the erosion of the silica between adjacent pairs of linea in valves not part of living cells at the time of collection. Compare also with *N. notha* Wallace.

DIMENSIONS (Illinois specimens): Length, 14–27 μm; width, 3–4 μm; striae, 16–22/10 μm.

71. **Navicula radiosa** var. **tenella** (Bréb. ex Kütz.) Grun. *Fig. 45b*.

CR: Patrick and Reimer (1966).

DESCRIPTION: Valves lanceolate with acute ends; axial area narrow; central area rounded to elongate-elliptic; one central stria on each side of the valve often almost reaching the central nodule; striae radiate, becoming convergent at the ends; lineae visible with oblique light.

DIMENSIONS: Length, 18–65 μm; width, 5–7 μm; striae, 15–18/10 μm.

72. **Navicula salinarum** Grun. var. **salinarum** *Fig. 52c*.

CR: Patrick and Reimer (1966).

DESCRIPTION: Valves elliptic-lanceolate with short, rostrate ends; axial area narrow; central area fairly large and irregular in shape; striae radiate, becoming parallel or convergent at the ends and often alternating longer and shorter around the central area; lineae more or less distinct.

DIMENSIONS: Length, 23–41 μm; width, 8–12 μm; striae, 14–18/10 μm.

73. **Navicula salinarum** var. **intermedia** (Grun.) Cl. *Fig. 52b.*

CR: Patrick and Reimer (1966).

DESCRIPTION: Valves elliptic-lanceolate with attenuate rostrate-capitate ends; axial area narrow; central area irregular in shape and moderately large; striae radiate, becoming parallel or convergent at the ends and in all cases more or less distinctly lineate. The striae alternate longer and shorter around the central area, and the centermost stria on each side frequently reaches almost to the central nodule. The attenuate ends are the principal difference between thhis variety and var. *salinarum*.

DIMENSIONS: Length, 25–50 μm; width, 7–10 μm; striae, 14–18/10 μm.

74. **Navicula sanctaecrucis** Oestr. var. **sanctaecrucis** *Fig. 44b.*

CR: Patrick and Reimer (1966).

DESCRIPTION: Valves elliptic-lanceolate with rather narrow, rostrate-capitate ends; axial area narrow; central area small, formed by slight shortening of a few striae; striae radiate and punctate; pseudosepta present at both valve ends.

DIMENSIONS: Length, 25–30 μm; width, 8.5–9.5 μm; striae, 15–16/10 μm at midvalve, becoming 19–20/10 μm toward the ends.

75. **Navicula scutelloides** W. Sm. ex Greg. var. **scutelloides** *Fig. 40d.*

CR: Patrick and Reimer (1966).

DESCRIPTION: Valves almost circular in outline; axial area narrow; central area fairly small; striae radiate and coarsely punctate. The striae alternate longer and shorter around the central area.

DIMENSIONS: Length, 10–30 μm; width, 8–20 μm; striae, 7–14/10 μm; puncta, 10–16/10 μm.

76. Navicula seminuloides Hust. var. seminuloides *Fig. 37k, l.*

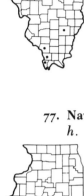

CR: Hustedt (1962).

DESCRIPTION: Valves elliptic with broadly rounded ends; axial area narrow; central area irregular in shape; striae radiate throughout; puncta not visible with the light microscope.

DIMENSIONS: Length, 9.2–14.5 μm; width, 3–6 μm; striae, 20–24/10 μm.

77. Navicula seminulum Grun. var. seminulum *Fig. 37a–c, g, h.*

CR: Hustedt (1962).

DESCRIPTION: Valve linear-elliptic with broadly rounded ends; axial area narrow; central area formed by the regular or irregular shortening of a few striae; striae radiate. Under the electron microscope the striae can be seen to be doubly punctate, and, although this feature cannot be observed directly with the light microscope, the double punctation gives the striae a somewhat blurred or fuzzy appearance. *N. seminulum* varies considerably in shape. I do not, therefore, put much faith in the varieties established on the basis of slight variations of shape alone.

DIMENSIONS: Length, 3–18 μm; width, 2–4.5 μm; striae, 18–21/10 μm.

78. Navicula seminulum var. radiosa Hust. *Fig. 37n.*

CR: Hustedt (1962).

DESCRIPTION: Valves elliptic-lanceolate with narrow, rounded ends; axial area narrow; central area indistinct; striae strongly radiate at midvalve, moderately radiate elsewhere. I have found this taxon so far only in Jackson Co.

DIMENSIONS: Length, around 14 μm; width, around 3.5 μm; striae, around 20/10 μm.

79. Navicula subadnata Hust. var. **subadnata** *Fig. 38b.*

CR: Hustedt (1962).

DESCRIPTION: Valves lanceolate with acute ends; central-axial area a broad lanceolate space; raphe lying in a narrow axial band of different thickness from the rest of the valve; striae strongly radiate throughout and not visibly punctate under the light microscope.

DIMENSIONS: Length, 7–13 μm; width, 3.3–4.3 μm; striae, 18–22/10 μm measured near midvalve.

80. Navicula subarvensis Hust. var. **subarvensis** *Fig. 35f.*

CR: Hustedt (1961).

DESCRIPTION: Valves linear with straight to weakly convex margins narrowing to form rostrate ends; axial area narrow; central area rounded or indistinct; striae radiate. The striae are fine and faint and may be observable only around the central area unless strongly obliqued light is used.

DIMENSIONS: Length, 10–13 μm; width, 3–5 μm; striae, 30/10 μm at midvalve, becoming 36–40/10 μm at the ends.

81. Navicula subhamulata Grun. in V.H. var. **subhamulata** *Fig. 39d.*

CR: Hustedt (1961).

DESCRIPTION: Valves linear-elliptic with broadly rounded ends; axial area narrow; central area no wider than axial area; striae parallel to weakly radiate near midvalve, becoming more distinctly radiate at the ends; puncta not visible with the light microscope. Each stria appears to be divided into two zones of differing thickness, giving one the impression that the striation on each side of the axis is crossed by a broad longitudinal band.

DIMENSIONS: Length, 14–20 μm; width, 4–7 μm; striae, 26/10 μm at midvalve, becoming 30/10 μm at the ends.

82. Navicula subminiscula Mang. var. **subminiscula** *Fig. 37p*.

CR: Hustedt (1962).

DESCRIPTION: Valves lanceolate; axial area narrow; central area no wider than the axial area; striae radiate throughout and not visibly punctate under the light microscope.

DIMENSIONS: Length, 8–12 μm; width, 3.0–4.5 μm; striae, 20–26/10 μm.

83. Navicula subsulcata Hust. var. **subsulcata** *Fig. 39g*.

CR: Hustedt in A.S.A. Taf. 402/59.

DESCRIPTION: Valves elliptic to elliptic-lanceolate with rounded ends; axial area narrow near the poles but widening toward midvalve and, with the central area, forming a moderately broad, lanceolate space; striae radiate throughout. Each stria is broken into two dashlike units, the gaps so aligned that they form thin longitudinal lines. In Hustedt's illustration, the gaps are near the margins. My few specimens are not entirely typical, having what appears to be a clear space along the margin. The valve structure was, however, indistinct and difficult to resolve, so it is possible that I may have failed to see the outer edges of the striae.

DIMENSIONS: Length, 15–22 μm; width, 5–7 μm; striae, 22–24/10 μm.

84. Navicula subtilissima Cl. var. **subtilissima** *Fig. 35e*.

CR: Patrick and Reimer (1966).

DESCRIPTION: Valves linear with narrow, rostrate or rostrate-capitate ends; raphe lying in a narrow axial band of different thickness from the rest of the valve; striae typically not visible with the light microscope except, if strongly obliqued light is used, near the central area.

DIMENSIONS: Length, 19–32 μm; width, 3.5–5.0 μm.

85. Navicula symmetrica Patr. var. **symmetrica** *Fig. 47c.*

CR: Patrick and Reimer (1966).

DESCRIPTION: Valves elliptic-lanceolate with subacute ends; axial area narrow; central area rounded to elongate-elliptic; striae lineate and distinctly radiate all the way to the ends.

DIMENSIONS: Length, 28–35 μm; width, 5–7 μm; striae, 15–17/10 μm.

86. Navicula tantula Hust. var. **tantula** *Fig. 37m.*

CR: Hustedt (1962).

DESCRIPTION: Valves linear with broadly rounded ends; axial area narrow; central area butterfly-shaped and almost reaching the valve margins; striae radiate throughout; puncta not visible with the light microscope.

DIMENSIONS: Length, 7–15 μm; width, 2–3 μm; striae, around 30/10 μm.

87. Navicula tenera Hust. var. **tenera** *Fig. 43b, c.*

CR: Patrick and Reimer (1966).

DESCRIPTION: Valves elliptic with rather blunt ends; striae short and distinctly punctate. The area between the striae and the raphe is clear, with the exception that the raphe branches on one side of the valve are bordered by a longitudinal row of puncta. When both valves are present, careful focusing is required since the longitudinal row on one valve is diagonally opposite to the row on the other valve. If both valves are simultaneously in focus, one may misidentify this species as *N. auriculata,* which has rows of puncta along both sides of each raphe branch.

DIMENSIONS: Length, 11–14 μm; width, 5–7 μm; striae, 16–18/10 μm.

88. Navicula terminata Hust. var. **terminata** *Fig. 42f.*

CF: Hustedt (1966).

DESCRIPTION: Valves elliptic-lanceolate with subacute ends; axial area moderately narrow; central area transversely expanded, almost reaching the valve margins; striae radiate and punctate; stigma present on one side of the central area.

DIMENSIONS: Length, 22–64 μm; width, 8–15 μm; striae, 20/10 μm at midvalve, becoming 24/10 μm at the ends.

89. Navicula terrestris Petersen var. **terrestris** *Fig. 44c.*

CR: Lund (1946).

DESCRIPTION: Valves linear-elliptic with broadly rounded ends; axial area moderately broad; central area longitudinally elliptic; both proximal raphe fissures directed sharply (almost at 90 degree angles) to the same side of the valve; striae crossed on each side of the valve by a thin longitudinal line near the axial area and parallel to it. Puncta indistinct or not visible.

DIMENSIONS: Length, 13–44 μm; width, 4–8 μm; striae, around 20/10 μm at midvalve (mine were as coarse as 14/10 μm), becoming 25–30/10 μm elsewhere.

90. Navicula terrestris var. **relicta** f. **triundulata** Lund *Fig. 44f.*

CR: Lund (1946).

DESCRIPTION: Valves similar in appearance to var. *terrestris* with the exception that the margins are weakly triundulate and the ends slightly set off subcapitate.

DIMENSIONS: Length, 14–27 μm; width, 5–6 μm; striae, as for var. *terrestris*.

91. **Navicula texana** Patr. var. **texana** *Fig. 42e*.

CR: Patrick and Reimer (1966).

DESCRIPTION: Valve margins straight to weakly convex at midvalve, narrowing fairly abruptly to form rostrate-capitate ends; axial area narrow; central area irregular in outline; striae radiate and distinctly punctate; striae around central area sometimes alternating longer and shorter; an isolated punctum found at the tip of one terminal stria.

DIMENSIONS: Length, 18–22 μm; width, 5–7 μm; striae, 18–21/10 μm.

92. **Navicula tridentula** Krasske var. **tridentula** *Fig. 35d*.

CR: Hustedt (1961).

DESCRIPTION: Valve margins triundulate; valve ends rostrate; raphe lying in a narrow axial band of different thickness from adjacent parts of the valve.

DIMENSIONS: Length, 12–19 μm; width, 3.5–4 μm; striae, not visible with the light microscope.

93. **Navicula tripunctata** O. F. Muell. var. **tripunctata** *Fig. 47a*.

CR: Patrick and Reimer (1966).

DESCRIPTION: Valves linear elliptic-lanceolate with subacute ends; axial area narrow; central area rounded, sometimes appearing almost square; striae weakly radiate near midvalve, becoming convergent at the ends; striae more or less distinctly lineate.

DIMENSIONS: Length, 33–60 μm; width, 6–10 μm; striae, 11–12/10 μm.

94. **Navicula tripunctata** var. **schizonemoides** (V.H.) Patr. *Figs. 47b, d; 50b.*

CR: Patrick and Reimer (1966).

DESCRIPTION: Valves lanceolate to elliptic-lanceolate with subacute ends; axial area narrow; central area irregularly rounded to longitudinally elliptic; striae moderately radiate near midvalve, becoming slightly convergent at the ends and more or less distinctly lineate throughout.

DIMENSIONS: Length, 24–60 μm; width, 6–11 μm; striae, 11–16/10 μm.

95. **Navicula tuscula** Ehr. var. **tuscula** *Fig. 40b.*

CR: Patrick and Reimer (1966).

DESCRIPTION: Valves elliptic with moderately protracted, narrow, rostrate-capitate ends; axial area narrow; central area transversely widened but not reaching the margins; striae doubly punctate. Each stria is broken into dashlike units for about half its length, becoming unbroken toward the valve margin. The units are so arranged that undulating longitudinal clear spaces are formed. A single isolated punctum may be seen in the central area.

DIMENSIONS: Length, 12–70 μm; width, 7–22 μm; striae, 12–14/10 μm.

96. **Navicula vaucheriae** Peters. var. **vaucheriae** *Fig. 38f, i.*

CR: Hustedt (1961).

DESCRIPTION: Valves lanceolate to elliptic-lanceolate with subacute to rounded ends; axial area narrow; central area no wider than the axial area; striae weakly radiate to parallel and somewhat more widely spaced near midvalve than at the ends, the density changing gradually; puncta not visible with the light microscope.

DIMENSIONS: Length, 8.5–15.0 μm; width, 4–6 μm; striae, around 23/10 μm on the average.

97. Navicula ventosa Hust. var. ventosa *Fig. 37-o*.

CR: Hustedt (1962).

DESCRIPTION: Valve elliptic-lanceolate with rounded ends; axial area narrow; central area no wider than axial area; striae radiate throughout; puncta not visible with the light microscope.

DIMENSIONS: Length, 6–8 μm; width, 2.5–3.5 μm; striae, 28/10 μm.

98. Navicula viridula (Kütz.) Kütz. emend. V.H. var. **viridula** *Fig. 46c*.

CR: Patrick and Reimer (1966).

DESCRIPTION: Valves linear elliptic-lanceolate with broadly subrostrate ends; axial area moderately narrow; central area moderately wide but not reaching the margins; striae radiate and somewhat undulate, becoming convergent at the ends; striae usually distinctly lineate. Some of the named varieties of this species may prove to be connected with the nominate variety by intergrades.

DIMENSIONS: Length, 35–80 μm; width, 10–15 μm; striae, 8–10/10 μm.

99. Navicula viridula var. **avenacea** (Bréb ex Grun.) V.H. *Fig. 45d*.

CR: Patrick and Reimer (1966).

DESCRIPTION: Valves lanceolate with somewhat blunt, almost truncate, ends; striae somewhat finer and less undulate than in var. *viridula;* other particulars as for var. viridula.

DIMENSIONS: Length, 30–60 μm; width, 8–10 μm; striae, 10–12/10 μm at midvalve, becoming 14/10 μm toward the ends.

100. **Navicula viridula** var. **linearis** Hust. *Fig. 46a*.

CR: Patrick and Reimer (1966).

DESCRIPTION: Valves linear with truncate-cuneate ends; valve margins nearly parallel at midvalve; other particulars of structure as for var. *viridula*.

DIMENSIONS: Length, 65–100 μm; width, 11–12 μm; striae, 8–10/10 μm.

101. **Navicula viridula** var. **rostellata** (Kütz.) Cl. *Fig. 46b*.

CR: Patrick and Reimer (1966).

DESCRIPTION: Valves broadly elliptic-lanceolate with narrow, attenuate, rostrate-capitate ends; other particulars of structure as for var. *viridula*.

DIMENSIONS: Length, 33–65 μm; width, 8–11 μm; striae, 9–13/10 μm.

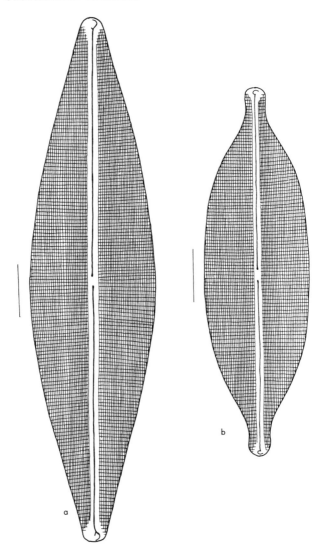

Fig. 31. a. *Navicula cuspidata* v. *cuspidata*. **b.** *N. cuspidata* v. *ambigua*. (Scale lines equal 10 micrometers).

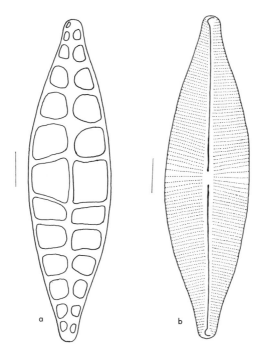

Fig. 32. a. *Navicula cuspidata* v. *cuspidata* (craticular plate). ***b.*** *N. cuspidata* v. *cuspidata* (atypical valve). (Scale lines equal 10 micrometers).

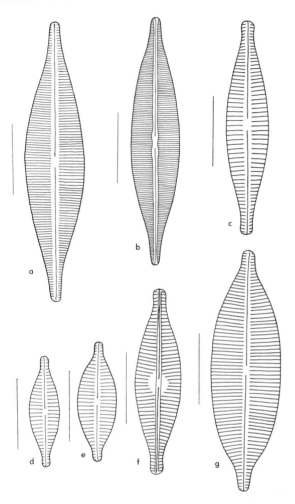

Fig. 33. *a. Navicula halophila* f. *tenuirostris. b. N. belliatula* v. *belliatula. c–e. N. gregaria* v. *gregaria. f. N. cryptocephala* v. *cryptocephala*(?). *g. N. accomoda* v. *accomoda.* (Scale lines equal 10 micrometers.)

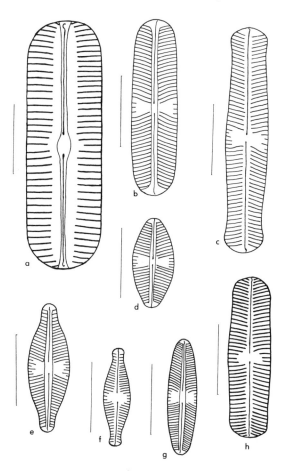

Fig. 34. a. Navicula americana v. *americana. b. N. pupula* v. *rectangularis. c. N. pupula* v. *pupula. d. N. pupula* v. *elliptica. e, f. N. pupula* f. *rostrata. g. N. pupula* v. *elliptica. h. N. laevissima* v. *laevissima.* (Scale lines equal 10 micrometers.)

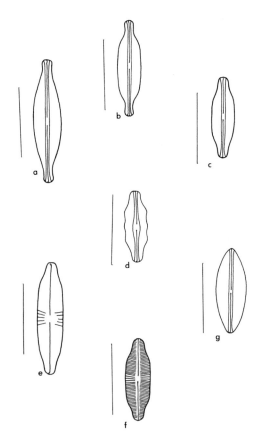

Fig. 35. *a. Navicula gysingensis* v. *gysingensis*. *b, c. N. arvensis* v. *arvensis*. *d. N. tridentula* v. *tridentula*. *e. N. subtilissima* v. *subtilissima*. *f. N. subarvensis* v. *subarvensis*. *g. N. indifferens* v. *indifferens*. (Scale lines equal 10 micrometers.)

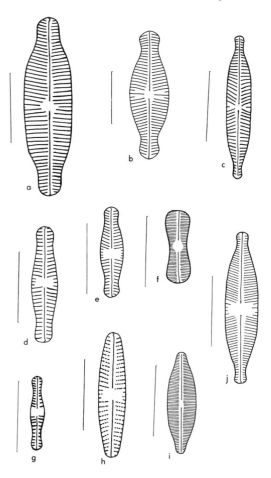

Fig. 36. a. *Navicula protracta* v. *protracta*. **b.** *N. laterostrata* v. *laterostrata*. **c.** *N. bicephala* v. *bicephala*. **d, e.** *N. disjuncta* v. *disjuncta*. **f.** *N. contenta* v. *biceps*. **g.** *N. hassiaca* v. *hassiaca*. **h.** *N. annexa* v. *annexa*. **i.** *N. miniscula* v. *miniscula*. **j.** *N. brockmannii* v. *brockmannii*. (Scale lines equal 10 micrometers.)

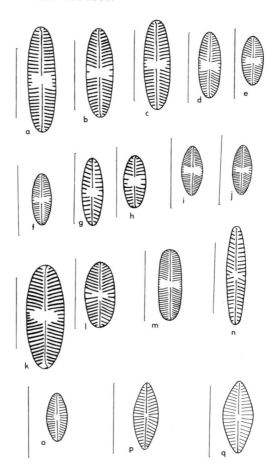

Fig. 37. *a–c. Navicula seminulum* v. *seminulum*. *d–f. N. minima* v. *minima*. *g, h. N. seminulum* v. *seminulum*. *i, j. N. nigrii* v. *nigrii*. *k, l. N. seminuloides* v. *seminuloides*. *m. N. tantula* v. *tantula*. *n. N. Seminulum* v. *radiosa*. *o. N. ventosa* v. *ventosa*. *p. N. subminiscula* v. *subminiscula*. *q. N. frugalis* v. *frugalis*. (Scale lines equal 10 micrometers.)

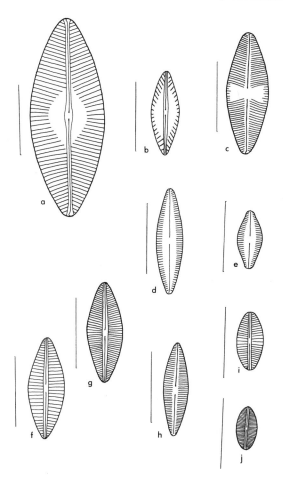

Fig. 38. a. Navicula bacilloides v. bacilloides. b. N. subadnata v. subadnata. c. N. lapidosa v. lapidosa. d, e. N. krasskei v. krasskei. f. N. vaucheriae v. vaucheriae. g. N. fluens v. fluens. h. N. biconica v. biconica. i. N. vaucheriae v. vaucheriae. j. N. muralis v. muralis. (Scale lines equal 10 micrometers.)

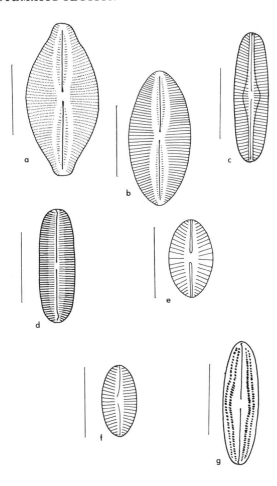

Fig. 39. *a, b. Navicula pygmaea* v. *pygmaea. c. N. monoculata* v. *monoculata. d. N. subhamulata* v. *subhamulata. e. N. excelsa* v. *excelsa. f. N. brevissima* v. *brevissima. g. N. subsulcata* v. *subsulcata.* (Scale lines equal 10 micrometers.)

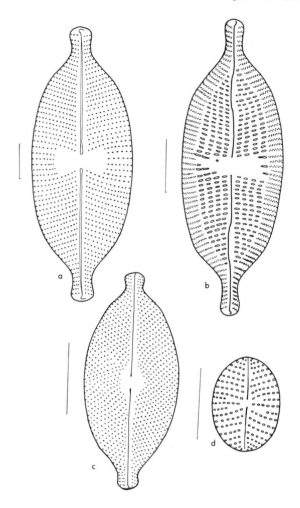

Fig. 40. a. Navicula amphibola v. amphibola. **b.** N. tuscula v. tuscula. **c.** N. placenta v. placenta. **d.** N. scutelloides v. scutelloides. (Scale lines equal 10 micrometers.)

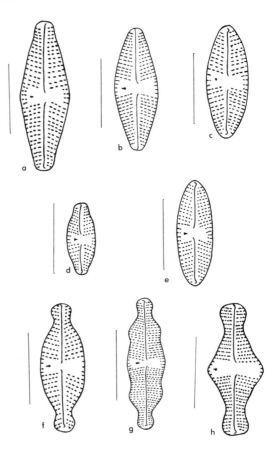

Fig. 41. *a–e. Navicula mutica* v. *mutica. f. N. mutica* v. *mutica*(?). *g. N. mutica* v. *nivalis. h. N. heufleriana* v. *heufleriana.* (Scale lines equal 10 micrometers.)

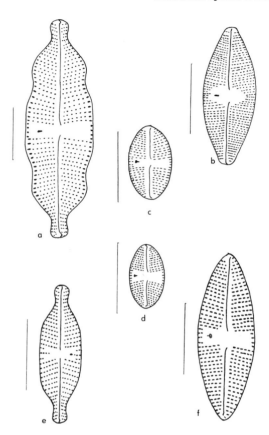

Fig. 42. a. Navicula charlatii v. *charlatii. b–d. N. muticoides* v. *muticoides. e.*
N. texana v. *texana. f. N. terminata* v. *terminata.* (Scale lines equal 10 micrometers.)

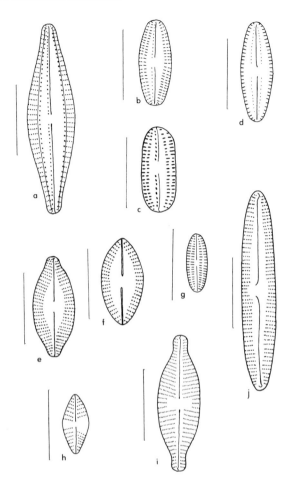

Fig. 43. a. *Navicula circumtexta* v. *circumtexta*. **b, c.** *N. tenera* v. *tenera*. **d.** *N. auriculata* v. *auriculata*. **e, f.** *N. confervacea* v. *confervacea*. **g.** *N. insociabilis* v. *insociabilis*. **h.** *N. ingenua* v. *ingenua*. **i.** *N. grimmei* v. *grimmei*. **j.** *N. monmouthiana-stodderi* v. *monmouthiana-stodderi*. (Scale lines equal 10 micrometers.)

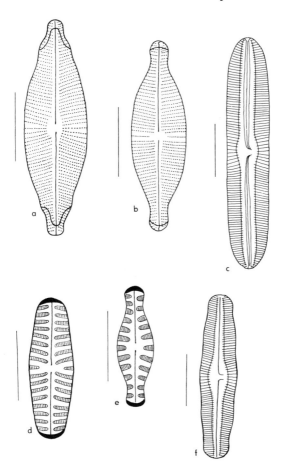

Fig. 44. a. Navicula integra v. integra. **b.** N. sanctaecrucis v. sanctaecrucis. **c.** N. terrestris v. terrestris. **d.** N. capitata v. hungarica. **e.** N. capitata v. capitata. **f.** N. terrestris v. relicta f. triundulata. (Scale lines equal 10 micrometers.)

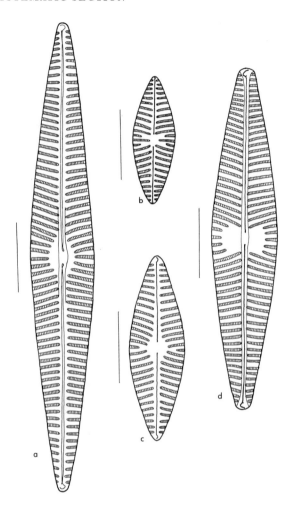

Fig. 45. *a. Navicula radiosa* v. *radiosa*. *b. N. radiosa* v. *tenella*. *c. N. menisculus* v. *menisculus*. *d. N. viridula* v. *avenacea*. (Scale lines equal 10 micrometers.)

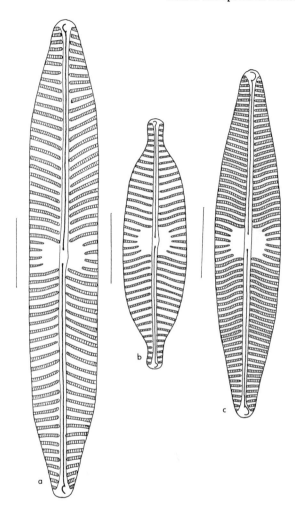

Fig. 46. a. Navicula viridula v. *linearis. b. N. viridula* v. *rostellata. c. N. viridula* v. *viridula.* (Scale lines equal 10 micrometers.)

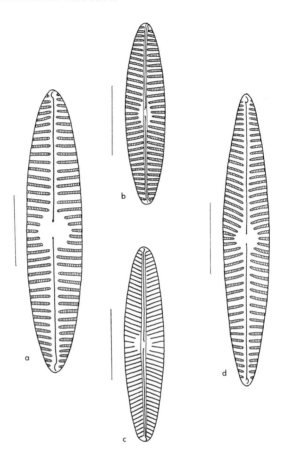

Fig. 47. a. Navicula tripunctata v. tripunctata. **b.** N. tripunctata v. schizonemoides. **c.** N. symmetrica v. symmetrica. **d.** N. tripunctata v. schizonemoides. (Scale lines equal 10 micrometers.)

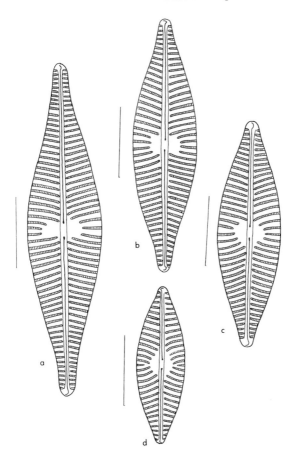

Fig. 48. a–d. Navicula lanceolata v. *lanceolata*. (Scale lines equal 10 micrometers.)

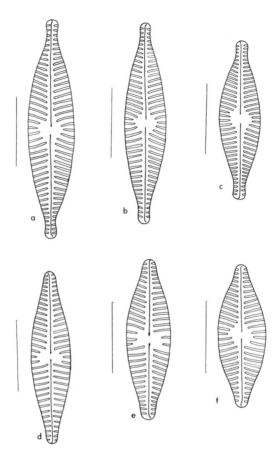

Fig. 49. a–f. *Navicula cryptocephala* v. *cryptocephala*. (Scale lines equal 10 micrometers.)

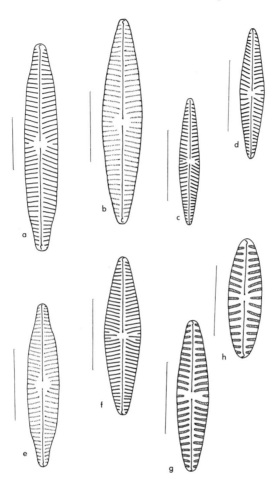

Fig. 50. *a. Navicula angusta* v. *angusta*. *b. N. tripunctata* v. *schizonemoides*. *c, d. N. radiosa* v. *minutissima*. *e. N. notha* v. *notha*. *f. N. radiosa tennelloides*(?). *g. N. heufleri* v. *heufleri*. *h. N. cincta* v. *cincta*. (Scale lines equal 10 micrometers.)

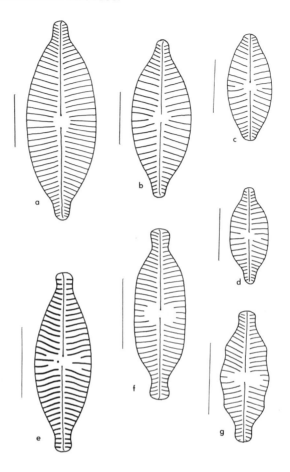

Fig. 51. *a–d. Navicula gastrum* v. *gastrum. e. N. decussis* v. *decussis. f. N. elginensis* v. *elginensis. g. N. elginensis* v. *neglecta.* (Scale lines equal 10 micrometers.)

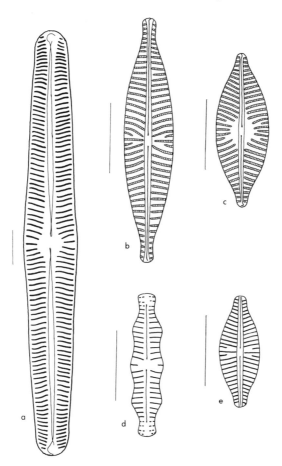

Fig. 52. a. Navicula oblonga v. oblonga. **b.** N. salinarum v. intermedia. **c.** N. salinarum v. salinarum. **d.** N. ignota v. ignota. **e.** N. hambergii v. hambergii. (Scale lines equal 10 micrometers.)

28. *Caloneis* Cleve 1894

Caloneis belongs to the order Naviculales of class Pennatibacillariophyceae. As with other genera in the order, valve markings are bilaterally symmetric and both valves bear a fully developed two-branch raphe that is not part of a keel-type or canal-type raphe structure.

Caloneis is distinguished from most of the other genera in the order by the nature of the striae. Instead of a row of simple or complex puncta in the valve surface, the stria in *Caloneis* is in fact a narrow tubular structure perforated by numerous fine pores that cannot be resolved with the light microscope. The tubular stria (in many species) is open to the inside of the valve by one fairly large pore. These openings are at about the same place on each stria, thus giving rise to the perception of a longitudinal line or band crossing the striae on both sides of the axial area. Such lines are, however, not evident in all species, especially the smaller ones.

The striae of *Pinnularia* are similar in structure to those of *Caloneis* and it is possible that *Caloneis* should in fact be considered a taxonomic section of *Pinnularia*. It is usually easy, though, to separate the two for practical purposes. The striae of *Caloneis* are almost always thin and are about the same width as the spaces between them. *Pinnularia* striae are usually broad and are frequently wider than the spaces between. *Caloneis* striae also are usually denser than those of *Pinnularia* (typically over 16/10 μm in *Caloneis*, but often considerably less in *Pinnularia*).

KEY TO THE TAXA OF *Caloneis* IN ILLINOIS

1. Valve margins triundulate; ends usually broadly rounded, sometimes rostrate _____ 2
1. Valve margins weakly to strongly convex; distinctive ends present or absent _____ 6
   2. Valves with distinct, short-rostrate ends _____
   _____ 12. *C. lewisii* var. *inflata*
   2. Valve ends rounded, not set off from the rest of the valve _____ 3
3. Valves with lunate markings (sometime faint) in the central area __ 4
3. Lunate marking absent from the central area _____ 5
   4. Valve margins deeply undulate; striae usually strongly radiate around the central area _____ 11. *C. lewisii*
   4. Valve margins weakly to moderately undulate; width 6–7 μm; striae moderately radiate around the central area __ 13. *C. limosa*

5. Margins weakly to moderately undulate; width 6–7 μm; length under 25 μm; central area reaching both margins _____ _____ 14. *C. ventricosa* var. *minuta*

5. Margins moderately undulate; width 7–10 μm; length 30–42 μm; central area rounded, sometimes reaching one or both margins _____ _____ 15. *C. ventricosa* var. *truncatula*

6. Terminal striae convergent _____ 7

6. Terminal striae radiate _____ 11

7. Valves capitate or subcapitate _____ 8

7. Valves narrowly lanceolate without distinctive ends; central area a narrow to elongate fascia _____ 9. *C. lagerstedtii*

8. Valves subcapitate _____ 9

8. Valves distinctly capitate _____ 10

9. Striae 16–18/10 μm; width 6–15 μm; longitudinal lines quite distinct _____ 1. *C. aequitorialis*

9. Striae 20–24/10 μm; width 5–6 μm; longitudinal lines not particularly distinct _____ 7. *C. hultenii*

10. Valves broad and elliptic with narrow ends; each stria seemingly thicker near the margin than near the axis, the zones sharply separated; central-axial area broadly lanceolate __ 2. *C. amphisbaena*

10. Valves linear-elliptic with rather broad, capitate ends; striae not divided into zones; central area reaching both margins _____ _____ 10. *C. lenzii*

11. Central area reaching both margins _____ 12

11. Central area rather narrow and elongate-elliptic _____ _____ 3. *C. bacillaris* var. *thermalis*

12. Striae 34–38/10 μm and frequently resolvable only with oblique light _____ 8. *C. hyalina*

12. Striae 26–30/10 μm and typically easy to resolve _____ 13

13. Valves linear with broadly rounded ends _____ 14

13. Valves lanceolate _____ 5. *C. bacillum* var. *angusta*

14. Axial area beginning to flare at a point halfway from the end to the central area _____ 6. *C. bacillum* f. *fonticola*

14. Axial area flaring just before the central area ____ 4. *C. bacillum*

1. Caloneis aequatorialis Hust. var. aequatorialis *Fig. 55b*.

CR: Manguin (1962).

DESCRIPTION: Valves linear with somewhat convex sides and broad, blunt ends; axial area lanceolate, opening suddenly into a central area that reaches both margins; striae radiate and rather thick; striation may continue around the poles; striae crossed by a moderately broad longitudinal band on each side of the cell.

DIMENSIONS: Length, 26–28 μm (mine reached 38 μm); width, 6–15 μm; striae, 16–18/10 μm (mine reached 20/10 μm).

2. Caloneis amphisbaena (Bory) Cl. var. amphisbaena *Fig. 53f*.

CR: Patrick and Reimer (1966).

DESCRIPTION: Valves broadly elliptic-lanceolate with narrow rostrate-capitate ends; axial area widening gradually into the lanceolate central area; striae radiate, becoming convergent at the ends; striae crossed by a broad band on both sides of the axial area.

DIMENSIONS: Length, 36–80 μm; width, 20–30 μm; striae, 15–18/10 μm (mine were as low as 13/10 μm).

3. Caloneis bacillaris var. thermalis (Grun.) A. Cl. *Fig. 54e–h*.

CR: Patrick and Reimer (1966).

DESCRIPTION: Valves linear-elliptic with rounded ends; axial area fairly narrow, opening into a rounded central area; striae moderately radiate; striae not crossed by very distinctive lateral lines.

DIMENSIONS: Length, 34–55 μm (mine were as short as 17 μm); width, 6–10 μm; striae, 18–23/10 μm.

4. Caloneis bacillum (Grun.) Cl. var. **bacillum** *Fig. 54b*.

CR: Hustedt (1930); Patrick and Reimer (1966).

DESCRIPTION: Valves linear-elliptic with broadly rounded ends; axial area widening gradually at first, then flaring fairly suddenly just before the central area; central area reaching both valve margins; striae weakly radiate throughout; longitudinal lines may cross the striae near the margin, but as these lines are weakly expressed in most specimens, I have omitted them from the drawing. Many varieties and forms of this species have been described, but many seem to have overlapping characters. I am inclined to believe that *C. bacillum* is simply a highly variable species.

DIMENSIONS: Length, 15–45 μm; width, 4–9 μm; striae, 21–30/10 μm.

5. Caloneis bacillum var. **angusta** A. Mayer *Fig. 54c, d*.

CR: Cleve-Euler (1955).

DESCRIPTION: Valves elliptic-lanceolate with subacute ends; axial area variable, sometimes flaring before the central area, sometimes not; central area reaching both valve margins; longitudinal lines not evident; striae radiate throughout.

DIMENSIONS: Length, 15–28 μm; width, 4–5 μm; striae, 21–25/10 μm.

6. Caloneis bacillum f. **fonticola** (Grun.) Mayer *Fig. 54a*.

CR: Mayer (1940).

DESCRIPTION: Valves linear-elliptic with rounded to broadly subacute ends; axial area beginning to flare about halfway from the ends to the central area, which reaches both margins; striae weakly to moderately radiate and not crossed by obvious longitudinal lines.

DIMENSIONS: Length, about 19 μm (mine varied from 15–25 μm); width, about 6 μm (mine varied from 4–6 μm); striae, about 23/10 μm (mine varied from 23–25/10 μm).

7. **Caloneis hultenii** Peters. var. **hultenii** *Fig. 55a.*

CR: Petersen (1946).

DESCRIPTION: Valves linear with weakly convex margins and ends that are weakly subcapitate; axial area narrow; central area widened to both margins; striae nearly parallel and continuing all around the polar regions; striae not crossed by prominent lines or bands.

DIMENSIONS: Length, 25–34 μm; width, 5–6 μm; striae, 20–24/10 μm.

8. **Caloneis hyalina** Hust. var. **hyalina** *Fig. 54i.*

CR: Patrick and Reimer (1966).

DESCRIPTION: Valves elliptic-lanceolate without distinctive ends; axial area narrow; central area widened to both margins; striae faint and almost parallel; striae not crossed by distinctive lines or bands.

DIMENSIONS: Length, 12–28 μm; width, 4–5 μm; striae, 34–38/10 μm.

9. **Caloneis lagerstedtii** Choln. var. **lagerstedtii** *Fig. 55d, e.*

CR: Cholnoky (1957).

DESCRIPTION: Valves lanceolate with acute to subacute ends; axial area narrow; raphe branches somewhat arched; central area reaching both valve margins; striae nearly parallel in most of the valve but becoming slightly convergent at the ends.

DIMENSIONS: Length, 14–26 μm; width, 4–6 μm; striae, 24–26/10 μm.

10. Caloneis lenzii Krasske var **lenzii** *Fig. 55c.*

CR: Krasske (1951).

DESCRIPTION: Valves linear with distinctly capitate ends; axial area widening gradually before opening into the central area, which reaches both margins; striae nearly parallel except at the ends where they are convergent; striae not crossed by distinctive lines or bands.

DIMENSIONS: Length, 29–30 μm; width, 5.8–6.5 μm; striae, 22–26/10 μm.

11. Caloneis **lewisii** Patr. var. **lewisii** *Fig. 53d.*

CR: Patrick and Reimer (1966).

DESCRIPTION: Valve margin triundulate, the polar undulations of equal width to the central one; axial area narrow; central area large and rounded, with two large lunate marks that face each other across the central nodule; striae radiate.

DIMENSIONS: Length, 27–42 μm; width, 8–11 μm; striae, 18–20/10 μm.

12. Caloneis **lewisii** var. **inflata** (Schultze) Patr. *Fig. 53a.*

CR: Patrick and Reimer (1966).

DESCRIPTION: Valve margin triundulate, all three undulations of approximately equal magnitude; ends protracted and narrowly rostrate; axial area narrow; central area large and rounded; lunate marks present or absent in central area.

DIMENSIONS: Length, 28–40 μm; width, 9–13 μm; striae, 18–20/10 μm.

13. Caloneis limosa (Kütz.) Patr. var. limosa *Fig. 53c.*

CR: Patrick and Reimer (1966).

DESCRIPTION: Valves linear to linear-elliptic with broadly rounded ends and margins that are weakly to strongly undulate; axial area moderately broad; central area rounded. In the central area is a pair of lunate marks that face across the central nodule like facing parentheses. These marks are usually well developed, but in some cases are not. If the pattern holds that I have seen with other species having such marks, specimens may well be found that lack the marks altogether. Likewise, a number of related taxa with typically enclosed central areas may have populations in which the central area reaches one or both margins, and I expect this with *C. limosa* also. The undulation of the valve margins of *C. limosa* is usually distinct, but in some specimens is not. In these cases, the form almost approaches that of *C. alpestris* (Grun.) Cl., which is usually more finely striate and is reported to have truly straight margins.

DIMENSIONS: Length, 22–56 μm; width, 8–14 μm; striae, 16–20/10 μm.

14. Caloneis ventricosa var. minuta (Grun.) Patr. *Fig. 53e.*

CR: Patrick and Reimer (1966).

DESCRIPTION: Valves more or less linear with triundulate margins and broadly subacute ends; axial area fairly narrow; central area reaching both margins.

DIMENSIONS: Length, 22–36 μm; width, 6–7 μm; striae, 21–22/10 μm.

15. Caloneis ventricosa var. **truncatula** (Grun.) Meist. *Fig. 53b.*

CR: Patrick and Reimer (1966).

DESCRIPTION: Valve margins weakly triundulate; ends broadly subacute; axial area narrow; central area rounded or reaching one or both margins.

DIMENSIONS: Length, 30–42 μm; width, 7–10 μm; striae, 20–22/10 μm.

29. *Pinnularia* Ehrenberg 1843, nom. cons.

Pinnularia belongs to the order Naviculales and, like all genera of the order, is characterized by the presence of a true raphe on both valves of each cell and by bilateral symmetry of valve markings. Cells of *Pinnularia* may be seen in both valve and girdle views. Cells are rectangular when seen in girdle view, but are not sufficiently distinctive in that view to allow identification beyond genus.

Valves are symmetric to both long and transverse axes. Both of these axes are straight. Valve markings consist of striaelike markings that have been shown to have different structure from true striae. Each of these pseudostriae is in fact a tubule that can be seen with the electron microscope to be finely perforate and that may be open to the interior of the valve by a narrow or wide opening or foramen. These openings can be seen with the light microscope and are aligned in such a way as to form longitudinal bands. Such bands are not visible in all species, and it appears to be the case in some instances (and it may be true in all) that this is because the striae are open to the valve interior all along their length.

Pinnularia is very similar to *Caloneis* and perhaps the two genera should be united. Those species which are called *Pinnularia* generally have fewer than 15/10 μm striae, and the striae generally appear robust, not slender. In general, cells that exceed 100 μm in length are almost surely *Pinnulariae*, but this is a practical, not a diagnostic, character. Differentiation of the genera is best learned by experience.

KEY TO Pinnularia CONFIRMED FOR ILLINOIS

1. Striae crossed by faint or distinct longitudinal bands on both sides of the valve _____ 2
1. Striae not crossed by such bands _____ 4
   2. Raphe complex _____ 3
   2. Raphe bandlike _____ 21. *P. maior* var. *transversa*
3. Striae very coarse, no more than 5/10 μm _____ 26. *P. streptoraphe*
3. Striae moderately coarse, 6–12/10 μm _____ 28. *P. viridis*
   4. Valves with ends set off by constructions or expansions _____ 18
   4. Valves with blunt, rounded, or acute ends that are not set off by constrictions or expansions _____ 5
5. Raphe complex _____ 28. *P. viridis*
5. Raphe linelike or bandlike _____ 6
   6. Each stria with an apparently thickened cap at the end next to the axial area _____ 17. *P. instita*
   6. Striae without such caps _____ 7
7. Striae 6/10 μm or fewer _____ 8
7. Striae 7/10 μm or more _____ 10
   8. Valve ends almost square _____ 16. *P. dubitabilis*
   8. Valve ends rounded to truncate-cuneate _____ 9
9. Valve ends rounded _____ 8. *P. borealis*
9. Valve ends truncate-cuneate _____ 9. *P. borealis* var. *congolensis*
   10. Distal raphe ends long and bayonet-shaped; lunate marks typically present in the central area _____ 25. *P. stomatophora*
   10. Distal raphe ends indistinct, comma-shaped, or hooked; no lunate marks in the central area _____ 11
11. Axial and central areas with a granular or "speckled" appearance throughout _____ 3. *P. acrosphaeria*
11. Axial and central areas devoid of granular or speckled markings __ 12
   12. Striae parallel near midvalve _____ 13
   12. Striae radiate near midvalve _____ 14
13. Valves 50 μm long or more _____ 5. *P. acuminata*
13. Valves under 40 μm long _____ 13. *P. castor.*
   14. Valves more or less linear with truncate-cuneate ends _____ 15
   14. Valves elliptic-lanceolate with subacute ends _____ 16
15. Valves typically under 5 μm wide; raphe filiform __ 18. *P. intermedia*
15. Valves typically over 7 μm wide; raphe narrowly bandlike _____
   _____ 24. *P. microstauron*
   16. Valve face arched; striae radiate throughout ___ 20. *P. leptosoma*
   16. Valve face more or less flat; striae convergent or sometimes parallel at the ends _____ 17

17. Valves 40 μm long or longer _____ 11. *P. brebissonii*
17. Valves less than 40 μm long _____ 12. *P. brebissonii* var. *diminuta*
  18. Central and axial areas broad, with a granular or "speckled" appearance throughout _____ 19
  18. Central and axial areas variously marked, but not granular throughout _____ 20
19. Valves almost linear, the ends set off by very weak constrictions _____ _____ 3. *P. acrosphaeria*
19. Valve margins strongly inflated at midvalve; ends distinctly capitate __ _____ 4. *P. acrosphaeria* var. *turgidula*
  20. Valve margins triundulate; ends capitate _____ 21
  20. Valve margins straight, convex, or weakly concave at midvalve; ends rostrate to capitate _____ 23
21. Large valves, 15 μm wide or wider _____ 19. *P. legumen*
21. Small to moderate sized valves, not exceeding 11 μm in width __ 22
  22. Small valves, typically under 30 μm long and under 6 μm wide __ _____ 7. *P. biceps* f. *petersenii*
  22. Moderate sized valves, typically over 30 μm long and over 6 μm wide _____ 23. *P. mesolepta*
23. Distal raphe ends long and bayonet-shaped; lunate marks sometimes present in the central area _____ 24
23. Distal raphe ends hooked, comma-shaped, or indistinct; lunate marks lacking _____ 26
  24. Valves almost linear with subrostrate ends; lunate marks sometimes present in the central area _____ 25. *P. stomatophora*
  24. Valve margins distinctly convex at midvalve; ends capitate to rostrate; lunate marks lacking _____ 25
25. Valves around 40 μm long; ends rostrate _____ 14. *P. caudata*
25. Valves usually over 45 μm long; ends capitate __ 22. *P. mesogongyla*
  26. Valve ends distinctly capitate _____ 27
  26. Valve ends rostrate to subcapitate _____ 30
27. Striae strongly radiate at midvalve, becoming strongly convergent toward the ends _____ 15. *P. divergentissima*
27. Striae moderately radiate over most of the valve, becoming convergent at the ends _____ 28
  28. Axial area beginning to flare around halfway between the ends and the central area _____ 10. *P. braunii* var. *amphicephala*
  28. Axial area flaring just before the central area _____ 29
29. Valves 6 μm wide or less and 30 μm long or less _____ _____ 7. *P. biceps* f. *petersenii*
29. Valves 7 μm wide or wider and 38 μm long or longer ___ 6. *P. biceps*

30. Valve ends rostrate _____ 31
30. Valve ends subcapitate _____ 32
31. Valves generally over 40 μm long; striae changing direction gradually from radiate to convergent at the ends _ 2. *P. abaujensis* var. *rostrata*
31. Valves generally under 40 μm long; striae changing direction more or less abruptly _____ 18. *P. intermedia*
32. Valves linear-lanceolate with broadly subcapitate ends; length generally over 50 μm _____ 1. *P. abaujensis*
32. Valves more or less linear with moderately wide subcapitate ends; length generally under 50 μm _____ 27. *P. subcapitata*

1. Pinnularia abaujensis (Pant.) Ross var. **abaujensis** *Fig. 57a*.

CR: Patrick and Reimer (1966).
DESCRIPTION: Valves linear-lanceolate with capitate to subcapitate ends. Axial area widening gradually toward the central area, which may be enclosed or which may reach one or both valve margins; axial raphe filiform, the proximal ends curving slightly toward one side of the valve and the distal ends hook-shaped; striae radiate, becoming convergent toward the ends. As with many other species, short striae continue all around the ends. Longitudinal lines or bands not visible.
DIMENSIONS: Length, 50–140 μm; width, 7–13 μm; striae, 9–13/10 μm.

2. Pinnularia abaujensis var. **rostrata** (Patr.) Patr. *Fig. 57b(?)*, *d–f*.

CR: Patrick and Reimer (1966).
DESCRIPTION: Valves linear, typically with weakly convex margins and with rather broad, rostrate ends; raphe similar to var. *abaujensis*; striae weakly radiate, becoming convergent toward the ends; short striae visible all the way around the ends. I have observed more variability in this taxon than Patrick and Reimer report. The axial area in some specimens flared just before the axial area while in others, such as the borderline specimen shown in *Fig. 57b*, the axial area began to widen very near the ends. Patrick and Reimer report that the central area is enclosed. I found it to have the same variability as in var. *abaujensis* and as often as not reaching both margins. It is quite possible that I was dealing with a complex of taxa, some described, some not, but all intergrading in form.

DIMENSIONS: Length, 65–80 μm (mine were as short as 40 μm); width, 8–11 μm; striae, 8–11/10 μm.

3. **Pinnularia acrosphaeria** W. Sm. var. **acrosphaeria** *Fig. 60b*.

CR: Patrick and Reimer (1966).

DESCRIPTION: Valves linear with broadly rounded ends; margins usually slightly convex near midvalve; axial and central areas forming a broad linear space, this combined space marked by irregularly placed ghostlike dots or warts; striae parallel to weakly radiate near midvalve, becoming more strongly radiate at the ends; raphe filamentous to slightly bandlike, but not complex.

DIMENSIONS: Length, 30–180 μm; width, 8–20 μm; striae, 6–14/10 μm.

4. **Pinnularia acrosphaeria** var. **turgidula** Grun. ex Cl. *Fig. 60a*.

CR: Patrick and Reimer (1966).

DESCRIPTION: Valve margins strongly convex at midvalve, then narrowing considerably before expanding to form the broadly capitate ends; central and axial areas forming a space similar in shape to the margin and marked by the same sort of ghostlike marks as in the nominate variety.

DIMENSIONS: Length, 48–70 μm; width, 12–15 μm; striae, 10–12/10 μm.

5. **Pinnularia acuminata** W. Sm. var. **acuminata** *Fig. 56f*.

CR: Patrick and Reimer (1966).

DESCRIPTION: Valves linear with straight margins and cuneate ends; axial area very broad; central area scarcely if at all wider than the axial area; raphe filiform, with hook-shaped distal ends; striae nearly parallel over most of the valve; striae not crossed by longitudinal lines or bands.

DIMENSIONS: Length, 55–90 μm; width, 11–18 μm; striae, 8–10/10 μm.

6. Pinnularia biceps Greg. var. **biceps** *Fig. 58b.*

CR: Patrick and Reimer (1966).

DESCRIPTION: Valves linear with distinctly capitate to rostrate-capitate ends that are narrower than the maximum valve width; valve margins typically straight, sometimes slightly concave at midvalve; axial area widening gradually toward the central area, which may be enclosed (rounded or rhombic) or may reach one or both valve margins; striae radiate, becoming convergent toward the ends; longitudinal lines or bands not evident. I am almost convinced that this taxon intergrades with *P. mesolepta* (Ehr.) W. Sm. by a form-series with increasing undulation of the margins.

DIMENSIONS: Length, 38–80 μm; width, 7–16 μm; striae, 9–14/10 μm.

7. Pinnularia biceps f. **petersenii** Ross *Fig. 58e, f.*

CR: Patrick and Reimer (1966).

DESCRIPTION: This taxon has valves of similar shape and appearance to var. *biceps*, but with different dimensions and with margins that vary from straight to triundulate.

DIMENSIONS: Length, 22–30 μm; width, 4.5–6 μm; striae, 12–15/10 μm.

8. Pinnularia borealis Ehr. var. **borealis** *Fig. 60e.*

CR: Patrick and Reimer (1966).

DESCRIPTION: Valves linear-elliptic with convex margins; axial area moderately broad; central area rounded to indistinct; raphe filiform; striae nearly parallel and very robust in appearance, not crossed by longitudinal lines.

DIMENSIONS: Length, 28–110 μm; width, 7–18 μm; striae, 4–6/10 μm.

9. Pinnularia borealis var. **congolensis** Zanon *Fig. 60c.*

CR: Hustedt (1949).

DESCRIPTION: Valves linear with cuneate ends; axial area moderately narrow; central area rounded; raphe filiform; striae weakly radiate except at the ends where they may be weakly convergent; striae robust and not crossed by longitudinal lines.

DIMENSIONS (my specimen): Length, 42 μm; width, 10 μm; striae, 6/10 μm.

10. Pinnularia braunii var. **amphicephala** (A. Mayer) Hust. *Fig. 58a.*

CR: Patrick and Reimer (1966).

DESCRIPTION: Valves more or less linear with straight to convex margins and narrow, capitate ends; axial area narrow near the poles but widening, gradually at first, then strongly as it nears the central area; central area reaching both margins. The overall appearance of the central and axial areas is of a broadly lanceolate space. In this respect it differs from *P. biceps* in which the axial area is narrower near the central area and therefore opens suddenly into the central area. Raphe filiform to narrowly bandlike, with comma-shaped distal ends and proximal ends deflected slightly to one side of the valve; longitudinal lateral bands not evident; striae moderately radiate, becoming convergent at the ends of the valve.

DIMENSIONS: Length, 45–55 μm; width, 7–8 μm; striae, 11–14/10 μm.

11. Pinnularia brebissonii (Kuetz.) Rabh. var. **brebissonii** *Fig. 56a.*

CR: Patrick and Reimer (1966).

DESCRIPTION: Valves linear-lanceolate with subacute to broadly rounded ends; axial area widening gradually toward the central area; central area rounded or rhombic or reaching one or both margins as a fascia; raphe bandlike but not complex; striae radiate, becoming convergent at the ends; striae not crossed by longitudinal bands.

DIMENSIONS: Length, 40–60 μm; width, 9–11 μm; striae, 10–14/10 μm.

12. Pinnularia brebissonii var. **diminuta** (Grun.) Cl. *Fig. 56b, d.*

CR: Patrick and Reimer (1966).

DESCRIPTION: Valves elliptic-lanceolate without distinctive ends; axial area widening significantly just before the central area, which is a fascia reaching both margins (in my specimen). The variation of the central area described for the nominate variety probably applies to this one as well. Striae radiate, becoming convergent at the ends; striae not crossed by longitudinal bands.

DIMENSIONS: Length, 20–33 μm; width, 5–8 μm; striae, 10–14/10 μm.

13. Pinnularia castor Hohn and Hellerman var. **castor** *Fig. 56e.*

CR: Hohn and Hellerman (1963).

DESCRIPTION: Valves linear-elliptic with rounded ends; axial area fairly broad; central area, according to the original description, slightly wider than the axial area (my specimens included examples that had central areas reaching one or both margins as well); striae parallel, becoming slightly convergent at the ends; raphe filiform and slightly undulate.

DIMENSIONS: Length, 31–35 μm; width, 7.0–7.5 μm; striae, around 12/10 μm.

14. Pinnularia caudata (Boyer) Patr. var. **caudata** *Fig. 59b*.

CR: Patrick and Reimer (1966).

DESCRIPTION: Valves elliptic-lanceolate with protracted subrostrate ends; axial area moderately wide; central area rounded, not reaching margins (at least in my specimens); raphe narrowly bandlike to filiform with bayonet-shaped distal ends and proximal ends slightly deflected to one side of the valve; longitudinal lateral bands not evident; striae moderately to weakly radiate, becoming convergent at the ends.

DIMENSIONS: Length, 40–45 μm; width, 9.0–11.5 μm; striae, 11–13/10 μm.

15. Pinnularia divergentissima (Grun.) Cl. var. **divergentissima** *Fig. 58c*.

CR: Patrick and Reimer (1966).

DESCRIPTION: Valves narrow and linear with moderately convex margins and rostrate to subcapitate ends; axial area narrow; central area reaching both margins; striae strongly radiate near midvalve, becoming strongly convergent before the ends.

DIMENSIONS: Length, 28–40 μm; width, 5–7 μm; striae, 12–14/10 μm (mine were only 9/10 μm near midvalve).

16. Pinnularia dubitabilis Hust. var. **dubitabilis** *Fig. 60d*.

CR: Hustedt (1949).

DESCRIPTION: Valves linear with straight margins and almost square ends; axial area moderately broad; central area reaching both margins; striae thick and so positioned that a stria on one side of the valve is sometimes not exactly opposite one on the other; raphe branches linelike and somewhat arched; terminal fissures distinct. This species is allied to *P. borealis* Ehr., but differs in the shape of the ends.

DIMENSIONS: Length, 20–40 μm; width, 5–6 μm; striae, 4–5/10 μm.

17. Pinnularia instita Hohn & Hellerman var. **instita** *Fig. 61b.*

CR: Hohn and Hellerman (1963).

DESCRIPTION: Valves linear with straight to slightly convex margins and rounded ends not set off from the rest of the valve; axial area moderately wide; central area closed and rounded or reaching one or both margins. This taxon is very similar in appearance to *P. viridis*, but has a broad bandlike raphe that is not complex. Foramina are apparently as long as the striae, so no lateral longitudinal band appears. Hohn and Hellerman note the presence of caps on the ends of the striae next to the axial area. I have seen these in some but not all specimens. I think it probable that these caps mark the ends of the very wide foramina. Striae parallel to weakly radiate, becoming convergent at the ends.

DIMENSIONS: Length, 50–70 μm; width, 8–13 μm; striae, 10–12/10 μm.

18. Pinnularia intermedia (Lagerst.) Cl. var. **intermedia** *Fig. 6of, g.*

CR: Lund (1946).

SYNONYM: *Pinnularia obscura* Krasske.

DESCRIPTION: Valves linear with weakly convex margins and subrostrate to rostrate ends; axial area moderately narrow; central area elongate and reaching both valve margins; striae radiate near midvalve, becoming strongly and fairly suddenly convergent near the ends; striae not crossed by longitudinal lines; raphe filiform.

DIMENSIONS (based on Patrick and Reimer, 1966): Length, 12–42 μm; width, 3–8 μm; striae, 7–15/10 μm.

19. Pinnularia legumen (Ehr.) Ehr. var. **legumen** *Fig. 59c.*

CR: Patrick and Reimer (1966).

DESCRIPTION: Valves linear with undulate margins and broadly capitate or subcapitate ends; axial area moderate in width; central area rounded, not reaching the valve margins (at least in the present specimens); in the central area a pair of lunate marks facing each other like parentheses across the central nodule (it is my experience that such marks vary in degree of expression and there are cases where they are very weak or absent altogether); raphe bandlike but not complex; proximal raphe ends straight; distal raphe ends sickle-shaped; lateral longitudinal bands not visible.

DIMENSIONS: Length, 60–130 μm; width, 15–23 μm; striae, 8–12/10 μm.

20. Pinnularia leptosoma Grun. var. **leptosoma** *Fig. 56c.*

CR: Hustedt (1930).

DESCRIPTION: Valves linear-lanceolate with subacute ends; valve surface somewhat arched; valves fairly heavily silicified, giving rise to the impression that the valve margins are very thick; valve margins may be slightly concave at the central area; axial area moderately narrow; central area reaching both valve margins; striae radiate throughout; raphe filiform with proximal ends slightly deflected toward one side of the valve.

DIMENSIONS: Length, 31–40 μm; width, 4–6 μm; striae, 15–17/10 μm near midvalve.

21. Pinnularia maior var. **transversa** (A.S.) Cl. *Fig. 61a.*

CR: Patrick and Reimer (1966).

DESCRIPTION: Valves linear; margins convex around central area, becoming weakly concave to straight before swelling once again at the ends; striae radiate at the central area, parallel between central area and ends, and convergent at the ends; axial area rather broad; central area somewhat wider than axial area on one side; raphe broad and bandlike (but not complex), with comma-shaped distal ends; lateral longitudinal bands visible.

DIMENSIONS: Length, 170–220 μm; width, 17–20 μm; striae, 8–9/10 μm.

22. Pinnularia mesogongyla Ehr. var. **mesogongyla** *Fig. 59a.*

CR: Hustedt (1930); Patrick and Reimer (1966).

DESCRIPTION: Valve margins convex, widening again to form the capitate ends; axial area moderately wide; central area broad and rounded; raphe bandlike with bayonet-shaped distal ends; striae radiate, becoming convergent near the ends, not crossed by longitudinal bands.

DIMENSIONS: Length, 46–80 μm; width, 8–15/10 μm; striae, 9–13/10 μm.

23. Pinnularia mesolepta (Ehr.) W. Sm. var. **mesolepta** *Fig. 58d.*

CR: Hustedt (1930).

DESCRIPTION: Valve margins distinctly but shallowly triundulate; ends capitate; axial area widening somewhat toward the central area, the central area either forming a continuation of the axial area or, more usually, reaching one or both valve margins as a fascia; striae radiate, becoming convergent at the ends, not crossed by longitudinal lines. Patrick and Reimer (1966) have noted the similarity of the present taxon to *P. biceps* Greg. in all respects but the margins.

DIMENSIONS: Length, 30–65 μm; width, 9–11 μm (mine were as narrow as 6 μm); striae, 10–14/10 μm.

24. Pinnularia microstauron (Ehr.) Cl. var. **microstauron** *Fig. 56g.*

CR: Patrick and Reimer (1966).

DESCRIPTION: Valves linear with more or less parallel margins and truncate-cuneate ends; axial area moderately narrow, flaring just before the central area, which may be enclosed (rounded to rhombic) or which may reach one or both margins; striae radiate, becoming convergent toward the ends; longitudinal bands lacking; raphe narrowly bandlike, the distal ends comma-shaped.

DIMENSIONS: Length, 25–90 μm; width, 7–11 μm; striae, 7–11/10 μm.

25. **Pinnularia** **stomatophora** (Grun.) Cl. var. **stomatophora**
Fig. 59d, e.

CR: Patrick and Reimer (1966).

DESCRIPTION: Valves linear with straight to very weakly undulate margins and rounded to subrostrate ends; axial area moderately wide, flaring before the central area, which reaches both valve margins; striae radiate, becoming convergent toward the ends; raphe nearly filamentous, the distal ends bayonet-shaped; longitudinal lines or bands absent. One characteristic that has been used to define this species is the presence of a pair of lunate marks on opposite sides of the central nodule. These marks may be strongly or weakly expressed. In some cases they are reduced to a few ghostlike speckles and may be essentially absent. I do not recognize the species *P. substomatophora* Hust., which has been separated on the basis of the absence of lunate marks.

DIMENSIONS: Length, 55–110 μm; width, 8–11 μm; striae, 11–14/ 10 μm.

26. **Pinnularia streptoraphe** Cl. var. **streptoraphe** *Fig. 62a.*

CR: Patrick and Reimer (1966).

DESCRIPTION: Valves linear with nearly straight margins and broadly rounded ends; axial area fairly wide; central area about the same width as the axial area, but slightly bulged on one side; raphe complex with comma-shaped distal ends and proximal ends that are slightly deflected to one side of the valve; striae parallel to weakly radiate, becoming convergent near the ends; longitudinal bands distinct. This species is similar to *P. viridis* (Nitz.) Ehr., but has coarser striae.

DIMENSIONS: Length, 135–260 μm; width, 20–35 μm; striae, around 5/10 μm.

27. **Pinnularia subcapitata** Greg. var. **subcapitata** *Figs. 57c; 60h, i.*

CR: Hustedt (1949); Patrick and Reimer (1966).

DESCRIPTION: Valves linear with straight to weakly convex margins and subcapitate ends; axial area moderately narrow, flaring slightly to distinctly before the central area; central area rounded and enclosed or reaching one or both valve margins.

DIMENSIONS: Length, 24–50 μm; width, 4–6 μm; striae, 12–13/10 μm (mine reached 15/10 μm.)

28. **Pinnularia viridis** (Nitz.) Ehr. var. **viridis** *Fig. 62b, c.*

CR: Patrick and Reimer (1966).

DESCRIPTION: Valves linear to linear-elliptic with rounded to broadly subacute ends; axial area varies in width from about 1/5 to ¼ of the valve width; central area usually only slightly wider than the axial area and sometimes developed on one side of the valve only; raphe complex; striae radiate to parallel at midvalve, becoming convergent at the ends; striae in most cases crossed by a longitudinal band on each side of the valve as shown in *Fig. 62b*, but in some cases (*Fig. 62c*) such bands not visible. I am not certain of the limits of variability of this species. Patrick and Reimer have retained several varieties based in part on the number and angle of the striae or the width of the axial area. I have, however, seen cells with one valve having radiate striae and the other valve fine, parallel striae, and other cells in which the one valve has a narrow axial area and the other a broad one. This leads me to doubt the stability of these characters. I have chosen not to separate the varieties in question and have instead expanded the circumscription of var. *viridis* to include them. There remains also the question of the raphe. Generally it is reported as complex and indeed is so in most specimens. I have seen, however, a number of specimens in which the raphe is more undulate than complex. These specimens are otherwise similar to the typical ones assigned to *P. viridis*. I am uncertain if they belong to *P. viridis* or to a different species.

DIMENSIONS: Length, 45–170 μm; width, 10–30 μm; striae, 6–12/10 μm (typically 8–10/10 μm).

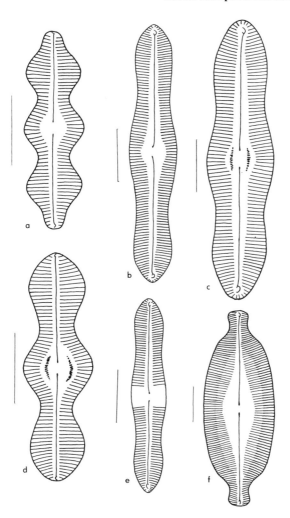

Fig. 53. *a. Caloneis lewisii* v. *inflata*. *b. C. ventricosa* v. *truncatula*. *c. C. limosa* v. *limosa*. *d. C. lewisii* v. *lewisii*. *e. C. ventricosa* v. *minuta*. *f. C. amphisbaena* v. *amphisbaena*. (Scale lines equal 10 micrometers.)

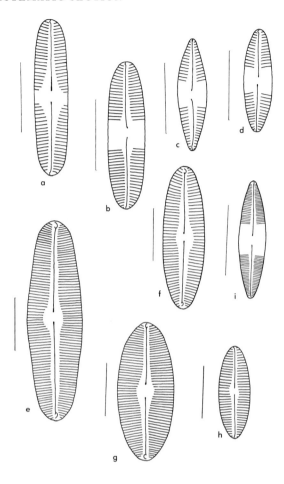

Fig. 54. a. *Caloneis bacillum* f. *fonticola*. **b.** *C. bacillum* v. *bacillum*. **c, d.** *C. bacillum* v. *angusta*. **e–h.** *C. bacillaris* v. *thermalis*. **i.** *C. hyalina* v. *hyalina*. (Scale lines equal 10 micrometers.)

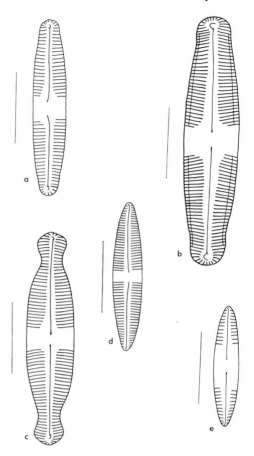

Fig. 55. a. *Caloneis hultenii* v. *hultenii*. **b.** *C. aequatorialis* v. *aequatorialis*. **c.** *C. lenzii* v. *lenzii*. **d, e.** *C. lagerstedtii* v. *lagerstedtii*. (Scale lines equal 10 micrometers.)

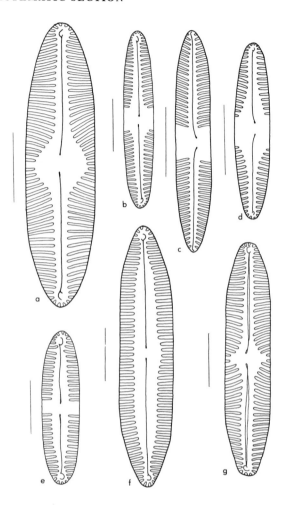

Fig. 56. a. *Pinnularia brebissonii* v. *brebissonii*. **b.** *P. brebissonii* v. *diminuta*. **c.** *P. leptosoma* v. *leptosoma*. **d.** *P. brebissonii* v. *diminuta*. **e.** *P. castor* v. *castor*. **f.** *P. acuminata* v. *acuminata*. **g.** *P. microstauron* v. *microstauron*. (Scale lines equal 10 micrometers.)

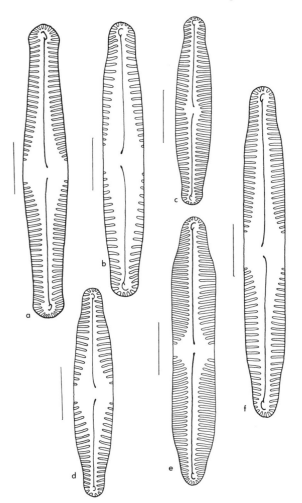

Fig. 57. a. *Pinnularia abaujensis* v. *abaujensis.* **b.** *P. abaujensis* v. *rostrata*(?). **c.**
P. subcapitata v. *subcapitata.* **d–f.** *P. abaujensis* v. *rostrata.* (Scale lines equal 10
micrometers.)

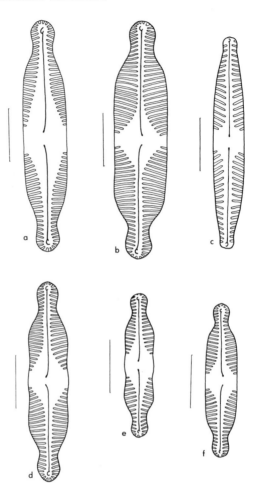

Fig. 58. a. Pinnularia braunii v. amphicephala. **b.** P. biceps v. biceps. **c.** P. divergentissima v. divergentissima. **d.** P. mesolepta v. mesolepta. **e, f.** P. biceps f. petersenii. (Scale lines equal 10 micrometers.)

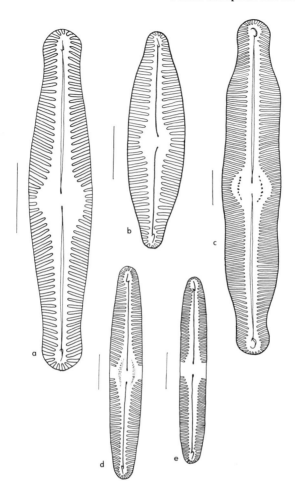

Fig. 59. a. *Pinnularia mesogongyla* v. *mesogongyla*. **b.** *P. caudata* v. *caudata*. **c.** *P. legumen* v. *legumen*. **d, e.** *P. stomatophora* v. *stomatophora*. (Scale lines equal 10 micrometers.)

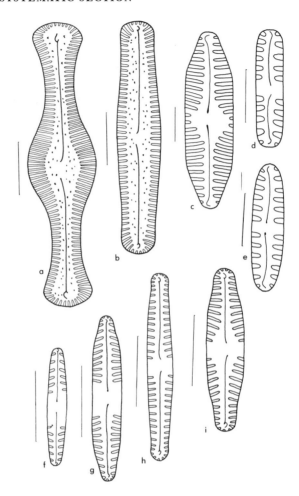

Fig. 60. a. *Pinnularia acrosphaeria* v. *turgidula.* **b.** *P. acrosphaeria* v. *acrosphaeria.* **c.** *P. borealis* v. *congolensis.* **d.** *P. dubitabilis* v. *dubitabilis.* **e.** *P. borealis* v. *borealis.* **f, g.** *P. intermedia* v. *intermedia.* **h, i.** *P. subcapitata* v. *subcapitata.* (Scale lines equal 10 micrometers.)

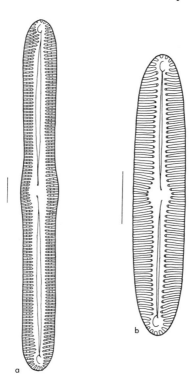

Fig. 61. a. Pinnularia maior v. *transversa. b. P. instita* v. *instita.* (Scale lines equal 10 micrometers.)

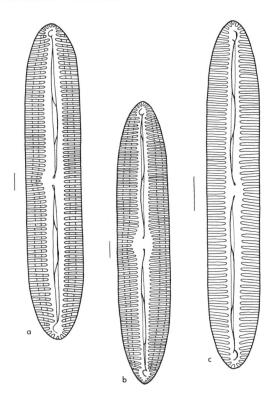

Fig. 62. a. *Pinnularia streptoraphe* v. *streptoraphe*. **b, c.** *P. viridis* v. *viridis*. (Scale lines equal 10 micrometers.)

30. *Plagiotropis* Pfitzer 1871

Plagiotropis belongs to the family Entomoneidaceae, order Naviculales, class Pennatibacillariophyceae. As with other genera in order Naviculales, a two-branched raphe is present on both valves of each cell.

Valves of *Plagiotropis* have a moderately tall, bilobed, longitudinal, central ridge or wing atop which the raphe is found. The wing base is more or less straight and lies alng the long axis of the valve. Valves can be seen in either valve or girdle view. In girdle view, both lobes of the wing of a valve of *Plagiotropis* can be seen in focus simultaneously. This is almost never the case in the closely related genus *Entomoneis*. There is but one species of *Plagiotropis* in Illinois and it is quite rare.

1. Plagiotropis lepidoptera var. **proboscidea** (Cl.) Reim. *Fig. 63a.*

CR: Patrick and Reimer (1975).

DESCRIPTION: Characters of the genus with these particulars: valves elliptic-lanceolate with narrow, rostrate, termini; axial area narrow; central area transversely rectangular or elliptic. The wing base is broad, but is narrower than the valve, so the junction line of the base with the rest of the valve is visible. My illustration is of a valve slightly tipped to one side. The apparent curvature of the raphe branches is caused in part by the curvature of the wing lobes.

DIMENSIONS: Length, 45–85 μm; width (valve view), 16–19 μm; striae, 16–18/10 μm.

31. *Entomoneis* Ehrenberg 1845

Entomoneis belongs to family Entomoneidaceae, order Naviculales, class Pennatibacillariophyceae. Like other genera in the Naviculales, each valve bears a fully developed two-branch raphe.

Entomoneis is characterized by the presence on each valve of a tall, longitudinal, bilobed wing atop which the raphe is found. The base of the wing is sigmoid. When seen in girdle view, as is almost always the case, the wing lobes can be seen to reach a low point near midvalve. Due to the geometry of the wing, it is almost impossible to see both lobes in focus simultaneously, even though it appears otherwise in the drawings. Identification of species is made on the basis of wing characters seen as the valve normally lies.

KEY TO SPECIES OF Entomoneis CONFIRMED FOR ILLINOIS

1. Lobes of wing flat-crested, the wing base undulate ____ 2. *E. ornata*
1. Lobes of wing rounded, the wing base not undulate _____ 2
    2. Striae on wing usually resolvable into puncta; irregularly placed dashlike thickenings usually present along the wing base _____ _____ 1. *E. alata*
    2. Striae on wing usually not distinctly punctate; wing base lacking dashlike thickenings _____ 3. *E. paludosa*

1. Entomoneis alata (Ehr.) Ehr. var. **alata** *Fig. 64b*.

CR: Patrick and Reimer (1975).

DESCRIPTION: Identification possible only when valve or cell is lying in girdle view; each valve with a large bilobed wing with rounded, not flat, wing crests; baseline of wing roughly parallel to wing crests and without undulations; valve and wing striate, the striae resolvable into fine puncta on the upper part of the wing; in addition to the striae there is a series of thickenings like large dots or dashes along the baseline of the wing. These are fewer in number than the striae and are irregular in placement.

DIMENSIONS: Length, 55–160 μm; width (height) of a single valve in girdle view, around 14.5 μm; striae, around 24/10 μm on valve body and somewhat coarser on the wing.

2. Entomoneis ornata (J. W. Bailey) Reim. var. **ornata** *Fig. 63b*.

CR: Patrick and Reimer (1975).

DESCRIPTION: Valves with the characteristics of the genus and these particulars: wing with flat-crested lobes; wing base undulate; wing punctate-striate.

DIMENSIONS: Length, 50–115 μm; cell width (girdle view), 28–42 μm; striae on wing, 18–24/10 μm.

3. **Entomoneis paludosa** (W. Sm.) Reim. var. **paludosa** *Fig. 64a.*

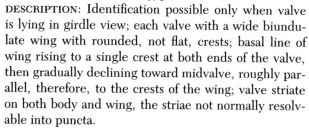

CR: Patrick and Reimer (1975).

DESCRIPTION: Identification possible only when valve is lying in girdle view; each valve with a wide biundulate wing with rounded, not flat, crests; basal line of wing rising to a single crest at both ends of the valve, then gradually declining toward midvalve, roughly parallel, therefore, to the crests of the wing; valve striate on both body and wing, the striae not normally resolvable into puncta.

DIMENSIONS: Length, 35–130 μm; width (height) of a single valve in girdle view, around 10 μm; striae, 20–24/10 μm.

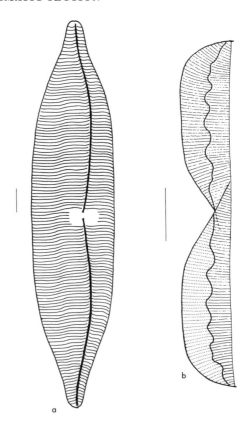

Fig. 63. a. *Plagiotropis lepidoptera* v. *proboscidea* (valve view). **b.** *Entomoneis ornata* v. *ornata* (one valve, girdle view). (Scale lines equal 10 micrometers.)

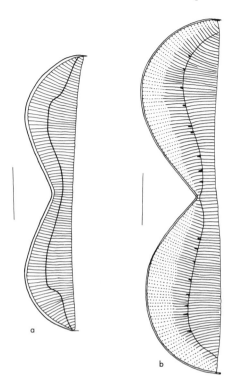

Fig. 64. a. *Entomoneis paludosa* v. *paludosa* (one valve, girdle view). **b.** *Entomoneis alata* v. *alata* (one valve, girdle view). (Scale lines equal 10 micrometers.)

32. *Cymbella* C. A. Agardh 1830

Cymbella belongs to family Cymbellaceae, order Naviculales, class Pennatibacillariophyceae. As with other genera in the Naviculales, both valves of each cell of *Cymbella* have a fully developed two-branch raphe that does not form part of a keel-type or canal-type raphe structure. Symmetry of valve markings is bilateral. The valve outline of *Cymbella* is symmetric to the transverse axis but not to the longitudinal axis. Most species have distinctly dorsiventral valves in which the dorsal margin is much more convex than the ventral margin. A few species approach symmetry to the long axis. The valves of a cell of *Cymbella* are more or less parallel, in contrast to the related genus *Amphora*, which has valves set at an angle to one another. Thus in *Cymbella* both girdle sides are about the same, while in *Amphora* the one side is longer in the pervalvar direction than the other.

Cymbella was previously divided into three genera (*Cocconema*, *Cymbella*, and *Encyonema*) based on growth habit (stalked cells, free-living, and tubular colonies, respectively), but this separation has been abandoned at the generic level.

KEY TO THE TAXA OF Cymbella IN ILLINOIS

1. One or more central striae tipped with isolated puncta or stigmata (careful observation sometimes necessary) _____ 2
1. Isolated puncta or stigmata absent _____ 13
   2. Isolated puncta or stigmata present on the dorsal side only ___ 3
   2. Isolated puncta or stigmata present on the ventral side only ___ 8
3. Valves almost straight, blunt-ended; ventral margin undulate _____
   -- 19. *C. sinuata*
3. Valve ends acute or subacute; ventral margins not undulate _____ 4
   4. Both dorsal and ventral margins convex; striae very coarsely punctate; stigma sometimes difficult to distinguish __ 21. *C. triangulum*
   4. Dorsal margin convex; ventral margin more or less straight; striae finely punctate or lineate (oblique light often required) _____ 5
5. Striae 9–12/10 μm at center, becoming 16–18/10 μm at the ends; width typically under 7 μm _____ 9. *C. lunata*
5. Striae 14–16/10 μm at the center, becoming 18–19/10 μm at the ends or, if 9–12/10 μm at the center, then width typically over 7 μm ___ 6
   6. Striae 14–16/10 μm at center; width 4.5–6.0 μm _ 11. *C. minuta*
   6. Striae 9–13/10 μm at center; width 7 μm or more _____ 7
7. Length 40–60 μm; width 9–12 μm; striae 9–11/10 μm _____
   ------------------------- 12. *C. minuta* var. *pseudogracilis*

7. Length 18–40 μm; width 7–9 μm; striae 11–13/10 μm at the center, becoming 16/10 μm at the ends _____ 13. *C. minuta* var. *silesiaca*
8. A single stout, slanting, ventral stigma present; valves robust in appearance; ends broadly rostrate _____ 22. *C. tumida*
8. One to several dotlike isolated puncta present on the ventral side _____ 9
9. Ventral margin weakly convex from pole to pole; two isolated puncta present; length typically under 40 μm _____ 23. *C. turgidula*
9. Ventral margin convex at midvalve, becoming concave before the ends _____ 10
10. Dorsal margin highly convex; ends not set off from the rest of the valve by constriction; isolated puncta 1–5 (typically over 3) _____ _____ 3. *C. cistula*
10. Dorsal margin slightly to distinctly reflexed at the ends (which are therefore weakly to distinctly rostrate) _____ 11
11. Isolated puncta typically 3; ends attenuate-rostrate ____ 8. *C. kappii*
11. Isolated puncta typically 1 or 2; ends weakly to distinctly rostrate _____ 12
12. Ends broad and blunt, only weakly set off; isolated puncta typically 1, sometimes 2; raphe forming an arc from pole to pole _____ _____ 5. *C. cymbiformis*
12. Ends narrow and usually distinctly rostrate; raphe more or less straight from pole to pole _____ 1. *C. affinis*
13. Valve ends rostrate or capitate _____ 14
13. Valve ends not set off from the rest of the valve by constriction of either margin _____ 20
14. Striae 20–25/10 μm at center, becoming 30/10 μm at the ends; ends narrow, rostrate-capitate; valves almost symmetrical to the long axis _____ 10. *C. microcephala*
14. Striae typically no more than 15/10 μm at center _____ 15
15. Valves robust, rather highly arched; raphe branches terminating short of the ends; valves striated around the ends; puncta distinct _____ _____ 15. *C. prostrata*
15. Raphe branches normally reaching the ends; at least a small terminal region free of striae _____ 16
16. Valve ends distinctly rostrate or capitate, formed by constriction of both margins _____ 17
16. Valve ends formed by constriction of one margin or, if of both margins, then constriction faint _____ 18
17. Striae distinctly punctate _____ 4. *C. cuspidata*
17. Striae resolvable into puncta or lineae only under strongly oblique light _____ 14. *C. naviculiformis*

18. Dorsal margin highly convex throughout; ventral margin convex at midvalve, becoming sharply concave near the ends (which are thus fairly narrow) _____ 16. *C. prostrata* var. *auerswaldii*

18. Valves more nearly symmetric to the long axis, the ventral margin convex or only slightly constricted at the ends _____ 19

19. Ends formed by constriction of the dorsal margin; margins weakly to moderately convex at midvalve _____ 20. *C. subaequalis*

19. Ends formed by weak constrictions of both margins; margins moderately to strongly convex at midvalve _____ 6. *C. hauckii*

20. Valves typically under 40 μm in length; striae resolvable into puncta only under strongly oblique light _____ 21

20. Valves typically over 40 μm long and distinctly punctate _____ 22

21. Striae 13/10 μm at center _____ 7. *C. hustedtii*

21. Striae 16/10 μm at center _____ 18. *C. pusilla*

22. Valves with very broad, rounded ends _____ 17. *C. proxima*

22. Valves tapering to moderately narrow, somewhat oblique-truncate ends _____ 2. *C. aspera*

1. Cymbella affinis Kuetz. var. **affinis** *Fig. 66e.*

CR: Patrick and Reimer (1975).

DESCRIPTION: Valves moderately to strongly dorsiventral; dorsal margin highly convex; ventral margin weakly convex; valve ends rostrate; axial area narrow; central area rounded and having one, sometimes more, isolated stigmata on the ventral side; raphe bandlike and curved; striae lineate.

DIMENSIONS: Length, 20–50 μm; width, 7–12 μm; striae (average), 9–11/10 μm.

2. Cymbella aspera (Ehr.) H. Peragallo var. **aspera** *Fig. 69a.*

CR: Patrick and Reimer (1975).

DESCRIPTION: Valves asymmetrically lanceolate, distinctly cymbelloid, the dorsal margin convex throughout; ventral margin swelling somewhat around central area and weakly concave between this swelling and the ends; valve ends moderately broad and more or less obliquely truncate to rounded-truncate; axial area fairly wide, swelling somewhat to form the central area; raphe broad and bandlike; distal raphe ends somewhat staple-shaped, the "points" of the staple aimed toward the dorsal margin; striae radiate throughout, the terminal striae no more radiate than elsewhere (unlike *C. lanceolata* (Ehr.) V.H., which has strongly radiate terminal striae); striae distinctly punctate throughout.

DIMENSIONS: Length, 70–200 μm; width, 20–30 μm; striae, 7–10/10 μm; puncta, 11–15/10 μm, sometimes somewhat coarser.

3. Cymbella cistula (Ehr.) Kirchn. var. **cistula** *Fig. 66b.*

CR: Patrick and Reimer (1975).

DESCRIPTION: Valves strongly dorsiventral; dorsal margin convex; ventral margin in part concave but often slightly convex at midvalve; raphe bandlike; axial area moderately broad; central area rounded with one to five isolated stigmata on the ventral side.

DIMENSIONS: Length, 40–120 μm; width, 15–25 μm; striae, 7–9/10 μm near midvalve, becoming 12/10 μm near the ends; puncta, 18–20/10 μm (mine were as few as 12/10 μm).

4. **Cymbella cuspidata** Kuetz. var. **cuspidata** *Fig. 68a.*

CR: Patrick and Reimer (1975).

DESCRIPTION: Valves weakly dorsiventral; dorsal margin highly convex; ventral margin convex but somewhat flattened at midvalve; ends rostrate; axial area moderately narrow; central area rounded; striae radiate and punctate; raphe narrowly bandlike, the proximal ends slightly hooked toward the ventral side.

DIMENSIONS: Length, 35–100 μm; width, 12–25 μm; striae, 8–12/10 μm at the center, becoming 15/10 μm at the ends; puncta (average), 20–22/10 μm.

5. **Cymbella cymbiformis** Ag. var. **cymbiformis** *Fig. 66a.*

CR: Patrick and Reimer (1975).

DESCRIPTION: Valves clearly cymbelloid, the dorsal margin convex throughout, the ventral margin convex near midvalve and concave elsewhere; ends fairly broad and rounded-truncate; axial area moderately wide; central area elongate, somewhat wider than axial area; raphe bandlike, the distal ends hooked toward the dorsal margin; striae radiate and punctate throughout. Patrick and Reimer have stated that one central stria on the ventral side is terminated by an isolated punctum.

I have found specimens that, though matching the illustrations in all other detail, have two isolated ventral puncta. This appears to cast some doubt on the stability of this character and, therefore, on the validity of var. *nonpunctata* Font. (which lacks any isolated puncta) as well.

DIMENSIONS: Length, 30–80 μm; width, 9–15 μm; striae, 8–9/10 μm at the center, becoming 12–14/10 μm at the ends; puncta, around 18–20/10 μm.

6. Cymbella hauckii V.H. var. **hauckii** *Fig. 68c.*

CR: Patrick and Reimer (1975).

DESCRIPTION: Valves having an asymmetric, lanceolate form with the dorsal margin distinctly more convex than the ventral margin; axial area narrow; central area rounded to linear-elliptic; raphe bandlike; distal raphe ends hooked toward the dorsal margin; striae radiate and punctate throughout, both striae and puncta becoming more numerous near the ends.

DIMENSIONS: Length, 39–80 μm; width, 13–18 μm; striae, 10/10 μm dorsal, 12/10 μm ventral at the center, becoming 16–17/10 μm at the ends; puncta, 16/10 μm at the center, 20–22/10 μm at the ends.

7. Cymbella hustedtii Krasske var. **hustedtii** *Fig. 68f.*

CR: Patrick and Reimer (1975).

DESCRIPTION: Dorsal margin convex; ventral margin weakly convex; ends not distinctive, axial area moderately narrow; central area no wider than axial area; raphe filiform, the branches somewhat arched; distal raphe fissures curved toward the dorsal margin; striae slender and indistinctly punctate.

DIMENSIONS: Length, 13–20 μm; width, 5–7 μm; striae, 13/10 μm at the center, 16/10 μm at the ends.

8. Cymbella kappii Cholnoky var. **kappii** *Fig. 66d.*

CR: Cholnoky (1956).

DESCRIPTION: Dorsal margin convex, reflexing slightly to form the rostrate ends; ventral margin convex at midvalve, becoming concave between midvalve and the ends; axial area narrow; central area somewhat wider than the axial area; raphe bandlike, the distal ends hooked toward the dorsal margin; striae radiate on the dorsal side and more nearly parallel on the ventral side; two or three ventral striae in the central area tipped with isolated puncta. Cholnoky notes the similarity of this taxon with both *C. tumidula* Grun. and *C. turgidula* Grun. It has the shape of one and the density of striae of the other. I suspect that it may prove to be part of the normal variation of one of them.

DIMENSIONS: Length, 25–40 μm; width, 8–11 μm; striae, around 10–11/10 μm at the midvalve, becoming somewhat finer toward the ends.

9. Cymbella lunata W. Sm. var. lunata *Fig. 65e.*

CR: Patrick and Reimer (1975).

DESCRIPTION: Valves cymbelloid, the dorsal margin moderately convex, the ventral margin straight or very slightly convex; ends drawn out and somewhat hooked toward the ventral side; axial area narrow; central area scarcely wider than axial area; centermost dorsal stria with an isolated punctum; raphe filiform, the elongate distal ends parallel to dorsal margin; proximal raphe ends close together and slightly deflected toward the dorsal margin.

DIMENSIONS: Length, 20–51 μm; width, 4.5–7.0 μm; striae, 9–12/10 μm at the center, becoming 16–18/10 μm at the ends.

10. Cymbella microcephala Grun. var. microcephala *Fig. 68d.*

CR: Patrick and Reimer (1975).

DESCRIPTION: Valves very weakly dorsiventral; both margins convex, the dorsal only slightly more convex than the ventral; ends narrow and rostrate-capitate; axial area narrow; central area varying from indistinct to somewhat asymmetric-rounded; raphe filiform, the proximal ends curved somewhat toward the dorsal margin; striae neither lineate nor punctate.

DIMENSIONS: Length, 10–20 μm; width, 2.5–4.0 μm; striae, 20–25/10 μm near the center, becoming 30/10 μm at the ends.

11. Cymbella minuta Hilse ex Rabh. var. **minuta** *Fig. 65d.*

CR: Patrick and Reimer (1975).

DESCRIPTION: Valves distinctly dorsiventral; dorsal margin distinctly convex; ventral margin straight or slightly concave or convex; ends acute but not set off by constriction; axial area and central area forming a moderately wide band that is much closer to the ventral than to the dorsal margin; striae punctate-lineate; middle dorsal stria with an indistinct isolated punctum.

DIMENSIONS: Length, 9–28 μm; width, 4.5–6.0 μm; striae, 14–16/10 μm at the center, 18–19/10 μm at the ends.

12. Cymbella minuta var. **pseudogracilis** (Choln.) Reim. *Fig. 65a.*

CR: Patrick and Reimer (1975).

DESCRIPTION: Valves distinctly dorsiventral; dorsal margin convex; ventral margin nearly straight; ends acute; axial and central areas forming a moderately broad band near the ventral margin; raphe filiform with terminal fissures very long and parallel to the dorsal margin; striae distinctly punctate-lineate; medial stria on the dorsal side with an indistinct to distinct isolated punctum.

DIMENSIONS: Length, 40–60 μm; width, 9–12 μm; striae, 9–11/10 μm.

13. Cymbella minuta var. **silesiaca** (Bleisch ex Rabh.) Reim. *Fig. 65b.*

CR: Patrick and Reimer (1975).

DESCRIPTION: Valves similar in shape to v. *minuta* with the exception that the ends appear somewhat more rounded; nature of axial and central areas and raphe similar to the nominate variety.

DIMENSIONS: Length, 18–40 μm; width, 7–9 μm; striae, 11–13/10 μm at the center, becoming 16/10 μm at the ends.

14. Cymbella naviculiformis Auers. var. naviculiformis *Fig. 68e*.

CR: Patrick and Reimer (1975).

DESCRIPTION: Valves naviculoid with an asymmetric linear-elliptic form and narrow rostrate or rostrate-capitate ends; dorsal margins convex, ventral margins weakly convex or straight; striae radiate throughout and more widely spaced at midvalve than at the ends; axial area narrow; central area rounded and rather large; striae punctate, but the puncta so close together that they cannot easily be resolved, the striae therefore appearing like lines; isolated puncta absent; raphe with distal ends hooked toward the dorsal margin; proximal ends curved somewhat toward the ventral margin.

DIMENSIONS: Length, 30–50 μm; width, 9–13 μm; striae, 12–14/10 μm near the center, becoming 18/10 μm at the ends.

15. Cymbella prostrata (Berk.) Cl. var. prostrata *Fig. 67b*.

CR: Patrick and Reimer (1975).

DESCRIPTION: Valves dorsiventral with highly convex dorsal margin and weakly convex ventral margin; ends subrostrate; axial area moderately broad; central area rounded; raphe with distal ends in subterminal position, the terminal zone striated; striae distinctly punctate-lineate. The cells of this species have a very robust, heavily silicified appearance and the valve surface appears moderately arched.

DIMENSIONS: Length, 40–80 μm; width, 14–30 μm; striae, 7–9/10 μm at the center and 11/10 μm at ends; puncta, 16–20/10 μm.

16. Cymbella prostrata var. **auerswaldii** (Rabh.) Reim. *Fig. 67c.*

CR: Patrick and Reimer (1975).

DESCRIPTION: Valves strongly dorsiventral; dorsal margin highly convex; ventral margin slightly convex near midvalve and somewhat concave between the midvalve convexity and the ends; ends subrostrate to nondistinctive; axial area widening gradually toward midvalve; central area not distinct from axial area; raphe filiform and straight, the distal ends reaching the margin; ends directed toward the ventral margin.

DIMENSIONS: Length, 15–32 μm; width, 8–12 μm; striae, 10–12/10 μm near the center, becoming 14/10 μm at the ends; puncta, around 18/10 μm (as low as 16/10 μm in my specimens).

17. Cymbella proxima Reim. var. **proxima** *Fig. 69b.*

CR: Patrick and Reimer (1975).

DESCRIPTION: Valves cymbelloid with convex dorsal margins and ventral margins slightly swollen at midvalve and straight to concave elsewhere; ends very broad and rounded, not set off in any way from the rest of the valve; axial area moderately wide; central area slightly wider than axial area; isolated puncta absent (the V-shaped mark shown in the illustration is actually a shadow of the central nodule); raphe bandlike with distal ends sharply hooked toward the dorsal margin; striae radiate and punctate throughout.

DIMENSIONS: Length, 45–120 μm; width, 18–24 μm; striae, 7–8/10 μm at the center, to 11/10 μm at the ends; puncta, 14–15/10 μm.

18. **Cymbella pusilla** Grun. var. **pusilla** *Fig. 68 g.*

CR: Patrick and Reimer (1975).

DESCRIPTION: Valves moderately dorsiventral; dorsal margin moderately convex; ventral margin weakly convex; ends not distinctive; axial area close to the longitudinal midline of the valve; central area rounded-irregular and only slightly wider than the axial area; dorsal striae radiate and not distinctly punctate; centermost dorsal striae somewhat shorter than adjacent striae; raphe filiform with small terminal fissures.

DIMENSIONS: Length, 14–40 μm; width, 3.5–8.0 μm; striae, 16/10 μm at the center, becoming 20/10 μm at the ends.

19. **Cymbella sinuata** Greg. var. **sinuata** *Fig. 65f.*

CR: Patrick and Reimer (1975).

DESCRIPTION: Valves weakly dorsiventral; dorsal margin weakly convex; ventral margin undulate; ends blunt; axial area narrow and nearly medial; raphe filiform; striae radiate and punctate; central area reaching margin on ventral side and bearing a distinct isolated punctum near the central nodule.

DIMENSIONS: Length, 11–40 μm; width, 3.5–9.0 μm; striae, 9–14/10 μm.

20. **Cymbella subaequalis** Grun. var. **subaequalis** *Fig. 68b.*

CR: Patrick and Reimer (1975).

DESCRIPTION: Valves weakly dorsiventral; dorsal margin convex; ventral margin moderately convex; ends subrostrate; axial area nearly medial; central area elongate-elliptic; raphe bandlike; striae indistinctly punctate and distinctly radiate on both dorsal and ventral sides.

DIMENSIONS: Length, 30–45 μm; width, 7–9 μm; striae, 12–14/10 μm, becoming somewhat finer near the ends.

21. Cymbella triangulum (Ehr.) Cl. var. **triangulum** *Fig. 65c.*

CR: Patrick and Reimer (1975).

DESCRIPTION: Valves asymmetric-lanceolate, the dorsal margin highly convex, the ventral margin moderately convex and the ends subrostrate-acute; axial and central areas forming a moderately wide band closer to the ventral than to the dorsal margin; raphe narrowly bandlike; striae coarsely punctate.

DIMENSIONS: Length, 30–70 μm; width, 13–20 μm; striae, 8–11/10 μm.

22. Cymbella tumida (Bréb. ex Kuetz.) V.H. var. **tumida** *Fig. 67a.*

CR: Patrick and Reimer (1975).

DESCRIPTION: Valves strongly dorsiventral with a highly convex dorsal margin and a ventral margin that is weakly convex near the middle and weakly concave near the ends; ends broadly rostrate; axial area nearly on the longitudinal midline of the valve and curved from pole to pole; central area rhombic to rounded with a distinct stigma on the ventral side; raphe filiform with the distal ends turned toward the dorsal margin; striae radiate and coarsely punctate-lineate.

DIMENSIONS: Length, 35–80 μm; width, 12–18/10 μm; striae, 8–10/10 μm at the center, becoming 12–13/10 μm at the ends.

23. Cymbella turgidula Grun. var. **turgidula** *Fig. 66c.*

CR: Patrick and Reimer (1975).

DESCRIPTION: Valves moderately dorsiventral with a highly convex dorsal margin and a weakly convex ventral margin; ends blunt to subrostrate; axial area moderately wide and somewhat closer to the ventral than to the dorsal margin; central area no wider than the axial area and marked on the ventral side by two isolated puncta; raphe bandlike with distal ends small and directed toward the dorsal margin; striae distinctly punctate.

DIMENSIONS: Length, 28–50 μm; width, 10–15 μm (mine were as narrow as 9.2 μm); striae, 9–11/10 μm at the center, becoming 12–14/10 μm at the ends.

33. *Amphora* Ehrenberg ex Kützing 1844

Amphora belongs to family Cymbellaceae, order Naviculales, class Pennatibacillariophyceae. As with other genera in the order Naviculales, both valves of *Amphora* cells bear a fully developed two-branch raphe. Symmetry of valve markings is bilateral. The valve outline of *Amphora* is symmetric to the transverse axis but not to the longitudinal axis. Unlike *Cymbella*, which has a similar type of symmetry, the valve planes of *Amphora* are not parallel, but are set at an angle so that one girdle side has a greater pervalvar length than the other. Moreover, one valve mantle is longer in the pervalvar direction than the other, so the valve has a "gabled" cross section. Complete frustules are usually seen in girdle view and valves are usually seen as somewhat tipped due to the unequal height of the dorsal and ventral valve mantles. The degree of this tip or "cant" affects the apparent shape. Since the keys and descriptions that follow are based on isolated valves in normal ("canted") lie, it is important to consider all features and not to rely on the apparent shape alone when making an identification.

KEY TO THE TAXA OF Amphora IN ILLINOIS

1. Striae broken into coarse, dashlike units _____ 2
1. Striae punctate or linelike, but not broken into dashlike units _____ 3
  2. Central group of dorsal striae interrupted by a gap or clear space __
  _____ 3. *A ovalis* var. *affinis*
  2. Central group of dorsal striae reaching the axial area without
  interruption _____ 2. *A ovalis* var. *ovalis*
3. Striae faint, 27–32/10 μm at the center of the dorsal side and often invisible at the ends _____ 6. *A. submontana*
3. Striae not exceeding 25/10 μm (usually less) on the dorsal side; striae usually distinct _____ 4
  4. Central area present and reaching the margins on both the ventral and dorsal sides; striae finely punctate _____ 4. *A. perpusilla*
  4. Central area (if present) not reaching the margin on the dorsal side
  _____ 5
5. Central area not wider than the axial area on the dorsal side _____ 6
5. Central area on dorsal side formed by shortening of the central group of striae (which in addition have a different appearance than the adjacent striae); most striae on the dorsal side finely punctate; ventral striae very short and faint _____ 5. *A. sabiniana*

6. Dorsal striae distinctly punctate; dorsal margin convex; ends weakly set off on the ventral side; ventral striae very short or not seen ___
   _____ 7. *A veneta*

6. Dorsal striae usually not visibly punctate; dorsal margin sharply reflexed at the ends, the ends therefore rostrate-capitate; ventral striae short and interrupted by a gap at midvalve _____
   _____ 1. *A. acutiuscula*

1. **Amphora acutiuscula** Kütz. var. **acutiuscula** *Fig. 70g*.

CR: Patrick and Reimer (1975).

DESCRIPTION: Dorsal margin convex, becoming reflexed at the ends; ventral margin weakly convex at midvalve, becoming concave before the ends; ends somewhat elongate, rostrate-capitate; raphe filiform and lying close to the dorsal striae; axial area wider on the ventral side than the dorsal; central area no wider than the axial area (on the dorsal side), the central area on the ventral side apparently consisting in a gap in the already short striation on the ventral margin; dorsal striae fairly coarse, the centermost striae somewhat more widely spaced than elsewhere; ventral striae fine and faint throughout.

DIMENSIONS: Length, 30–60 μm; width (valve), 6–9 μm; striae (dorsal), 10–12/10 μm at the center, 15–16/10 μm at the ends; striae (ventral), 18–19/10 μm.

2. **Amphora ovalis** (Kütz.) Kütz. var. **ovalis** *Fig. 70a*.

CR: Patrick and Reimer (1975).

DESCRIPTION: Dorsal margin convex; ventral margin straight to weakly concave; valve ends acute to subacute; axial area narrow; raphe branches bow-shaped; striae broken into dashlike puncta; each ventral stria composed of a single row of puncta and each dorsal stria of several rows; dorsal striae not interrupted by an enclosed clear space at midvalve. This last character is the principal difference between var. *ovalis* and var. *affinis* (Kütz.) V.H., which is far more common in Illinois.

DIMENSIONS: Length, 35–85 μm; width (valve), 9–17 μm; striae, 10–12/10 μm.

3. **Amphora ovalis** var. **affinis** (Kütz.) V.H. ex De T. *Fig. 70b.*

CR: Patrick and Reimer (1975).

DESCRIPTION: Dorsal margin strongly convex; ventral margin straight to weakly concave; valve ends acute to subacute; axial area narrow; raphe branches filiform and bow-shaped; dorsal striae broken into dashlike puncta; ventral striae consisting of a single row of dashlike puncta close to the raphe, with a clear area between this row of puncta and the ventral margin; dorsal striation with a pronounced gap at midvalve due to the absence of several puncta in three or four central striae.

DIMENSIONS: Length, 28–80 μm; width (valve), 7–12 μm; striae, 11–14/10 μm.

4. **Amphora perpusilla** (Grun.) Grun. var. **perpusilla** *Fig. 70c.*

CR: Patrick and Reimer (1975).

DESCRIPTION: Dorsal margin convex; ventral margin straight; ends acute to subacute; dorsal striae radiate and punctate; ventral striae punctate, radiate near midvalve, becoming convergent at the ends; axial area narrow; central area a fascia reaching both margins; raphe branches fairly straight.

DIMENSIONS: Length, 4–17 μm; width (valve), 2–3 μm; striae, 16–24/10 μm.

5. **Amphora sabiniana** Reim. var. **sabiniana** *Fig. 70f.*

CR: Patrick and Reimer (1975).

DESCRIPTION: Dorsal valve margin convex, recurved at the ends; ventral margin slightly convex at midvalve and concave between the central swelling and the ends; valve ends protracted and rostrate-capitate; axial area widening toward midvalve; central area formed on the dorsal side by a slight shortening of the thickened central group of striae; central area not developed on the ventral side; dorsal striae punctate; ventral striae short and dashlike; raphe branches lying close to the dorsal striae.

DIMENSIONS: Length, 22–45 μm; width (valve), 6–7.5 μm; dorsal striae, 14/10 μm at midvalve, becoming 18–20/10 μm at the ends;

ventral striae, 16/10 μm at midvalve, becoming 32/10 μm at the ends.

6. Amphora submontana Hust. var. **submontana** *Fig. 70d, h.*

CR: Patrick and Reimer (1975).

DESCRIPTION: Dorsal margin convex; ventral margin slightly swollen at midvalve, becoming straight to slightly concave between the central swelling and the ends; ends slightly protracted; axial area moderately wide; raphe branches lying close to the dorsal side of the axial area; central area formed on the dorsal side by shortening of the central group of striae; central area not developed on the ventral side; puncta not visible with the light microscope.

DIMENSIONS: Length, 15–25 μm; width (valve), 3–5 μm; dorsal striae, 27–32/10 μm at midvalve and 36/10 μm or more at the ends; ventral striae, around 32/10 μm at midvalve, becoming 40/10 μm at the ends.

7. Amphora veneta Kutz. var. **veneta** *Fig. 70e.*

CR: Patrick and Reimer (1975).

DESCRIPTION: Valves with convex dorsal margins and straight to weakly concave ventral margins; valve ends rounded and not set off by constrictions or expansions; ventral striae short (in my specimens, not seen); dorsal striae radiate and punctate; the spacing of dorsal striae gradually decreasing from midvalve to the ends; axial area rather wide; central area not developed on either side.

DIMENSIONS: Length, 10–45 μm; width (valve), 4–6 μm; dorsal striae, 14–20/10 μm at midvalve, becoming 26/10 μm near the ends; ventral striae, if visible, 24–26/10 μm.

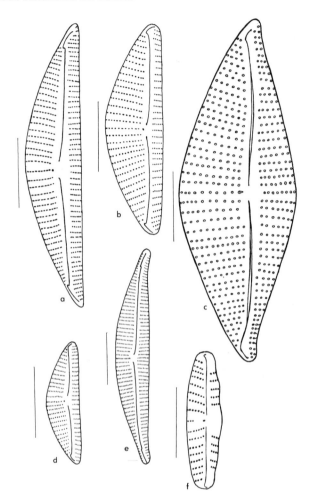

Fig. 65. a. *Cymbella minuta* v. *pseudogracilis*. **b.** *C. minuta* v. *silesiaca*. **c.** *C. triangulum* v. *triangulum*. **d.** *C. minuta* v. *minuta*. **e.** *C. lunata* v. *lunata*. **f.** *C. sinuata* v. *sinuata*. (Scale lines equal 10 micrometers.)

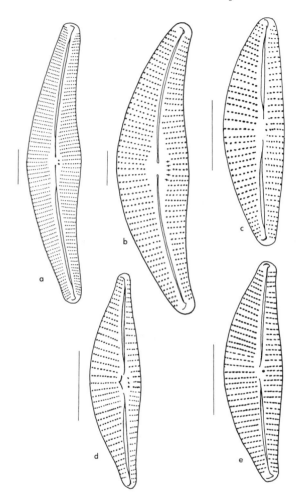

Fig. 66. a. *Cymbella cymbiformis* v. *cymbiformis*. **b.** *C. cistula* v. *cistula*. **c.** *C. tur-gidula* v. *turgidula*. **d.** *C. kappii* v. *kappii*. **e.** *C. affinis* v. *affinis*. (Scale lines equal 10 micrometers.)

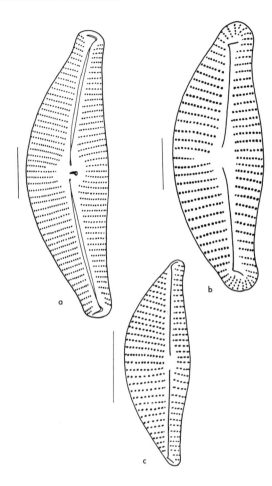

Fig. 67. a. *Cymbella tumida* v. *tumida*. **b.** *C. prostrata* v. *prostrata*. **c.** *C. prostrata* v. *auerswaldii*. (Scale lines equal 10 micrometers.)

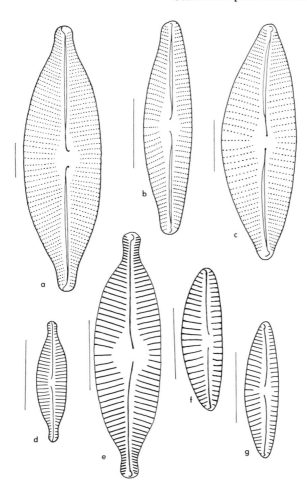

Fig. 68. a. *Cymbella cuspidata* v. *cuspidata*. **b.** *C. subaequalis* v. *subaequalis*. **c.** *C. hauckii* v. *hauckii*. **d.** *C. microcephala* v. *microcephala*. **e.** *C. naviculiformis* v. *naviculiformis*. **f.** *C. hustedtii* v. *hustedtii*. **g.** *C. pusilla* v. *pusilla*. (Scale lines equal 10 micrometers.)

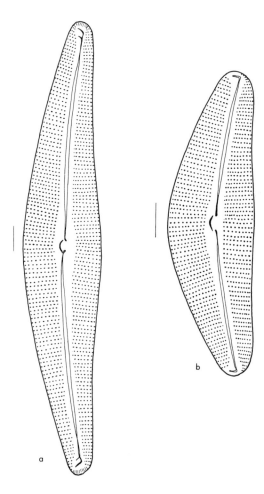

Fig. 69. a. *Cymbella aspera* v. *aspera*. **b.** *C. proxima* v. *proxima*. (Scale lines equal 10 micrometers.)

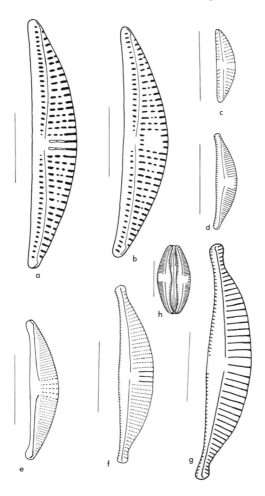

Fig. 70. a. *Amphora ovalis* v. *ovalis*. **b.** A. *ovalis* v. *affinis*. **c.** A. *perpusilla* v. *perpusilla*. **d.** A. *submontana* v. *submontana*. **e.** A. *veneta* v. *veneta*. **f.** A. *sabiniana* v. *sabiniana*. **g.** A. *acutiuscula* v. *acutiuscula*. **h.** A. *submontana* v. *submontana* (cell). (Scale lines equal 10 micrometers.)

34. *Gomphonema* Ehrenberg 1831 nom. cons., non Agardh 1824

Gomphonema belongs to family Gomphonemaceae, order Naviculales, class Pennatibacillariophyceae. As with other genera in order Naviculales, both valves of *Gomphonema* cells have a fully developed two-branched raphe that is not associated with a keel-type or canal-type raphe structure. Symmetry of valve markings is bilateral. The valve outline of *Gomphonema* is symmetric to the longitudinal axis but not to the transverse axis. Most species have valves that are distinctly divided into "head" and "foot" ends. A few species have valves almost symmetric to the transverse axis. Some species are fairly consistent in valve shape, but many are extremely variable in this regard. I have declined to recognize a number of varieties of some common species when these varieties have been separated on the basis of shape alone.

KEY TO THE TAXA OF Gomphonema IN ILLINOIS

1. Valves with one or more isolated puncta in the central area _____ 1
1. Valves lacking isolated puncta _____ 2
  2. Striae short, the axial and central areas forming a broad space ___ _____ 1. *G. abbreviatum*
  2. Striae fairly long. the axial area narrow _____ 3
3. Valves with a broad, rounded head end _____ 15. *G. olivaceum*
3. Valves with the head end rostrate or subrostrate _____ _____ 16. *G. olivaceum* var. *calcarea*
  4. More than one central stria tipped by an isolated punctum ___ 5
  4. Only one isolated punctum present in the central area _____ 6
5. Valve ends rostrate to subrostrate _____ _____ abnormal forms of 5. *G. angustatum*
5. Valve ends rounded, the head end rather broad; four central striae (two on each side of the valve) tipped by an isolated punctum (these cells otherwise similar to *G. olivaceum* in shape) ____ 14. *G. olivaceoides*
  6. Valve margins constricted near the head end, the valves therefore rostrate or capitate _____ 7
  6. Valve margins either not constricted or shallowly constricted beginning some distance from the head end _____ 17
7. Valve ends simply capitate (that is, there is no apiculate tip to the head) _____ 8
7. Valve ends rostrate, apiculate, or apiculate-rostrate _____ 11
  8. Valve margins nearly parallel at midvalve, sometimes weakly un-

dulate; both head and foot ends subcapitate to capitate _ _ _ _ _ _ _ _
_ 7. *G. angustatum* var. *productum*

8. Valve margins distinctly convex at midvalve _ _ _ _ _ _ _ _ _ _ _ _ _ _ 9

9. Valves with broadly convex margins and very narrow, distinctly capitate ends _ 19. *G. sphaerophorum*

9. Valves moderately convex at midvalve, the head set off by a moderately deep constriction but expanding to a width often more than one-half the maximum valve width _ 10

10. Valve head rounded; valves generally appearing rather narrow (5–8 μm) for their length _ 21. *G. subtile*

10. Valve head rather broadly truncate; valves typically 8–14 μm wide (sometimes as narrow as 6 μm) _ _ _ _ _ _ _ _ _ _ _ _ 23. *G. truncatum*

11. Valve ends broadly capitate, the head with an apiculate rostrate tip _ _
_ 2. *G. acuminatum*

11. Valve ends rostrate or apiculate _ 12

12. Valve ends narrowly rostrate to acuminate _ _ _ _ _ _ _ _ _ _ _ _ _ _ 13

12. Valve ends broadly rostrate, comparatively weakly set off from the rest of the valve _ 16

13. Maximum valve breadth confined to the region around the central nodule _ 14

13. Maximum valve breadth nearer the head end or prolonged for some distance from the central nodule toward the head _ _ _ _ _ _ _ _ _ _ _ 15

14. Valve margins convex from the foot pole almost to the head end, the ends therefore short-rostrate _ _ _ _ _ _ _ _ _ _ _ 13. *G. grunowii*

14. Valve margins forming a complex curve, straight to weakly concave from foot pole to midvalve, convex at midvalve, then again weakly concave, becoming again convex near the head, followed by an abrupt concave narrowing to form the acuminate or acuminate-rostrate end _ 25. *G. turris*

15. Valve margins almost straight in the head end, narrowing sharply and suddenly to form the apiculate tip _ _ _ _ _ _ 10. *G. augur* var. *gautieri*

15. Valve margins convex in the head end, narrowing sharply but not suddenly before the acuminate-rostrate end _
_ 9. *G. augur*

16. Raphe filiform; striae generally around 14–16/10 μm _ _ _ _ _ _ _ _ _ _
_ 17. *G. parvulum*

16. Raphe narrowly bandlike; striae generally under 14/10 μm _ _ _ _ _
_ _ _ _ _ _ _ _ _ _ _ _ _ _ _ _ 5, 6, 7, 8. *G. angustatum* and varieties

17. Striae rather short, the central and axial areas forming a broad space similar in shape to the valve margins; valves almost symmetric to the transverse axis _ 11. *G. clevei*

17. Axial area comparatively narrow _____ 18
   18. Maximum valve width close to the head end; isolated punctum
      close to the central nodule _____ 22. *G. tergestinum*
   18. Maximum valve breadth near midvalve _____ 19
19. Head end usually distinctly broader than the foot end, the difference
   in width of the two halves of the valve emphasized in many cases by
   strong concavity of the margins in the foot half of the valve _____ 20
19. Valves almost naviculoid or, if distinctly asymmetric to the transverse
   axis, the asymmetry expressed primarily in the length of the two cell
   halves _____ 21
   20. Head portion of valve very broad; valves generally wider than 10
      μm _____ 24. *G. truncatum* var. *capitatum*
   20. Head portion of the valve moderately wide, but the head end itself
      not conspicuously broad; valve width usually under 10 μm _____
      _____ 20. *G. subclavatum*
21. Valves generally under 20 μm long and under 4 μm wide; head end
   only slightly broader than foot end ___ 18. *G. parvulum* var. *aequalis*
21. Valves usually over 20 μm long and usually wider than 4 μm ____ 22
   22. Valves nearly symmetric to the transverse axis; ends acute to sub-
      acute in most cases _____ 12. *G. gracile*
   22. Valves with the head end slightly wider than the foot end, and the
      distance from the foot end to the central nodule usually longer than
      from the head end to the central nodule_____
      _____ 3. *G. affine* and 4. var. *insigne*

1. **Gomphonema abbreviatum** Ag. var. **abbreviatum** *Fig. 77a–d.*

CR: Patrick and Reimer (1975).
DESCRIPTION: Maximum valve breadth typically close to the head pole, sometimes somewhat closer to mid-valve; head pole nondistinctive; foot pole nondistinctive to subcapitate; valve margin typically a smooth curve, rarely undulate; central-axial area a broad space, the striae therefore short; isolated puncta absent; striae finely and indistinctly punctate (oblique light usually required); raphe filiform.
DIMENSIONS: Length, 7.5–34.0 μm; width, 2.5–6.0 μm; striae, 14–22/10 μm.

2. Gomphonema acuminatum Ehr. var. **acuminatum** *Fig. 73a.*

CR: Patrick and Reimer (1975).

DESCRIPTION: Valve margin inflated near midvalve, narrowing before the head end; head end broadly capitate with a short, narrow, rostrate tip; valve margins in the foot portion tapering from the midvalve inflation to a nondistinctive end; maximum valve breadth usually at the head end just before the rostrate tip; axial area narrow; raphe narrowly bandlike; central area elongate-elliptic; one central stria with an isolated punctum; striae distinctly punctate.

DIMENSIONS: Length, 30–85 μm; width, 8–11 μm; striae, 7–11/10 μm.

3. Gomphonema affine Kuetz. var. **affine** *Fig. 74a.*

CR: Patrick and Reimer (1975).

DESCRIPTION: Maximum valve breadth near midvalve; valves asymmetrically lanceolate, the head end slightly less narrow than the foot end; striae radiate and punctate; axial area moderately narrow; central area formed by shortening of a few striae; an isolated punctum present in the central area; raphe bandlike.

DIMENSIONS: Length, 30–75 μm; width, 7–11 μm; striae, 10–13/10 μm.

4. Gomphonema affine var. **insigne** (Greg.) Andrews *Fig. 74b.*

CR: Patrick and Reimer (1975).

DESCRIPTION: Valves asymmetric-lanceolate with convex margins and subacute ends in the head end of the valve and straight to concave margns and narrow ends in the foot end of the valve; axial area moderately broad; central area small and rounded; one central stria with an isolated punctum; striae weakly radiate and punctate.

DIMENSIONS: Length, 30–60 μm; width, 7–11 μm; striae, 7–12/10 μm (mine reached 14/10 μm in one specimen).

5. Gomphonema angustatum (Kuetz.) Rabh. var. **angustatum**
Fig. 75a.

CR: Patrick and Reimer (1975).

DESCRIPTION: Valves with greatest breadth near mid-valve; margins convex over most of their length, reflexing slightly to form broadly rostrate head and foot poles; axial area narrow; central area formed by shortening of a few striae; central area with a single isolated punctum; striae radiate and punctate; raphe narrowly bandlike to nearly filiform.

DIMENSIONS: Length, 12–45 μm; width, 5–9 μm; striae, 9–13/10 μm.

6. Gomphonema angustatum var. **obtusatum** (Kütz.) Grun.
Fig. 75d.

CR: Patrick and Reimer (1975).

DESCRIPTION: Valve with greatest width near head end; head end broadly rostrate; valve margins tapering from maximum width to foot pole without noticeable constriction; margins nearly straight; foot pole slightly bulbous and distinctly narrower than the head pole; raphe narrowly bandlike; axial area narrow; central area formed by the shortening of one central stria (the area appears larger, however, since the central group of striae is more widely spaced than other striae); one central stria with an isolated punctum. I think it quite possible that this taxon intergrades with var. *angustatum*.

DIMENSIONS: Length, 25–36 μm; width, 7–9 μm; striae, 9–12/10 μm.

7. Gomphonema angustatum var. **productum** Grun. *Fig. 75c.*

CR: Patrick and Reimer (1975).

DESCRIPTION: Valves with maximum breadth about halfway from the central nodule to the head end; valve margins straight to weakly undulate; head and foot poles capitate; axial area narrow; central area small; central area with a single isolated punctum; striae moderately to weakly radiate and faintly lineate; raphe filiform.

DIMENSIONS: Length, 13–48 μm; width, 4–6 μm; striae, 10–11/10 μm at the center, 13/10 μm at the ends.

8. Gomphonema angustatum var. **sarcophagus** (Greg.) Grun. *Fig. 75b.*

CR: Patrick and Reimer (1975).

DESCRIPTION: Valves nearly linear; head end slightly wider than foot end; head and foot poles rostrate; axial area narrow; central area formed by shortening of a few striae and distinguished by a slightly greater spacing of striae; one central stria with an isolated punctum; striae radiate and punctate; raphe narrowly bandlike.

DIMENSIONS: Length, 12–45 μm; width, 5–9 μm; striae, 7–13/10 μm.

9. Gomphonema augur Ehr. var. **augur** *Fig. 72c.*

CR: Patrick and Reimer (1975).

DESCRIPTION: Valve margins more or less linear in the tapered foot portion of the valve, becoming convex in the head portion and narrowing abruptly to form a short, narrow, rostrate end; maximum valve breadth near the head end; axial area narrow; central area small and rounded; one central stria with an isolated punctum; striae radiate and punctate.

DIMENSIONS: Length, 17–50 μm; width, 9–13 μm; striae, 11–15/10 μm.

10. Gomphonema augur var. **gautieri** V.H. *Fig. 72b.*

CR: Hustedt (1930).

DESCRIPTION: Similar to var. *augur* but typically larger and with head end valve margins having a prolonged linear region before narrowing to the short, rostrate end instead of having the convex-margined region characteristic of var. *augur*.

DIMENSIONS: Length, 40–70 μm; width, 12–15 μm; striae, 12–15/10 μm.

11. Gomphonema clevei Fricke var. clevei *Fig. 77f.*

CR: Patrick and Reimer (1975).

DESCRIPTION: Head portion of valve slightly broader than foot portion; ends typically nondistinctive but sometimes slightly protracted; axial area rather broad, central area not distinguished from the axial area; striae short, with puncta visible under oblique light; a single isolated punctum visible near the central nodule; raphe filiform.

DIMENSIONS: Length, 17–37 μm; width, 5–8 μm (my specimens were as narrow as 4 μm); striae, 10/10 μm near the central nodule, becoming 13–14/10 μm at the ends.

12. Gomphonema gracile Ehr. emend. V.H. var. gracile *Fig. 74c, d.*

CR: Patrick and Reimer (1975).

DESCRIPTION: Valves lanceolate and almost symmetric to the transverse axis; ends narrow and typically acute; axial area narrow; central area indistinct; one central stria terminated with an isolated punctum; striae radiate and indistinctly punctate; raphe more or less bandlike in large specimens but tending toward filamentous in smaller ones. *G. gracile* is best distinguished from similar species by its near symmetry to the transverse axis and by the narrow, acute ends.

DIMENSIONS: Length, 20–90 μm; width, 3–11 μm; striae, 11–16/10 μm.

13. Gomphonema grunowii Patr. var. grunowii *Fig. 76b.*

CR: Patrick and Reimer (1975).

DESCRIPTION: Valves with greatest breadth near midvalve; valve margins convex in the head portion; head end attenuate-rostrate; valve margins in the foot portion weakly convex, becoming slightly concave near the end; axial area narrow; central area small; one central stria terminated by an isolated punctum; raphe narrowly bandlike. *G. grunowii* is not quite so symmetric to the transverse axis as *G. gracile*, nor does it have the broadly subacute ends of *G. affine*.

DIMENSIONS: Length, 22–56 μm; width, 7–10 μm;

striae, 12–13/10 μm near the central area, becoming 16/10 μm at the ends.

14. Gomphonema olivaceoides Hust. var. olivaceoides *Fig.* 77*e*.

CR: Patrick and Reimer (1975).

DESCRIPTION: Valves with greatest breadth near the broadly rounded head; valve margins tapering from the head to the fairly narrow foot end; central area formed by shortening several striae on each side of the valve, the four striae defining the margins of the central area each terminated with an isolated punctum; axial area narrow; striae radiate and indistinctly punctate. I am not entirely convinced that the present specimens were not merely abnormalities of the much more common *G. olivaceum*, which, however, lacks the isolated puncta.

DIMENSIONS: Length, 18–35 μm; width, 5–6 μm; striae, 10–12/10 μm, becoming finer near the ends.

15. Gomphonema olivaceum (Lyngb.) Kuetz. var. olivaceum *Fig.* 77*g*.

CR: Patrick and Reimer (1975).

DESCRIPTION: Valves clavate with narrow foot pole and broadly rounded head pole; axial area narrow; central area characterized by wider spacing of the striae, a butterfly shape, one or more shortened striae, and an absence of an isolated punctum; striae radiate and punctate, becoming more closely spaced near the poles.

DIMENSIONS: Length, 15–40 μm; width, 5–10 μm; striae, 11–14/10 μm.

16. Gomphonema olivaceum var. **calcarea** (Cl.) Cl. *Fig. 77h.*

CR: Patrick and Reimer (1975).

DESCRIPTION: Valves similar in form and structure to var. *olivaceum* with the exception that the head end is subrostrate rather than broadly rounded. This taxon has been found so far only in Carroll County.

DIMENSIONS: Length, 22–47 μm; width, 5–8 μm; striae, 10–14/10 μm.

17. Gomphonema parvulum (Kütz.) Kütz. var. **parvulum** *Fig. 76c, e–g.*

CR: Wallace and Patrick (1950).

DESCRIPTION: Valve outline highly variable; margins more or less convex, sometimes undulate, narrowing to form rostrate ends at the head and rostrate to subrostrate ends at the foot in the majority of cases; in some cases valves asymmetric-lanceolate without distinctive ends; in all cases, axial area narrow, central area small to indistinct with one stria bearing an isolated punctum, the striae weakly radiate to parallel and indistinctly punctate; raphe filiform, never bandlike.

DIMENSIONS: Length, 12–36 μm; width, 4–8 μm; striae, 12–19/10 μm.

18. Gomphonema parvulum var. **aequalis** A. Mayer *Fig. 76d.*

CR: A. Mayer (1928).

DESCRIPTION: Valves lanceolate, almost symmetrical to the transverse axis; greatest valve breadth near midvalve; valve margins in head part nearly straight; margins in foot part becoming weakly concave near the end; head end blunt, fairly narrow, rounded, not conspicuously set off from the rest of the valve, sometimes weakly bulbous; foot end rounded, narrower than head end, not set off from the rest of the valve; axial area narrow; central area small, formed by shortening of a single stria; one central stria with an isolated punctum; striae radiate and indistinctly puncate. These specimens are similar in form to *G. dichotomum* Kütz., but are much smaller.

DIMENSIONS: (my specimens): Length, 11–20 mm; width, 3.3–4.0 mm; striae, 11–15/10 mm.

19. Gomphonema sphaerophorum Ehr. var. **sphaerophorum** *Fig. 73b*.

CR: Patrick and Reimer (1975).

DESCRIPTION: Valves with greatest width near midvalve; margins of head portion of valve strongly convex, narrowing sharply before the narrow, distinctly capitate end; margins of foot portion of valve convex near the central area, becoming straight and finally concave before the end; axial area moderately narrow; central area small; one central stria terminated by an isolated punctum; raphe bandlike. This species is similar in form to *G. augur*, which, however, has a narrowly rostrate head end.

DIMENSIONS: Length, 30–47 mm; width, 7–11 mm; striae, 11–16/10 mm.

20. Gomphonema subclavatum (Grun.) Grun. var. **subclavatum** *Fig. 73c–e*.

CR: Cleve-Euler (1955); Patrick and Reimer (1975).

DESCRIPTION: Valves clavate with straight to slightly convex margins in the head portion and concave margins in the foot portion; maximum valve breadth near midvalve; ends rounded; central area formed by the shortening of one to several striae on each side of the valve; one central stria terminated with an isolated punctum; striae radiate and distinctly punctate; raphe narrowly bandlike to nearly filiform.

DIMENSIONS: Length, 25–70 mm; width, 7–10 mm; striae, 11–14/10 mm.

21. **Gomphonema subtile** Ehr. var. **subtile** *Fig. 76a.*

CR: Patrick and Reimer (1975).

DESCRIPTION: Valve margins convex at midvalve; margin in the head portion narrowing before the distinctly capitate head end; margin in the foot portion becoming somewhat concave before the end; greatest valve breadth near midvalve; axial area widening from the ends toward midvalve; central area not distinguished from the axial area; one isolated punctum near the central nodule; striae finely punctate and radiate throughout; raphe filiform.

DIMENSIONS: Length, 33–50 mm; width, 5–8 mm; striae, around 12/10 mm near midvalve, becoming 14/10 mm at the ends.

22. **Gomphonema tergestinum** (Grun.) Fricke var. **tergestinum** *Fig. 76h.*

CR: Patrick and Reimer (1975).

DESCRIPTION: Valves asymmetric-elliptic-lanceolate with the greatest breadth near the head end; head end broadly rounded; foot end narrow and rounded; neither end set off by constriction or expansion; axial area narrow; central area formed by shortening of a single stria; an isolated punctum present close to the central nodule; striae radiate and punctate.

DIMENSIONS: Length, 12–55 mm; width, 4–12 mm; striae, 12–14/10 mm.

23. **Gomphonema truncatum** Ehr. var. **truncatum** *Fig. 71c.*

CR: Patrick and Reimer (1975).

DESCRIPTION: Valve margins convex at midvalve; margins in head portion narrowing before expanding to form the broad, flat-capitate end; margins in the foot portion narrowing to a nondistinctive terminus; axial area narrow; central area rounded to irregular; one central stria terminated by an isolated punctum; striae radiate and punctate.

DIMENSIONS: Length, 26–65 mm; width, 6–14 mm; striae, 10–12/10 mm.

24. Gomphonema truncatum var. **capitatum** (Ehr.) Patr. *Fig.* 71a, b.

CR: Patrick and Reimer (1975).

DESCRIPTION: Valves with maximum breadth near midvalve; margins in head portion narrowing gradually before the broadly rounded terminus; margins in the foot portion narrowing more strongly and in part concave; axial area narrow; central area irregular; one central stria with an isolated punctum; striae radiate and punctate; raphe bandlike.

DIMENSIONS: Length, 16–65 mm; width, 6–13 mm; striae, 10–12/10 mm.

25. Gomphonema turris Ehr. var. **turris** *Fig. 72a*.

CR: Patrick and Reimer (1975).

DESCRIPTION: Valve margins of the foot end somewhat concave, narrowing to a subacute terminus; valve margin of the head end convex near midvalve, changing angle fairly abruptly close to the end, then narrowing considerably to the narrow, rostrate tip; axial area moderately narrow; central area small; one central stria with an isolated punctum; striae radiate and punctate.

DIMENSIONS: Length, 40–60 mm; width, about 13 mm; striae, 8–10/10 mm at the center and 12/10 mm at the ends.

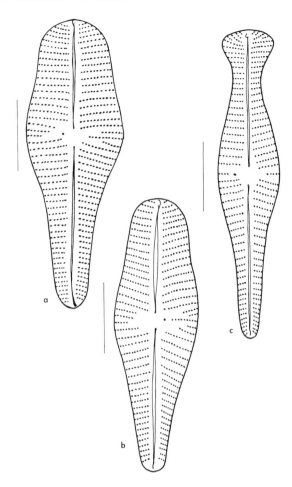

Fig. 71. a, b. *Gomphonema truncatum* v. *capitatum. c. G. truncatum* v. *truncatum.* (Scale lines equal 10 micrometers.)

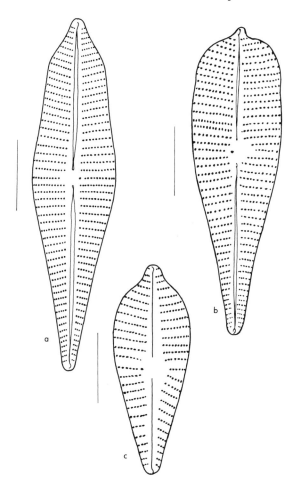

Fig. 72. a. *Gomphonema turris* v. turris. b. *G. augur* v. *gautieri*. c. *G. augur* v. *augur*. (Scale lines equal 10 micrometers.)

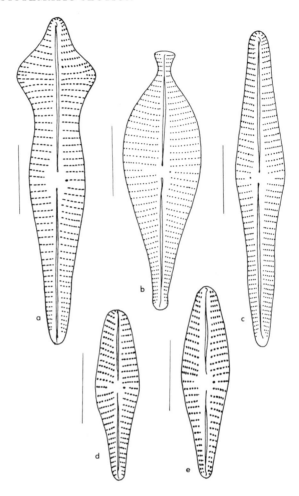

Fig. 73. *a. Gomphonema acuminatum* v. *acuminatum. b. G. sphaerophorum* v. *sphaerophorum. c–e. G. subclavatum* v. *subclavatum.* (Scale lines equal 10 micrometers.)

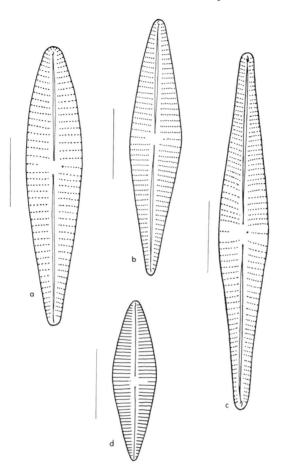

Fig. 74. a. *Gomphonema affine* v. *affine*. **b.** *G. affine* v. *insigne*. **c, d.** *G. gracile* v. *gracile*. (Scale lines equal 10 micrometers.)

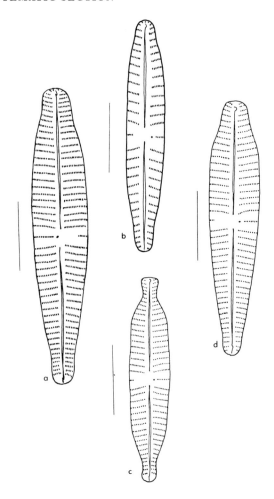

Fig. 75. a. *Gomphonema angustatum* v. *angustatum*. **b.** *G. angustatum* v. *sarcophagus*. **c.** *G. angustatum* v. *productum*. **d.** *G. angustatum* v. *obtusatum*. (Scale lines equal 10 micrometers.)

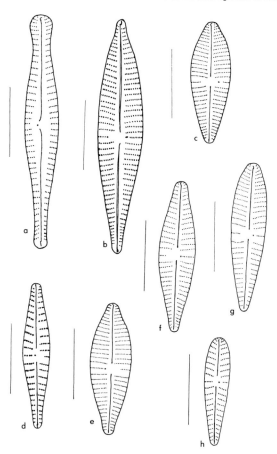

Fig. 76. *a. Gomphonema subtile* v. *subtile*. *b. G. grunowii* v. *grunowii*. *c. G. parvulum* v. *parvulum*. *d. G. parvulum* v. *aequalis*. *e–g. G. parvulum* v. *parvulum*. *h. G. tergestinum* v. *tergestinum*. (Scale lines equal 10 micrometers.)

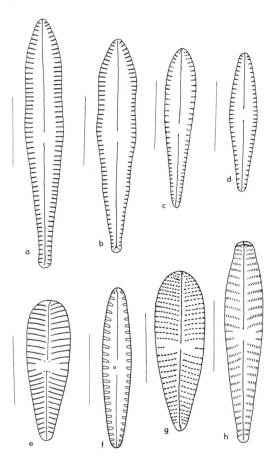

Fig. 77. a–d. *Gomphonema abbreviatum* v. *abbreviatum.* **e.** *G. olivaceoides* v. *olivaceoides.* **f.** *G. clevei* v. *clevei.* **g.** *G. olivaceum* v. *olivaceum.* **h.** *G. olivaceum* v. *calcarea.* (Scale lines equal 10 micrometers.)

35. *Denticula* Kützing 1844.

Denticula belongs to family Epithemiaceae, order Epithemiales, class Pennatibacillariophyceae. As with other genera in the order Epithemiales, both valves of each cell of *Denticula* have a tubular canal raphe structure sunken into the valve surface. Valve markings are bilaterally symmetric.

The valve outline of *Denticula* is symmetric to both the long and transverse axes. The canal raphe may run close to the longitudinal midline of the valve or in positions closer to the margin. In the one species I found, the canal lies along the extreme edge of the valve face and is hard to distinguish from the margin. Valve markings consist of transverse striae, which may or may not be visibly punctate, and thin or thick costae. Some species have internal plates or septa. Cells may be seen in valve or girdle view but can be identified only in valve view.

1. **Denticula tenuis** f. **diminuta** Mang. *Fig. 78d.*

CR: A.N.S.P. File, referring to Manguin (1964).

DESCRIPTION: Valves narrowly linear with subacute ends; canal raphe lying along extreme edge of valve; striae faint and not resolvable into puncta; costae distinct.

DIMENSIONS: Length, 12–16 μm; width, around 2.5 μm; striae, 22–24/10 μm costae, 6–9/10 μm.

36. *Epithemia* Brébisson 1838

Epithemia belongs to family Epithemiaceae, order Epithemiales, class Pennatibacillariophyceae. As with the other genera in the order Epithemiales, both valves of each cell have a tubular raphe structure sunken into the valve surface. Valve markings are bilaterally symmetric.

Valves of *Epithemia* are symmetric to the transverse axis but not to the longitudinal axis. The valve face is flat to slightly convex in cross section and both valve mantles are of equal height. The raphe canal is formed of two bow-shaped branches that meet at an acute angle at midvalve. In some species most of the canal is hidden on the valve mantle with only the terminal and/or central portions visible in valve view. This is of taxonomic significance. The valves are marked by transverse rows of complex pores called alveoli. Interspersed among the rows of alveoli are thickened transverse costae. Each valve may be associated with an internal plate or septum with large irregular openings but no other markings. The costae are prolonged onto the valve mantle, and for absolute confirmation of species identifications, it is frequently necessary to see or infer the shape of the ends of these costae in girdle view. In some species, the costae terminate in rounded thickenings that may mark the junction of their internal prolongations with the crossbars of the septum.

KEY TO THE TAXA OF Epithemia IN ILLINOIS

1. Crossbars of the septum very thick; each crossbar interrupted by a gap near the ventral margin; in girdle view the septum marked by a series of knoblike thickenings _____ 3. *E. argus*
1. Crossbars of the septum rather thin, usually not much wider than the costae; knoblike thickenings along the septum not evident when cells are observed in girdle view _____ 2
   2. Dorsal margin rather highly convex; raphe canal approaching the dorsal margin at midvalve _____ 4. *E. sorex*
   2. Dorsal margin moderately convex; raphe canal usually less than one-half way to the dorsal margin at midvalve _____ 3
3. Valves usually over 60 μm long; two or three rows of alveoli between each pair of costae; 7–9 rows of alveoli in 10 μm _____ 5. *E. turgida*
3. Valves usually 50 μm long or less; 3–7 rows of alveoli between each pair of costae; 11–14 rows of alveoli in 10 μm _____ 4
   4. Valve ends rostrate-capitate _____ 2. *E. adnata* var. *proboscidea*
   4. Valve ends subrostrate-truncate _____ 1. *E. adnata* var. *minor*

1. Epithemia adnata var. **minor** (Peragallo & Herib.) Patr. *Fig. 78c.*

CR: Patrick and Reimer (1975).

DESCRIPTION: Ventral margin nearly straight; dorsal margin convex, flattened near midvalve; ends short and subrostrate; raphe canal mostly hidden on the valve mantle, appearing near midvalve as a very short "V"; septum not massive (see var. *proboscidea* for explanation); distinguished from var. *proboscidea* by the shape of the ends and the small size.

DIMENSIONS: Length, 26–41 μm; width, 8–11 μm; striae, 12–14/10 μm; costae, 3–5/10 μm; usually 3–5 striae between each pair of costae.

2. Epithemia adnata var. **proboscidea** (Kütz.) Patr. *Fig. 78f.*

CR: Patrick and Reimer (1975).

DESCRIPTION: Ventral margin weakly concave to nearly straight; dorsal margin convex, somewhat flattened at midvalve and reflexed to form the rostrate or subcapitate ends; raphe canal V-shaped, the central nodule usually reaching less than halfway from the ventral margin to the dorsal; septum not massively developed, in girdle view the vertical ribs of the septum not tipped with knoblike thickenings (for an example of a massive septum, see *E. argus*).

DIMENSIONS: Length, 40–50 μm; width, 8–10 μm; striae, 11–14/10 μm; costae, 2.5–4/10 μm; usually 3–7 striae between each pair of costae.

3. Epithemia argus (Ehr.) Kütz. var. **argus** *Fig. 78a,b.*

CR: Patrick and Reimer (1975).

DESCRIPTION: Ventral margin weakly convex, reflexing slightly at the ends; dorsal margin convex, reflexed at the ends; valve ends short-rostrate or rostrate-capitate; raphe canal running along the ventral margin near the ends, curving toward the dorsal margin and forming the typical "V" shape at midvalve; each valve associated with a massive internal septum parallel to the valve; crossbars of the septum wider than the costae on the valve; in girdle view the ends of the vertical plate of the septum (one for each crossbar) tipped with knoblike thickenings; each crossbar of the septum is "broken" near the ventral margin. This gap can be seen as a focus feature. The septum must be seen for confirmation, but the dimensions are fairly distinctive along Illinois species.

DIMENSIONS: Length, 30–130 μm; width, 6–15 μm; striae, 10–12/10 μm; costae, 1–3/10 μm; usually there are 5–8 rows of striae between each pair of costae.

4. Epithemia sorex Kuetz. var. sorex *Fig. 78e.*

CR: Hustedt (1930).

DESCRIPTION: Dorsal margin highly convex; ventral margin slightly to moderately concave; ends rostrate-capitate; costae and "striae" radiate; ends of costae in girdle view not capitate; midpoint of canal raphe approaching dorsal margin.

DIMENSIONS: Length, 20–65 μm; width, 8–15 μm; "striae," 12–15/10 μm; costae, 5–7/10 μm.

5. Epithemia turgida (Ehr.) Kuetz. var. turgida *Fig. 78g.*

CR: Hustedt (1930).

DESCRIPTION: Dorsal margin moderately convex; ventral margin straight to weakly concave; ends broadly rostrate; costae and "striae" radiate; usually two or three rows of "striae" between costae; midpoint of raphe closer to ventral than to dorsal margin.

DIMENSIONS: Length, 60–220 μm; width, 15–18 μm; "striae," 7–9/10 μm; costae, 3–5/10 μm.

37. *Rhopalodia* O. Mueller 1897

Rhopalodia belongs to the order Epithemiales. Like the other genera in the order, the raphe is a slit in a tubular structure sunken in the valve surface. Due to the shape of the valve, the raphe is almost never seen.

If you could cut a valve of *Rhopalodia* in cross section, you would see that the section was gabled, that is, it has a peak or ridge like the roof peak of a house. This peak is located closer to one side of the section than the other (I have illustrated the appearance of the section in the Glossary under **Gabled**). The valve mantle on the side with the peak is higher than the mantle on the other side. Most of the valve surface is a slanted plane between the high ridge and the low mantle. Due to the shape of the valves, both whole cells and single valves lie on their sides. When seen in this way, the valves have a bracket shape, with the straight or concave side being the actual valve margin while the convex (or bracket-shaped) side is in actuality the ridge (the other true valve margin is obscured). I suggest that you see the Glossary for illustrations of these relations as well as the illustrations of species on *Fig. 79*. The canal raphe of *Rhopalodia* lies along the crest of the ridge. Valve markings consist of transverse costae between which are rows of faint striae composed of pores or alveoli.

KEY TO THE TAXA OF Rhopalodia IN ILLINOIS

1. Valve margin (ridge side) a simple arc from pole to pole _____ 2
1. Valve margin (ridge side) a complex curve, indented slightly at mid-valve; between midvalve and either end, the ridge side convex, then concave, then convex again near the end _____ 1. *R. gibba*
   2. Width of a single valve, 6–7 μm __2. *R. gibberula* var. *vanheurckii*
   2. Width of a single valve, 11–15 μm _____ 3. *R. musculus*

1. Rhopalodia gibba (Ehr.) O. Muell. var. **gibba** *Fig. 79a,b*.

CR: Hustedt (1930).

DESCRIPTION: Dorsal valve margin bracket-shaped with a convex midportion, becoming somewhat concave between middle and ends; dorsal margin sometimes slightly indented at midvalve; ventral margin more or less straight, curving slightly away from the dorsal margin at the ends; costae and "striae" more or less parallel; girdle sometimes marked by rows of fine pores.

DIMENSIONS: Length, 35–300 μm; width (single valve), 8–11 μm; "striae," 12–16/10 μm; costae, 6–8/10 μm.

2. Rhopalodia gibberula var. **vanheurckii** O. Muell. *Fig. 79d*.

CR: Patrick and Reimer (1975).

DESCRIPTION: Dorsal valve margin convex without undulation or indentation; ventral margin more or less straight; ends slightly rostrate; "striae" and costae weakly radiate.

DIMENSIONS: Length, 27–40 μm; width, 6–7 μm; "striae," 14–17/10 μm; costae, 2–8/10 μm.

3. **Rhopalodia musculus** (Kütz.) O. Muell. var. **musculus** *Fig.* 79c.

CR: Patrick and Reimer (1975).

DESCRIPTION: Dorsal valve margin convex with neither indentations nor changes in curvature; ventral margin straight with the exception that the ends are hooked slightly away from the dorsal margin to form weakly rostrate tips; costae and "striae" weakly radiate.

DIMENSIONS: Length, 30–80 μm; width of valve, 11–15 μm, "striae," 12–16/10 μm; costae, 3–5/10 μm.

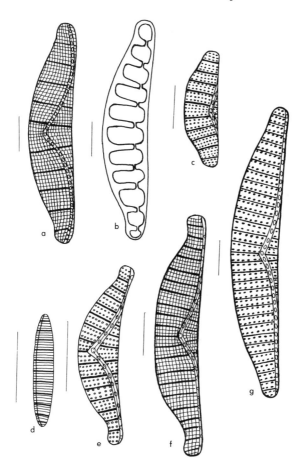

Fig. 78. a. Epithemia argus v. argus (valve). **b.** Epithemia argus v. argus (septum). **c.** Epithemia adnata v. minor (valve). **d.** Denticula tenuis f. diminuta (valve). **e.** Epithemia sorex v. sorex (valve). **f.** Epithemia adnata v. proboscidea (valve). **g.** Epithemia turgida v. turgida (valve). (Scale lines equal 10 micrometers.)

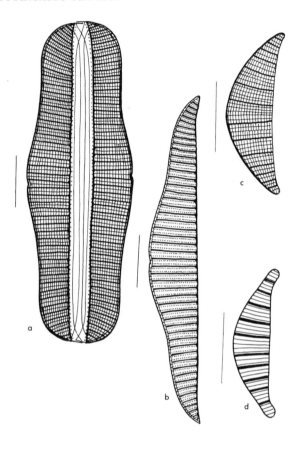

Fig. 79. a. *Rhopalodia gibba* v. *gibba* (cell, girdle view). **b.** *R. gibba* v. *gibba* (valve as normally seen). **c.** *R. musculus* v. *musculus* (valve as normally seen). **d.** *R. gibberula* v. *vanheurckii* (valve as normally seen). (Scale lines equal 10 micrometers.)

38. *Bacillaria* Gmelin 1788

Bacillaria belongs to family Bacillariaceae, order Bacillariales, class Pennatibacillariophyceae. As with other genera in the order, each valve of a *Bacillaria* cell has a single keel-type raphe structure. Symmetry of valve markings is bilateral. The keel is in general a low, longitudinal ridge that bears along its crest a tubular structure. The raphe is a thin slit in the outer surface of the tube and is normally invisible. The tube is partially open to the interior of the cell, and is bridged by silica crossbars or fibulae. The fibulae have been reported in older works as "keel puncta," and both their nature and density are of taxonomic significance. The structure of a typical keel is illustrated in the Glossary.

Bacillaria differs from *Nitzschia* and *Hantzschia* primarily on the basis of colony formation. Cells of *Bacillaria* are united into band-like colonies that continually change shape due to the individual movement of the cells. Even when the colony form is lost during sample preparation, however, one can almost always identify *Bacillaria* by its distinct striae and the central position of the keel.
There is but one species in Illinois.

1. **Bacillaria paradoxa** Gmelin var. **paradoxa** *Fig. 95f.*

CR: Hustedt (1930).
DESCRIPTION: Valves linear with acute, sometimes slightly protracted ends; keel narrow and close to the midline of the valve; striae parallel and indistinctly punctate.
DIMENSIONS: Length, 55–150 μm; width, 4–8 μm; striae, 20–25/10 μm; fibulae, 6–8/10 μm.

39. *Hantzschia* Grunow 1880

Hantzschia belongs to family Bacillariaceae, order Bacillariales, class Pennatibacillariophyceae. As with other genera in the order, both valves of each cell bear a single keel-type raphe structure. Valve markings are bilaterally symmetric.
Valves of *Hantzschia* are symmetric to the transverse axis, but are asymmetric in structure and form to the longitudinal axis. The

structural asymmetry arises from the marginal position of the keel. The valve outline itself is weakly asymmetric to the long axis. Striae are visible and punctate. The structure of the keel is essentially the same as the structure of keels in other genera in the order. I have illustrated the basic elements of keel structure in the Glossary and have provided some further explanation in the introduction of the genus *Bacillaria* above.

Hantzschia is distinguished from the related genus *Bacillaria* both in having a marginal, rather than central, keel and in the absence of the characteristic *Bacillaria* colony structure. *Hantzschia* differs from *Nitzschia* in the position of the keels. If you could see a cell of *Hantzschia* in cross section, you would find that the keels are both on the same side of the cell, unlike *Nitzschia*, which has keels either in subcentral position or on diagonally opposite valve margins. In the case of complete cells in valve view, this feature can be seen by careful focusing. The species of *Hantzschia* have a distinctive appearance and are few in number. I have found but one species with three varieties in Illinois. You should learn to recognize these taxa on sight, since the generic character will not be evident in the case of isolated valves.

KEY TO THE VARIETIES OF Hantzschia amphioxys IN ILLINOIS

1. Valves 20–100 μm long and 10 μm wide or less _____
_____ 1. *H. amphioxys* var. *amphioxys*
1. Valves typically over 100 μm in length and wider than 10 μm _____ 2
   2. Valve ends elongate-attenuate _____ 3. *H. amphioxys* var. *vivax*
   2. Valve margins converging abruptly to form short-rostrate or capitate ends _____ 2. *H. amphioxys* var. *major*

1. **Hantzschia amphioxys** (Ehr.) Grun. var. **amphioxys** *Fig. 80c,d.*

CR: Hustedt (1930).

DESCRIPTION: Valves with the characters of the genus; valve ends vary from rostrate to capitate; degree of dorsiventrality varies but is usually only moderate; keel puncta massive; striae variable in number; puncta very distinct or indistinct. This variety is separated from the others primarily on the basis of size.

DIMENSIONS: Length, 20–100 μm; width, 5–10 μm; striae, 13–20/10 μm (sometimes more); fibulae, 5–8/10 μm.

2. Hantzschia amphioxys var. **major** Grun. *Fig. 80a.*

CR: Hustedt (1930).

DESCRIPTION: Basic features similar to var. *amphioxys*, with the exception of size and the shape of the ends, which are more often capitate or rostrate-capitate than is the case with var. *amphioxys*.

DIMENSIONS: Length, 100–200 μm; width, 13–15 μm; striae and fibulae, similar to var. *amphioxys*.

3. Hantzschia amphioxys var. **vivax** (Hantz.) Grun. *Fig. 80b.*

CR: Hustedt (1930).

DESCRIPTION: Basic features similar to var. *amphioxys*, with the exception that the valve ends are generally long-attenuate and subcapitate-rostrate. In size, the cells are generally similar to var. *major*.

DIMENSIONS: Length, 100–210 μm; width, 11–12 μm; fibulae and striae, similar to var. *amphioxys*.

40. *Nitzschia* Hassall 1845

Nitzschia belongs to family Bacillariaceae, order Bacillariales, class Pennatibacillariophyceae and is by far the largest genus in the order. As with related genera, each valve of a cell of *Nitzschia* has a single keel-type raphe structure. I have illustrated the structure of a typical keel in the Glossary and have made some further remarks on it in the introduction to the genus *Bacillaria* above.

Nitzschia is distinguished from the related genera *Bacillaria* and *Hantzschia* in part on the basis of keel position and in part on the basis of colony structure. Most species of *Nitzschia* have keels located along or near the valve margins. In cells of this type, if you could see them in cross section, you would find that the keels are diagonally opposite, unlike in *Hantzschia,* which has both keels on the same side of the cell. Some species have central keels, like those of *Bacillaria,* but none have the peculiar colony structure of that genus.

Nitzschia species are distinguished with the light microscope on the basis of the position of the keel on the valve, the density, form, and arrangement of the fibulae, the shape and size of the valve, and the nature and density of striae, costae, and puncta. Careful observation is necessary, since the distinguishing characters are often subtle. Many species have an extraordinary variability of form, which may be deceptive. In some cases only a small, rather obscure, character is used to separate one species from another while the more obvious details are ignored.

I feel it fair to warn you that the taxonomy of this genus is in a state of flux. The approach I have taken is midway between the modern and the classical views and may prove to be unsuitable as more becomes known. There are those who believe that it may ultimately become impossible to deal with the lanceolate *Nitzschiae* (a large and ecologically important group) without using the electron microscope as a routine tool.

The key and descriptions that follow have been based as far as possible on characters visible with the light microscope. It is, unfortunately, necessary to stipulate that an excellent set of lenses is required. You will need to be able to discern—and measure—striae that may be as dense as 38–40/10 μm and surely as dense as 36/10 μm.

There have been (Kalinsky, 1973) over one thousand species, varieties, and forms described for this genus. I have included only fifty-five. While many of the "taxa" of *Nitzschia* in fact prove to be multiple descriptions of the same thing, there are many genuine species that I expect here but did not find. If the key and descriptions do not provide a good fit for what you have, do not hesitate to turn elsewhere!

KEY TO Nitzschia CONFIRMED FOR ILLINOIS

1. Valves sigmoid in valve view, girdle view, or both _____ 2
1. Longitudinal axis of valve more or less straight _____ 11
   2. Striae resolvable without oblique light or with moderate obliquing
    -- 3
   2. Striae not resolvable or resolvable only with strongly obliqued light
    -- 8
3. Striae arranged in both longitudinal and transverse systems _____
   --- 38. *N. sigma*
3. Striae transverse only _____ 4

4. Striae 20/10 μm or less _____ 5
4. Striae over 20/10 μm _____ 6
5. Valves under 6 μm wide _____ 28. *N. lorenziana* v. *subtilis*
5. Valves 8 μm wide or wider _____ 27. *N. lorenziana*
  6. Keel submarginal; valves sigmoid in girdle view and irregular in
    curvature in valve view _____ 39. *N. sigmoidea*
  6. Keel marginal; valves sigmoid in valve view _____ 7
7. Valves generally over 100 μm long _____ 52. *N. vermicularis*
7. Valves generally under 100 μm long _____ 16. *N. filiformis*
  8. Valve ends narrow and needlelike, curved in opposite directions _
    _____ 35. *N. reversa*
  8. Valve ends not distinctive _____ 9
9. Valves typically over 100 μm long _____ 52. *N. vermicularis*
9. Valves typically under 100 μm in length _____ 10
  10. Valve ends acute _____ 23. *N. ignorata*
  10. Valve ends rounded _____ 16. *N. filiformis*
11. Valves with needlelike ends that are distinctly set off from the rest of
    the valve _____ 12
11. Valves ends distinctive or nondistinctive but not needlelike _____ 13
  12. Valves 50 μm long or more _____ 2. *N. acicularis*
  12. Valves under 30 μm long _____ 5. *N. agnewii*
13. Valve ends hooked or scalpeliform _____ 14
13. Valve ends capitate, rostrate, attenuate or nondistinctive _____ 22
  14. Striae distinctly, though sometimes finely, punctate _____ 15
  14. Striae either not resolvable into puncta or resolvable into puncta
    only with strongly obliqued light _____ 16
15. Striae regularly punctate _____ 54. *N. vitrea*
15. Striae irregularly punctate _____ 55. *N. vitrea* v. *scaphiformis*
  16. Striae 30/10 μm or fewer _____ 17
  16. Striae over 30/10 μm _____ 20
17. Large valves, 160 μm long or longer; keel submarginal _____
    _____ 39. *N. sigmoidea*
17. Valves typically no longer than 100 μm, or if longer, then keel marginal
    _____ 18
  18. Keel submarginal in position, the center pair of fibulae no more
    widely spaced than the rest _____ 45. *N. subvitrea*
  18. Keel marginal, the center pair of fibulae more widely spaced than
    the rest _____ 19
19. Valves with nearly parallel margins at midvalve; length usually over 70
    μm and frequently over 100 μm; width, 5–6 μm ____ 26. *N. linearis*

19. Valves with margins slightly concave at midvalve; length under 100 μm
    and typically under 70 μm; width 5–10 μm _____ 50. *N. umbonata*
20. Valves appearing linear, the scalpeliform ends short and narrow __
    _____ 4. *N. adapta*
20. Valves with rather long ends, giving the cell a rhombic appearance
    or a subsigmoid appearance _____ 21
21. Ends curved, the appearance of the valve subsigmoid _____
    _____ 23. *N. ignorata*
21. Ends straight-scalpeliform and the valves appearing linear-rhombic __
    _____ 8. *N. brevissima*
    22. Keel central to submarginal _____ 23
    22. Keel marginal or indistinct _____ 27
23. Valves broad and linear-elliptic, the striae clearly visible and distinctly
    punctate; keel submarginal _____ 33. *N. plana*
23. Valves linear to lanceolate, the striae visible or invisible _____ 24
    24. Keel central _____ 25
    24. Keel subcentral to submarginal _____ 26
25. Striae distinct _____ *Bacillaria paradoxa* (See genus 38)
25. Striae not visible even with strongly obliqued light; valves linear ____
    _____ 20. *N. geitleri*
    26. Valves lanceolate without distinctive ends; length 100 μm or more
        _____ 3. *N. acula*
    26. Valves lanceolate with slightly protracted to rostrate ends; length
        70 μm or less _____ 15. *N. dissipata*
27. Keel structure formed by enlarged fibulae so arranged that the keel
    appears to be formed of a series of bubblelike "cells"; striae distinctly
    punctate _____ 28
27. Fibulae long or short but not producing the effect of bubblelike "cells";
    striae and puncta visible or invisible _____ 29
    28. Valves narrowly lanceolate _____ 41. *N. sinuata* var. *delognei*
    28. Valve margins strongly inflated at midvalve _____
        _____ 42. *N. sinuata* var. *tabellaria*
29. Fibulae equal in number to the striae _____ 30
29. Fibulae fewer in number than the striae _____ 34
    30. Tiny lanceolate valves under 3 μm in width; ends nondistinctive _
        _____ 10. *N. chasei*
    30. Valves with distinctive ends or, if ends not distinctive, then width
        greater than 3 μm _____ 31
31. Small valves, under 25 μm long; striae distinctly punctate; ends nar-
    rowly rostrate-capitate _____ 40. *N. silicula* var. *commutata*
31. Valves lanceolate or with rostrate ends _____ 32

32. Valves lanceolate, the ends very slightly protracted _____
    _____ 53. *N. vexans*
32. Valve ends distinctly rostrate _____ 33
33. Striae not visibly punctate but interrupted by a narrow clear space
    along the long axis of the valve _____ 13. *N. constricta*
33. Striae easily resolved into puncta and not interrupted by an axial clear
    space _____ 7. *N. angustata*
34. Valve outline elliptic to broadly elliptic-lanceolate, the margins
    usually convex, rarely concave; ends nondistinctive _____ 35
34. Valve outline linear to lanceolate; ends distinctive or nondistinctive
    _____ 39
35. Valves marked by punctate striae but not by thickened costae _____
    _____ 33. *N. plana*
35. Valves marked by costae and sometimes also by striae _____ 36
36. Valves marked by a narrow band of costae that do not reach either
    margin _____ 32. *N. perversa*
36. Valves marked by costae that reach one or both margins ____ 37
37. Costae reaching both margins; faint rows of striae visible between the
    costae _____ 46. *N. tryblionella*
37. Costae alone are visible _____ 38
38. Costae broad, arising from both margins and apparently meeting
    at a zigzag centerline _____ 48. *N. tryblionella* var. *victoriae*
38. Costae slender and reaching one or both margins but not appar-
    ently interrupted at midvalve by a zigzag centerline _____
    _____ 47. *N. tryblionella* var. *levidensis*
39. Striae interrupted by a narrow clear space along the longitudinal axis
    of the valve; margins straight to concave; ends truncate-cunceate to
    rostrate _____ 22. *N. hungarica*
39. Striae visible or invisible but, if visible, then not interrupted by a lon-
    gitudinal clear area _____ 40
40. Center pair of fibulae more widely spaced than adjacent pairs, or
    fibulae irregularly spaced _____ 41
40. Fibulae evenly spaced _____ 52
41. Striae 36/10 μm or more and not resolvable with most light micro-
    scopes even using oblique light _____ 42
41. Striae up to 36/10 μm and usually resolvable without difficulty using
    oblique light _____ 46
42. Fibulae elongate and undulate; ends narrowly rostrate _____
    _____ 34. *N. recta?* (borderline form)
42. Fibulae short _____ 43
43. Valves with narrow attenuate ends or valves linear-lanceolate ___ 44

43. Valves with short, rostrate or capitate ends _____ 45
44. Valves with narrow, attenuate ends _____ 43. *N. spiculoides*
44. Valves linear-lanceolate _____ 4. *N. adapta*
45. Valves with short, narrow, capitate ends _____ 37. *N. rufitorrentis*
45. Valves with narrow, rostrate, somewhat attenuate, ends _____
_____ 19. *N. gandersheimiensis*
46. Valves linear, with broadly rounded, nondistinctive, ends _____
_____ 16. *N. filiformis*
46. Valves lanceolate, or, if linear, then with distinctive ends ____ 47
47. Valves broadly linear, narrowing suddendly to short, apiculate, ends _
_____ 49. *N. umbilicata*
47. Valves narrowly linear to linear-lanceolate, and, if relatively broad,
then the ends rostrate-capitate _____ 48
48. Valve margins slightly concave near midvalve _____
_____ 19. *N. gandersheimiensis*
48. Valve margins straight or convex at midvalve _____ 49
49. Striae around 35/10 μm _____ 44. *N. sublinearis*
49. Striae 30/10 μm or less _____ 50
50. Valves 2–4 μm wide and 45 μm or less in length; striae usually
resolvable into puncta with strongly obliqued light _____
_____ 18. *N. frustulum*
50. Valves 5 μm wide or more _____ 51
51. Valves typically under 70 μm long and over 6 μm wide _____
_____ 50. *N. umbonata*
51. Valves over 70 μm long and no wider than 6 μm ____ 26. *N. linearis*
52. Striae visible, and the puncta distinct or resolvable with moderately oblique light _____ 53
52. Striae visible or invisible, but if visible, then the puncta resolvable,
if at all, only with strongly obliqued light _____ 55
53. Valve ends distinctly capitate or rostrate-capitate _____
_____ 30. *N. microcephala* var. *elegantula*
53. Valve ends truncate-cuneate to somewhat protracted _____ 54.
54. Fibulae prolonged as ribs that reach beyond midvalve _____
_____ 14. *N. denticula*
54. Fibulae usually short, and, if prolonged as ribs, then the ribs not
reaching midvalve _____ 6. *N. amphibia*
55. Striae distinct or easily resolved with oblique light _____ 63
55. Striae not resolvable or resolvable only with difficulty using strongly
obliqued light _____ 56
56. Valve margins constricted abruptly to form very narrow capitate or
rostrate-capitate ends _____ 57

56. Valves with various types of ends but, if capitate, then the margins gradually narrowing before the ends _____ 58
57. Valves under 20 μm in length _____ 29. *N. microcephala*
57. Valves 20 μm long or more _____ 1. *N. accommodata*
58. Valves narrowly lanceolate, under 3 μm wide, the margins tapering gradually from midvalve to the ends _____ 9. *N. caledonensis*
58. Valves linear or lanceolate but, if narrowly lanceolate, the ends attenuate _____ 59
59. Valves with a more or less linear midportion and long-attenuate ends _____ 21. *N. gracilis*
59. Valves linear or lanceolate but, if lanceolate, then the ends short-attenuate or nondistinctive _____ 60
60. Valves linear with acute ends; keel submarginal, the line of the raphe visible _____ 34. *N. recta*
60. Valves linear or lanceolate, but the keel marginal and the line of the raphe not evident _____ 61
61. Valves linear to linear-elliptic with blunt, sometimes slightly protracted, ends _____ 11. *N. communis*
61. Valves linear or lanceolate with acute to subcapitate ends _____ 62
62. Valves distinctly lanceolate with slightly protracted ends _____ _____ 25. *N. kuetzingiana*
62. Valves linear with somewhat protracted rostrate to rostrate-capitate ends _____ 31. *N. palea*
63. Striae 20/10 μm or less and length less than 15 μm _____ 64
63. Striae more than 20/10 μm _____ 65
64. Valves elliptic-lanceolate with acute ends _____ 10. *N. chasei*
64. Valves elliptic-lanceolate with rounded ends __ 51. *N. valdestriata*
65. Valves narrowly linear-lanceolate with acute, nonattenuate, ends ____ _____ 12. *N. confinis*
65. Valves linear or lanceolate but, if lanceolate, then the ends attenuate or subcapitate _____ 66
66. Valves more or less linear, typically over 4 μm wide and over 30 μm long _____ 24. *N. intermedia*
66. Valves lanceolate or, if linear, then 30 μm long or less _____ 67
67. Valves linear or lanceolate and striae 28–30/10 μm __ 17. *N. fonticola*
67. Valves lanceolate and striae 23–25/10 μm _____ 36. *N. romana*

1. Nitzschia accommodata Hust. var. **accommodata** *Fig. 90c.*

CR: Hustedt (1949).

DESCRIPTION: Valve margins parallel near midvalve; valve ends short, narrow, and rostrate-capitate to capitate; keel marginal; fibulae more or less evenly spaced.

DIMENSIONS: Length, 27–32 μm (mine were as short as 20 μm); width, 3.5–4.5 μm; striae around 36/10 μm and difficult to resolve; fibulae, 10–14/10 μm.

2. Nitzschia acicularis (Kütz.) W. Sm. var. **acicularis** *Fig. 92b.*

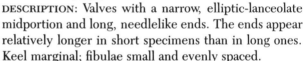

CR: Hustedt (1930).

DESCRIPTION: Valves with a narrow, elliptic-lanceolate midportion and long, needlelike ends. The ends appear relatively longer in short specimens than in long ones. Keel marginal; fibulae small and evenly spaced.

DIMENSIONS: Length, 50–150 μm; width at midvalve, 3–4 μm; striae not visible with the light microscope; fibulae, 17–20/10 μm.

3. Nitzschia acula Hantz. var. **acula** *Fig. 95b.*

CR: Hustedt (1930).

DESCRIPTION: Valves linear-lanceolate with acute, nondistinctive ends; keel subcentral in position; fibulae somewhat irregularly placed, but the center pair not more widely spaced than the rest.

DIMENSIONS: Length, 100–150 μm; width, 3.5–5.0 μm; striae not visible with the light microscope; fibulae, 5–8/10 μm.

4. Nitzschia adapta Hust. var. **adapta** *Fig. 93d.*

CR: Hustedt (1949).

DESCRIPTION: Valves linear to linear lanceolate; ends vary in appearance with the "lie" of the valve: tapered with short-scalpeliform tips in some "lies," almost capitate in others; keel marginal and indented at midvalve; center pair of fibulae more widely spaced than the rest.

DIMENSIONS: Length, 70–150 μm; width, around 3 μm; striae 34–45/10 μm and not resolvable with most light microscopes; fibulae, 11–14/10 μm.

5. Nitzschia agnewii Choln. var. **agnewii** *Fig. 92c.*

CR: Schoeman (1973).

DESCRIPTION: Valves narrowly lanceolate with protracted, almost needlelike ends; keel marginal; fibulae small and evenly spaced.

DIMENSIONS: Length, 15–21 μm; width at midvalve, around 2 μm; striae not visible with the light microscope; fibulae, 16–22/10 μm.

6. Nitzschia amphibia Grun. var. **amphibia** *Fig. 91b, c.*

CR: Hustedt (1930).

DESCRIPTION: Valves linear to lanceolate with cuneate to slightly protracted ends; keel marginal; fibulae not transversely protracted as ribs on the valve face or, if protracted, the ribs not reaching midvalve; striae coarse and distinctly punctate.

DIMENSIONS: Length, 12–50 μm; width, 3–5 μm; striae, 15–19/10 μm; fibulae, 7–9/10 μm.

7. Nitzschia angustata (W. Sm.) Grun. var. **angustata** *Fig. 83a, b.*

CR: Hustedt (1930).

DESCRIPTION: Valves linear with cuneate or, typically, rostrate ends; valve margins weakly convex, parallel, or concave at midvalve; valve surface with a weak longitudinal fold along the midline that gives the striae an undulate appearance; keel marginal and indistinct, the fibulae equal in number to the striae; puncta usually visible without oblique light in the Illinois specimens.

DIMENSIONS: Length, 25–110 μm; width, 5–10 μm; striae and fibulae, 12–18/10 μm.

8. Nitzschia brevissima (Lewis) Grun. var. **brevissima** *Fig. 96a, b.*

CR: Hustedt (1930), Lund (1946), in both reported as *N. parvula* Lewis; Lange-Bertalot & Simonsen (1978).

DESCRIPTION: Valves linear with rostrate to scalpeliform ends; valve margins frequently concave at midvalve; keel marginal; fibulae coarse and more widely spaced at midvalve than elsewhere.

DIMENSIONS: Length, 20–70 μm; width, 3–6 μm; fibulae, 5–9/10 μm; striae 30–40/10 μm, typically 36/10 μm or more and visible, if at all, only with strongly obliqued light.

9. Nitzschia caledonensis Schoeman var. **caledonensis** *Fig. 92d.*

CR: Schoeman (1973).

DESCRIPTION: Valves narrowly lanceolate, the margins tapering gradually to the ends; keel marginal; fibulae small and evenly spaced.

DIMENSIONS: Length, 25–30 μm; width, 1.7–2.5 μm; striae not visible with the light microscope; fibulae, 16–17/10 μm.

10. Nitzschia chasei Choln. var. **chasei** *Fig. 89g.*

CR: Archibald (1971).

DESCRIPTION: Valves narrowly elliptic-lanceolate with acute ends; keel marginal and indistinct. The original description indicated that the fibulae are equal in number to the striae and the illustration showed the keel as indistinct. Archibald reported that the fibulae are half as many as the striae. My specimens showed no distinctive fibulae and therefore seem to be more like those originally described.

DIMENSIONS: Length, 7–12 μm; width, 2.0–2.5 μm; striae, 16–20/10 μm; fibulae, according to Cholnoky, 16–20/10 μm or, according to Archibald, 8–10/10 μm.

11. Nitzschia communis Rabh. var. **communis** *Fig. 90d.*

CR: Hustedt (1930).

DESCRIPTION: Valves linear with weakly convex margins and broadly rounded, sometimes slightly protracted, ends; keel marginal; fibulae evenly spaced.

DIMENSIONS: Length, 20–40 μm; width, 4–5 μm; striae 30–38/10 μm (Illinois specimens had striae that were closer to 38/10 μm); fibulae, 10–14/10 μm.

12. Nitzschia confinis Hust. var. **confinis** *Fig. 88d–f.*

CR: Hustedt (1949).

DESCRIPTION: Valves lanceolate with acute ends; keel marginal; fibulae evenly spaced.

DIMENSIONS: Length, 20–55 μm; width, 2.0–2.5 μm (some of my specimens were around 3 μm wide); striae, 30–33/10 μm; fibulae, 13–15/10 μm.

13. Nitzschia constricta (Kütz.) Ralfs var. **constricta** *Fig. 81c.*

CR: Hustedt (1930), reported as *N. apiculata* Greg.

DESCRIPTION: Valves linear with rostrate ends and concave margins at midvalve; keel marginal and indistinct, the fibulae equal in number to the striae; striae not visibly punctate; striation interrupted by a narrow clear space along the longitudinal axis of the valve.

DIMENSIONS: Length, 20–50 μm; width, 5–8 μm; striae and fibulae, 17–20/10 μm.

14. Nitzschia denticula Grun. var. **denticula** *Fig. 91a.*

CR: Hustedt (1930).

DESCRIPTION: Valves lanceolate to linear-lanceolate; ends acute; keel marginal, the fibulae transversely protracted as distinct ribs on the valve face. The ribs diminish in thickness beyond midvalve and are indistinguishable near the margin opposite the keel. Striae distinct and fairly coarsely punctate.

DIMENSIONS: Length, 10–100 μm; width, 3–8 μm; striae, 14–20/10 μm; fibulae, 5–8/10 μm.

15. Nitzschia dissipata (Kütz.) Grun. var. **dissipata** *Fig. 95c–e.*

CR: Hustedt (1930).

DESCRIPTION: Valves lanceolate to linear-lanceolate with slightly protracted to subrostrate ends; keel most often subcentral, but sometimes submarginal; fibulae evenly spaced.

DIMENSIONS: Length, 15–70 μm; width, 4–7 μm; striae, around 40/10 μm, but typically not visible with most light microscopes; fibulae, 6–8/10 μm.

16. Nitzschia filiformis (W. Sm.) var. **filiformis** *Fig. 96d, e.*

CR: Hustedt (1930).

DESCRIPTION: Valves linear to moderately sigmoid with rounded to somewhat attenuate ends; keel marginal, the central pair of fibulae more widely spaced than the rest; striae fine but easily resolved.

DIMENSIONS: Length, 20–100 μm; width, 4–6 μm; striae, around 36/10 μm; fibulae, 8–11/10 μm.

17. Nitzschia fonticola Grun. var. **fonticola** *Fig. 87c, 88e.*

CR: Hustedt (1930, 1949).

DESCRIPTION: Valves typically lanceolate with somewhat attenuate ends; more rarely, valves linear with weakly convex margins and almost rostrate ends; keel marginal; fibulae evenly spaced; striae not visibly punctate.

DIMENSIONS: Length, 11–30 μm; width, 2.5–4.0 μm; striae, 28–30/10 μm; fibulae, 12–15/10 μm.

18. Nitzschia frustulum (Kütz.) Grun. var. **frustulum** *Fig. 88d.*

CR: Hustedt (1930); Lange-Bertalot & Simonsen (1978).

DESCRIPTION: Valves linear with cuneate or subrostrate ends. Shorter specimens may be elliptic-lanceolate with subacute ends. Keel marginal, the central pair of fibulae always more widely spaced than adjacent pairs; striae distinct and frequently distinctly punctate.

DIMENSIONS: Length, 3–45 μm; width, 2–4 μm; striae, 19–30/10 μm; fibulae, 10–16/10 μm.

19. Nitzschia gandersheimiensis Krasske var. **gandersheimiensis** *Fig. 88b, 90b.*

CR: Lange-Bertalot & Simonsen (1978).

DESCRIPTION: Valves linear to linear-lanceolate with protracted, often subcapitate, ends; keel marginal, the central pair of fibulae more widely spaced than the rest. Valve margins usually indented at midvalve. This taxon is quite variable both in appearance and dimensions. I recommend that you consult the critical reference, which contains numerous photographic illustrations.

DIMENSIONS: Length, 14–70 μm; width, 3–6 μm; striae, 23–42/10 μm; fibulae, 8–18/10 μm.

20. Nitzschia geitleri Hust. var. **geitleri** *Fig. 95a.*

CR: Hustedt (1959a).

DESCRIPTION: Valves narrowly linear with nearly parallel margins and broadly rounded ends; keel almost central; valves weakly sigmoid in girdle view. This large *Nitzschia* is distinctive in appearance, but may easily be overlooked because of its lightly silicified valves.

DIMENSIONS: Length, 160 to over 200 μm; width, 5–8 μm; striae not visible with the light microscope; fibulae, 5–7/10 μm.

21. Nitzschia gracilis Hantzsch var. **gracilis** *Fig. 93a, b.*

CR: Hustedt (1930).

DESCRIPTION: Valves linear to linear-lanceolate with narrow, attenuate ends; keel marginal; fibulae small and more or less evenly spaced, the central pair not more widely spaced than other pairs.

DIMENSIONS: Length, 45–110 μm; width, 2–4 μm; striae typically invisible under the light microscope; fibulae, 12–16/10 μm.

22. Nitzschia hungarica Grun. var. **hungarica** *Fig. 83c, d.*

CR: Hustedt (1930).

DESCRIPTION: Valves linear with cuneate to rostrate ends and margins that vary from straight to concave; keel marginal, the fibulae fewer than the striae and usually distinct; striation interrupted by a fairly narrow longitudinal clear space.

DIMENSIONS: Length, 20–110 μm; width, 5–9 μm; striae, 16–22/10 μm; fibulae, 7–11/10 μm.

23. Nitzschia ignorata Krasske var. **ignorata** *Fig. 96c.*

CR: Hustedt (1930).

DESCRIPTION: Valves moderately sigmoid with tapered ends; keel marginal, the central pair of fibulae more widely spaced than others; unlike the sigmoid examples of *N. filiformis*, the striae are difficult to resolve, even with strongly obliqued light.

DIMENSIONS: Length, 40–70 μm; width, 3–4 μm; striae 36/10 μm, perhaps somewhat more; fibulae, 8–11/10 μm.

24. Nitzschia intermedia Hantzsch var. **intermedia** *Figs. 87b; 88a, c.*

CR: Lange-Bertalot (1977).

DESCRIPTION: Valves linear to linear-lanceolate, tapering to ends that vary from rounded to capitate; keel marginal, the fibulae not more widely spaced at midvalve than elsewhere. As circumscribed by Lange-Bertalot, this taxon is extremely variable in appearance and dimensions. The critical reference is well illustrated with photographs. I advise its use.

DIMENSIONS: Length, 25–200 μm; width, 4–7 μm; striae, 21–33/10 μm; fibulae, 8–13/10 μm.

25. Nitzschia kuetzingiana Hilse var. **kuetzingiana** *Fig. 90h, i.*

CR: Hustedt (1930).

DESCRIPTION: Valves lanceolate with somewhat attenuate or attenuate-rostrate ends; keel marginal; fibulae small and evenly spaced.

DIMENSIONS: Length, 10–28 μm; width, 2.5–5.0 μm; striae, around 36/10 μm; fibulae, 14–18/10 μm.

26. Nitzschia linearis W. Sm. var. **linearis** *Fig. 86a.*

CR: Hustedt (1930).

DESCRIPTION: Valves linear, the margins nearly parallel at midvalve. The shape of the ends depends on the "lie." When the valve is not tipped, the ends are scalpeliform-rostrate. When the valve is somewhat tipped, the ends appear more rostrate-capitate. Keel marginal, the center pair of fibulae more widely spaced than the rest and the entire keel structure robust in appearance.

DIMENSIONS: Length, 70–180 μm; width, 5–6 μm; striae, 28–30/10 μm; fibulae, 8–13/10 μm.

27. Nitzschia lorenziana Grun. var. **lorenziana** *Fig. 97a.*

CR: Peragallo & Peragallo (1897–1908).

DESCRIPTION: Valves sigmoid-lanceolate; ends narrow and somewhat protracted; keel marginal, the fibulae evenly spaced; striae distinct: puncta distinct. This taxon is rare in Illinois.

DIMENSIONS: Length, 130–220 μm; width (my specimens), around 9 μm; striae, 13–14/10 μm at midvalve, approaching 20/10 μm at the ends (mine were around 20/10 μm throughout): fibulae, 6–7/10 μm (mine reached 10/10 μm).

28. Nitzschia lorenziana var. **subtilis** Grun. *Fig. 97b.*

CR: Hustedt (1930).

DESCRIPTION: Valves narrowly sigmoid-lanceolate; ends narrow and somewhat attenuate; keel marginal, the fibulae somewhat irregularly spaced.

DIMENSIONS: Length, 65–165 μm; width, 3–5 μm; striae, 16–19/10 μm; fibulae, 6–8/10 μm.

29. Nitzschia microcephala Grun. var. **microcephala** *Fig. 91i.*

CR: Hustedt (1930).

DESCRIPTION: Valves linear with straight to weakly convex margins that are suddenly constricted to form small capitate or rostrate-capitate ends; keel marginal, the fibulae small and evenly spaced.

DIMENSIONS: Length, 7–16 μm; width, around 3 μm; striae, 33–36/10 μm; fibulae, 12–13/10 μm.

30. Nitzschia microcephala var. **elegantula** (Grun.) V.H. *Fig. 91g.*

CR: Hohn & Hellerman (1963).

DESCRIPTION: Similar in form and structure to var. *microcephala*, but the striae distinct and distinctly punctate.

DIMENSIONS: Length, 14–17 μm; width, 2.5–3.0 μm; striae, 24–26/10 μm; fibulae, 12–13/10 μm.

31. Nitzschia palea (Kütz.) W. Sm. var. palea *Fig. 90a, f, g.*

CR: Hustedt (1930).

DESCRIPTION: Valves linear to linear-lanceolate with rostrate to rostrate-capitate ends; keel marginal, the fibulae evenly spaced.

DIMENSIONS: Length, 20–65 μm; width, 2.5–5.0 μm; striae, 32–40/10 μm and typically 36/10 μm or more; fibulae, 10–15/10 μm.

32. Nitzschia perversa Grun. var. perversa *Fig. 82d–f.*

CR: Cleve & Grunow (1880); Foged (1976).

DESCRIPTION: Valves elliptic with broadly subacute or rounded ends; costae distinct, forming a narrow band removed from the margins; keel indistinct. In some specimens I saw marginal structures that may have been fibulae, though they were not evident in all specimens. Striae, if visible at all, indistinct and resolvable only with strongly obliqued light; valve surface sometimes marked with irregularly placed speckles. *Fig. 82d* illustrates a specimen that had a faint band of very fine lines in the position normally occupied by the costae.

DIMENSIONS (my specimens): Length, 16–25 μm; width, 6–10 μm; costae, 12–20/10 μm (38/10 μm?); striae, if visible, around 40/10 μm; fibulae?, if visible, 7–9/10 μm.

33. Nitzschia plana W. Sm. var. plana *Fig. 81a.*

CR: Peragallo and Peragallo (1897–1908).

DESCRIPTION: Valves linear with margins weakly concave at midvalve; ends broadly rounded; keel submarginal; fibulae large and very distinct; striae distinct and punctate, the puncta regularly and closely spaced at the outer ends of the striae, but irregular in spacing at midvalve. The specimen I drew had a striae-free band along the margin opposite the keel.

DIMENSIONS: Length, 80–170 μm; width, around 18 μm in shorter specimens; striae, 18–20/10 μm; fibulae, 6–8/10 μm.

34. **Nitzschia recta** Hantzsch var. **recta** *Fig. 94a, b(?).*

CR: Hustedt (1930).

DESCRIPTION: Valves linear with nearly straight margins and somewhat attenuate ends; striae not visible even under strongly obliqued light; keel submarginal, the line of the raphe often faintly visible; fibulae somewhat irregularly spaced and moderately long. I have seen also some specimens, such as the one illustrated in *Fig. 94b,* that have the form and the elongate fibulae characteristic of *N. bremensis* Hust., but that have striae almost twice as fine as those reported. These specimens give one the impression that they are closer to *N. recta* than to *N. bremensis,* but I have not seen enough specimens to make a decision. I have retained this illustration only to show that such specimens exist. I hope you will find large enough populations to make a study of it, and preferably with the electron microscope.

DIMENSIONS: Length, 46–130 μm; width, 4.0–7.5 μm; fibulae, 5–9/10 μm.

35. **Nitzschia reversa** W. Sm. var. **reversa** *Fig. 92a.*

CR: Van Heurck (1880–83).

DESCRIPTION: Valves sigmoid, the fairly straight middle portion lanceolate, the curved ends narrow and almost needlelike; keel marginal, the center pair of fibulae more widely spaced than the rest.

DIMENSIONS (my specimens): Length, around 70 μm; width, around 4 μm; striae, not visible; fibulae, around 15/10 μm.

36. Nitzschia romana Grun. var. **romana** *Fig. 89a, b.*

CR: Hustedt (1930).
DESCRIPTION: Valves lanceolate with slightly protracted ends; keel marginal, the fibulae evenly spaced.
DIMENSIONS: Length, 20–45 μm; width, 4–5 μm; striae, 23–25/10 μm; fibulae, 11–13/10 μm.

37. Nitzschia rufitorrentis Choln. var. **rufitorrentis** *Fig. 90e.*

CR: Cholnoky (1960).
DESCRIPTION: Valves linear with nearly parallel margins that narrow abruptly to form short rostrate-capitate ends; keel marginal and slightly indented at midvalve; fibulae robust, the central pair perhaps more widely spaced than the rest.
DIMENSIONS: Length, 41–60 μm; width, 5–6 μm; striae not visible; fibulae, 5–10/10 μm.

38. Nitzschia sigma (Kütz.) W. Sm. var. **sigma** *Fig. 97c.*

CR: Hustedt (1930).
DESCRIPTION: Valves weakly to moderately sigmoid-lanceolate; keel marginal, the fibulae large and evenly spaced; striae distinct and punctate, the puncta so arranged as to form undulating longitudinal striae as well as the straight transverse ones.
DIMENSIONS: Length, 50–1000 μm (the very long cells have not been found in Illinois!); width, 4–15 μm; transverse striae, 22–30/10 μm; fibulae, 7–16/10 μm.

39. Nitzschia sigmoidea (Nitz.) W. Sm. var. **sigmoidea** *Fig. 98d.*

CR: Hustedt (1930).

DESCRIPTION: Valves distinctly sigmoid in girdle view and often seen in this view; valves nearly linear in valve view; ends scalpeliform. It should be noted that valves are typically not seen fully in valve view, so the linear valves may actually appear irregularly curved due to a combination of the shapes of the valve and girdle views. The keel is submarginal in position and the fibulae are massive.

DIMENSIONS: Length, 160–500 μm; width, 8–14 μm; striae, 23–26/10 μm; fibulae, 5–7/10 μm.

40. Nitzschia silicula var. **commutata** Reim. *Fig. 91h.*

CR: Reimer (1966).

DESCRIPTION: Valve margins weakly convex at mid-valve, narrowing abruptly to form rostrate-capitate ends; keel marginal and indistinct, the fibulae equal in number to the striae; striae distinctly punctate.

DIMENSIONS: Length, 15–23 μm; width, 4–5 μm; striae and fibulae, 14–17/10 μm.

41. Nitzschia sinuata var. **delognei** Lange-Bertalot *Fig. 91f.*

CR: Lange-Bertalot (1980).

DESCRIPTION: Valves narrowly lanceolate; keel marginal, the fibulae elongate and so arranged that there appears to be a series of bubblelike "cells" along the margin; striae distinctly punctate.

DIMENSIONS: Length, 10–22 μm; width, 3–4 μm; striae, 22–32/10 μm; fibulae, 6–8/10 μm.

42. Nitzschia sinuata var. tabellaria (Grun.) Grun. *Fig. 91e.*

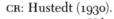

CR: Hustedt (1930).

DESCRIPTION: Valves lanceolate, the margins strongly inflated at midvalve and, in many specimens, weakly undulate toward the ends; keel marginal, the fibulae elongate and so arranged as to give the impression of a series of bubblelike "cells" along the margin; striae distinctly punctate.

DIMENSIONS: Length, 15–30 μm; width, 5–8 μm; striae, 18–20/10 μm; fibulae, 5–6/10 μm.

43. Nitzschia spiculoides Hust. var. **spiculoides** *Fig. 93c.*

CR: Hustedt (1949).

DESCRIPTION: Valves narrowly linear-lanceolate with narrow, attenuate ends; keel marginal, the center pair of fibulae more widely spaced than the rest.

DIMENSIONS: Length, 42–95 μm; width, 2.5–4.5 μm; striae, typically not visible; fibulae, 12–16/10 μm.

44. Nitzschia sublinearis Hust. var. **sublinearis** *Fig. 87a.*

CR: Hustedt (1930).

DESCRIPTION: Valves linear, tapering toward rostrate-capitate ends; keel marginal, the central pair of fibulae slightly more widely spaced than the rest; striae fine but easily resolved with oblique light.

DIMENSIONS: Length, 40–75 μm; width, 4–6 μm; striae, around 35/10 μm; fibulae, 13–15/10 μm.

45. Nitzschia subvitrea Hust. var. **subvitrea** *Fig. 84c*.

CR: Cholnoky (1958); A.N.S.P. File.

DESCRIPTION: Valves linear with scalpeliform ends; keel submarginal, the fibulae robust; striae distinct, but not resolvable into puncta with most microscopes.

DIMENSIONS: Length, 45–94 μm; width, 9–10 μm; striae, 28–29/10 μm; fibulae, 4–7/10 μm.

46. Nitzschia tryblionella Hantzsch var. **tryblionella** *Fig. 81b*.

CR: Hustedt (1930).

DESCRIPTION: Valves broadly elliptic-lanceolate without distinctive ends; keel marginal, the fibulae massive; keel margin slightly indented at midvalve; valve surface costate with several rows of finely punctate striae between each pair of costae.

DIMENSIONS: Length, 60–150 μm; width, 16–35 μm; striae, about 32/10 μm; costae, about 8–10 μm; fibulae, 7–9/10 μm.

47. Nitzschia tryblionella var. **levidensis** (W. Sm.) Grun. *Fig. 82a, b.*

CR: Hustedt (1930).

DESCRIPTION: Valves typically linear-elliptic with straight to convex margins; some specimens with concave margins; keel marginal, the fibulae more or less distinct; valve surface costate, the costae typically crossing the valve, but in some cases stopping short of the nonkeel margin. The valve surface is weakly folded, the long axis of the fold parallel to the long axis of the valve. This folding causes the costae to appear slightly undulate.

DIMENSIONS: Length, 18–66 μm; width, 9–14 μm; costae, 8–13/10 μm; fibulae, 7–11/10 μm.

48. Nitzschia tryblionella var. victoriae Grun. *Fig. 82c.*

CR: Hustedt (1930).

DESCRIPTION: Valves broadly elliptic with broadly subacute ends; keel marginal, the fibulae distinct; valve surface marked by broad costae that appear to be broken at midvalve. In part this appears to be an illusion brought about by folding of the valve surface.

DIMENSIONS: Length, 30–65 μm; width, 15–26 μm; costae, 5–8/10 μm; fibulae, around 6/10 μm.

49. Nitzschia umbilicata Hust. var. umbilicata *Fig. 83e.*

CR: Hustedt (1949).

DESCRIPTION: Valves linear with short, rostrate-apiculate ends; keel marginal, the central pair of fibulae more widely spaced than the rest; transverse costae distinct.

DIMENSIONS: Length, around 35 μm; width, around 8 μm; costae, 18–20/10 μm; fibulae, about 11/10 μm.

50. Nitzschia umbonata (Ehr.) Lange-Bertalot var. umbonata *Figs. 85a–c; 86b, c.*

CR: Lange-Bertalot (1978); Hustedt (1930), reported as *N. thermalis* Kütz. and *N. stagnorum* Rabh.

DESCRIPTION: Valves linear with tapered, sometimes substrate to subcapitate, ends that in some orientations have a hooked appearance; valve margins frequently concave at midvalve on one or both sides; keel marginal, the central pair of fibulae more widely spaced than the rest. As with several other of Lange-Bertalot's revisions, the new circumscription includes specimens of various appearance. I recommend consulting the critical reference.

DIMENSIONS (compiled from Hustedt, 1930): Length, 30–100 μm; width, 5–10 μm; striae, 25–28/10 μm; fibulae, 7–10/10 μm.

51. Nitzschia valdestriata Aleem & Hustedt var. **valdestriata**
Fig. 89c.

CR: Lange-Bertalot & Simonsen (1978).

DESCRIPTION: Valves linear to linear-elliptic with rounded to broadly subacute ends; striae distinct; puncta not visible. In many specimens the striae tend to run at a slight angle to the long axis at the ends, being slightly closer together at the keel margin than at the nonkeel margin.

DIMENSIONS: Length, 5–13 µm; width, 2.5–3 µm; striae, 16–19/10 µm; fibulae, 6–14/10 µm.

52. Nitzschia vermicularis (Kütz.) Grun. var. **vermicularis** *Fig.* 98a–c.

CR: Hustedt (1930).

DESCRIPTION: Valves sigmoid in valve view and weakly to distinctly sigmoid in girdle view. I have seen specimens close in appearance to my *Fig.* 98c on the Van Heurck exsiccata slide A–VH 15 at the Academy of Natural Sciences of Philadelphia. These specimens are more highly sigmoid in valve view and more coarsely striate than those usually illustrated. The keel is marginal with robust fibulae.

DIMENSIONS: Length, 90–250 µm; width, 4–7 µm; striae, 28–36/10 µm; fibulae, 7–12/10 µm.

53. Nitzschia vexans Grun. in V.H. var. **vexans** *Fig. 91d.*

CR: Van Heurck (1880–83), Pl. 57, Fig. 8.

DESCRIPTION: Valves lanceolate with slightly protracted ends; keel marginal, the fibulae equal in number to the striae; striae distinctly and fairly coarsely punctate.

DIMENSIONS: Length, around 20 µm; width, around 5 µm; striae and fibulae, around 15–16/10 µm.

54. **Nitzschia vitrea** Norman var. **vitrea** *Fig. 84a.*

CR: Hustedt (1930).

DESCRIPTION: Valves linear with scalpeliform ends; keel submarginal, the fibulae massive; striae finely and regularly punctate.

DIMENSIONS: Length, 39–200 μm; width, 6–13 μm; striae, 17–24/10 μm; fibulae, 4–7/10 μm.

55. **Nitzschia vitrea** var. **scaphiformis** Wisl. & Poretz. *Fig. 84b.*

CR: Reimer (1970).

DESCRIPTION: Valves linear with scalpeliform ends; keel submarginal, the fibulae massive; striae punctate, the puncta irregularly placed.

DIMENSIONS: Length, 28–51 μm; width, 6–7 μm; striae, 20–26/10 μm; fibulae, 5–6/10 μm.

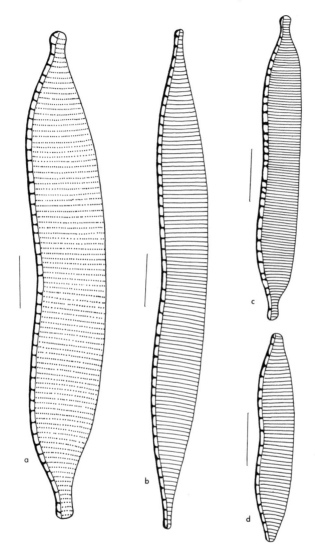

Fig. 80. a. Hantzschia amphioxys v. major. b. H. amphioxys v. vivax. c, d. H. amphioxys v. amphioxys. (Scale lines equal 10 micrometers.)

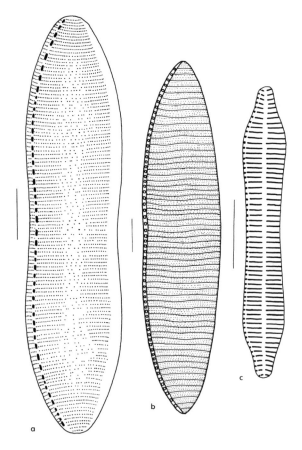

Fig. 81. a. *Nitzschia plana* v. *plana*. **b.** *N. tryblionella* v. *tryblionella*. **c.** *N. constricta* v. *constricta*. (Scale lines equal 10 micrometers.)

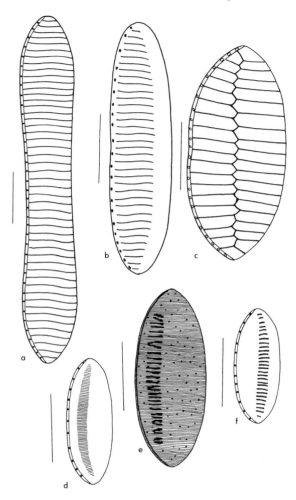

Fig. 82. *a, b.* *Nitzschia tryblionella* v. *levidensis*. *c.* *N. tryblionella* v. *victoriae*. *d–f.* *N. perversa* v. *perversa*. (Scale lines equal 10 micrometers.)

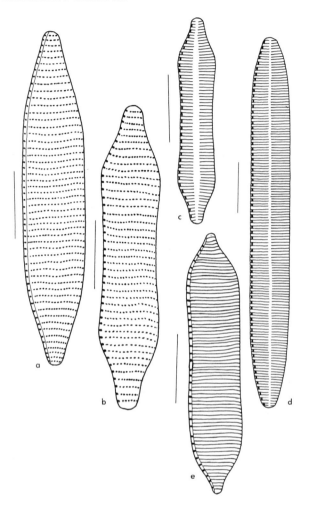

Fig. 83. *a, b. Nitzschia angustata* v. *angustata. c, d. N. hungarica* v. *hungarica. e. N. umbilicata* v. *umbilicata.* (Scale lines equal 10 micrometers.)

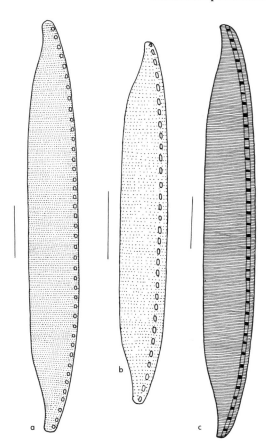

Fig. 84. a. *Nitzschia vitrea* v. *vitrea*. **b.** *N. vitrea* v. *scaphiformis*. **c.** *N. subvitrea* v. *subvitrea*. (Scale lines equal 10 micrometers.)

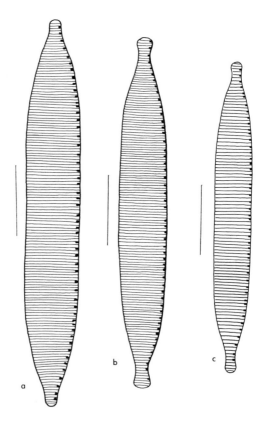

Fig. 85. a–c. *Nitzschia umbonata* v. *umbonata*. (Scale lines equal 10 micrometers.)

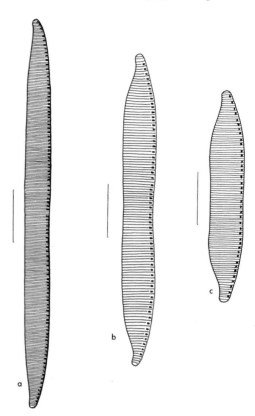

Fig. 86. a. *Nitzschia linearis* v. *linearis*. *b, c. N. umbonata* v. *umbonata*. (Scale lines equal 10 micrometers.)

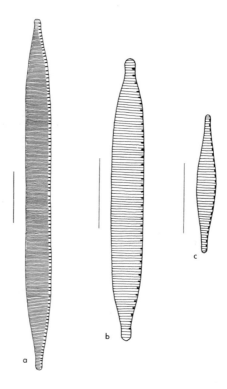

Fig. 87. a. *Nitzschia sublinearis* v. *sublinearis*. **b.** *N. intermedia* v. *intermedia*. **c.** *N. fonticola* v. *fonticola*. (Scale lines equal 10 micrometers.)

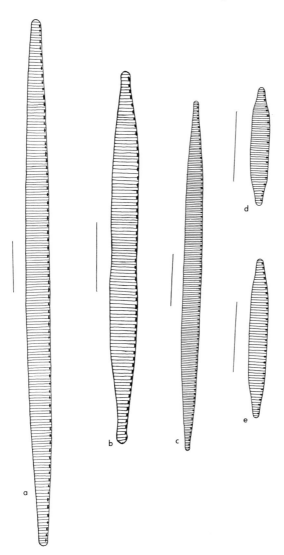

Fig. 88. a. *Nitzschia intermedia* v. *intermedia*. ***b.*** *N. gandersheimiensis* v. *gandersheimiensis*. ***c.*** *N. intermedia* v. *intermedia*. ***d.*** *N. frustulum* v. *frustulum*. ***e.*** *N. fonticola* v. *fonticola*. (Scale lines equal 10 micrometers.)

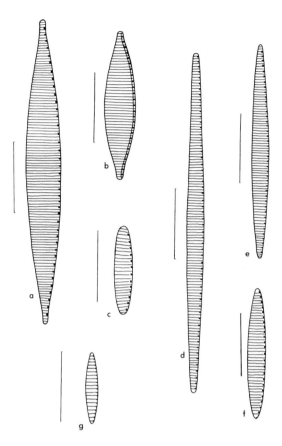

Fig. 89. a, b. *Nitzschia romana* v. *romana*. **c.** *N. valdestriata* v. *valdestriata*. **d–f.** *N. confinis* v. *confinis*. **g.** *N. chasei* v. *chasei*. (Scale lines equal 10 micrometers.)

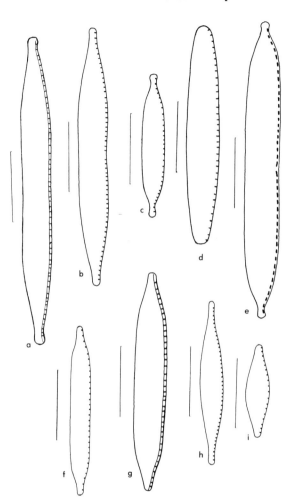

Fig. 90. a. *Nitzschia palea* v. *palea*. **b.** *N. gandersheimiensis* v. *gandersheimiensis*. *c.*
N. accommodata v. *accommodata*. **d.** *N. communis* v. *communis*. **e.** *N. rufitorrentis*
v. *rufitorrentis*. **f, g.** *N. palea* v. *palea*. **h, i.** *N. kuetzingiana* v. *kuetzingiana*. (Scale
lines equal 10 micrometers.)

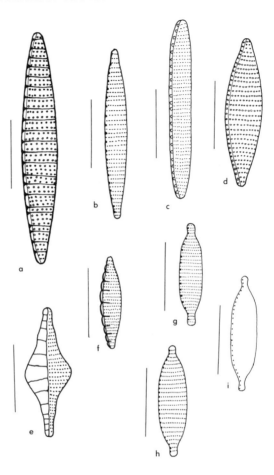

Fig. 91. a. Nitzschia denticula v. denticula. **b, c.** N. amphibia v. amphibia. **d.** N. vexans v. vexans. **e.** N. sinuata v. tabellaria. **f.** N. sinuata v. delognei. **g.** N. microcephala v. elegantula. **h.** N. silicula v. commutata. **i.** N. microcephala v. microcephala. (Scale lines equal 10 micrometers.)

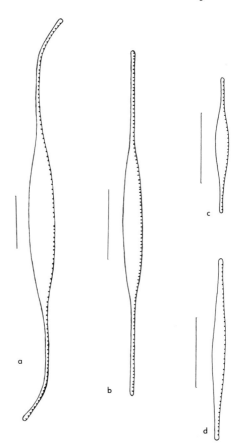

Fig. 92. a. *Nitzschia reversa* v. *reversa*. **b.** *N. acicularis* v. *acicularis*. **c.** *N. agnewii* v. *agnewii*. **d.** *N. caledonensis* v. *caledonensis*. (Scale lines equal 10 micrometers.)

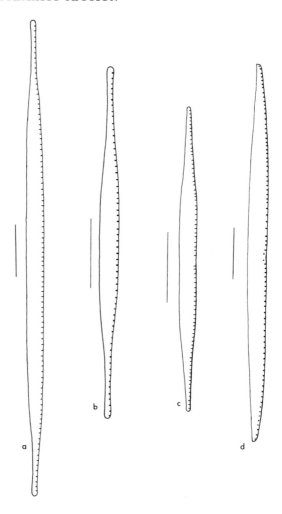

Fig. 93. *a, b. Nitzschia gracilis* v. *gracilis. c. N. spiculoides* v. *spiculoides. d. N. adapta* v. *adapta.* (Scale lines equal 10 micrometers.)

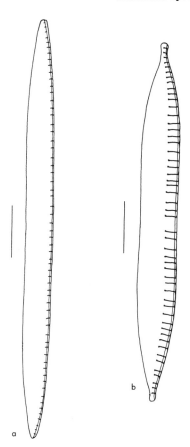

Fig. 94. a. Nitzschia recta v. *recta. b. N. recta* v. *recta*(?). (Scale lines equal 10 micrometers.)

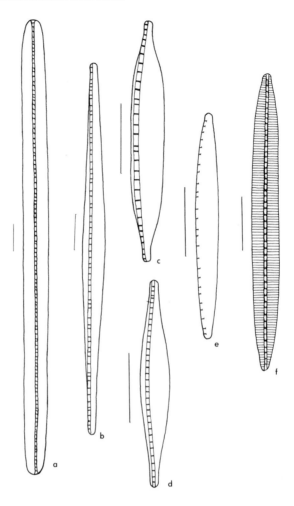

Fig. 95. a. *Nitzschia geitleri* v. *geitleri.* **b.** *Nitzschia acula* v. *acula.* **c–e.** *Nitzschia dissipata* v. *dissipata.* **f.** *Bacillaria paradoxa* v. *paradoxa.* (Scale lines equal 10 micrometers.)

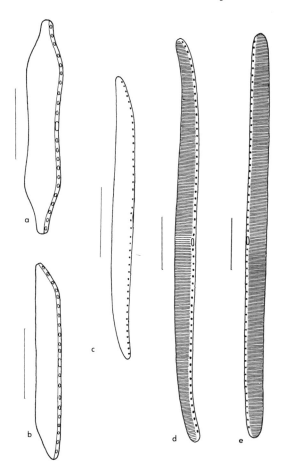

Fig. 96. a, b. *Nitzschia brevissima* v. *brevissima. c. N. ignorata* v. *ignorata. d, e. N. filiformis* v. *filiformis.* (Scale lines equal 10 micrometers.)

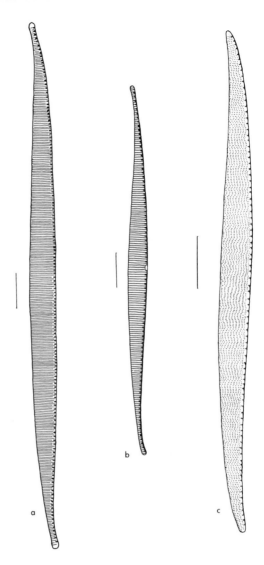

Fig. 97. a. *Nitzschia lorenziana* v. *lorenziana*. **b.** *N. lorenziana* v. *subtilis*. **c.** *N. sigma* v. *sigma*. (Scale lines equal 10 micrometers.)

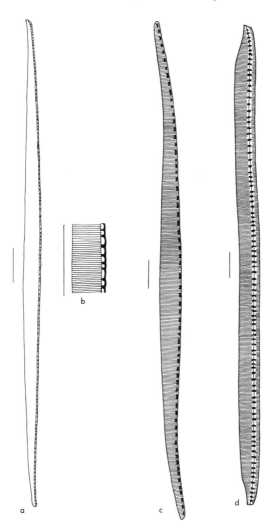

Fig. 98. a. *Nitzschia vermicularis* v. *vermicularis*. **b.** *N. vermicularis* v. *vermicularis* (detail). **c.** *N. vermicularis* v. *vermicularis*. **d.** *N. sigmoidea* v. *sigmoidea*. (Scale lines equal 10 micrometers.)

41. Cymatopleura William Smith 1851

This genus is a member of the order Surirellales and, like all genera in the order, is characterized by bilateral symmetry of markings on the valve and by the presence of two keel-type raphes on each valve, one running along each valve margin. The tubular raphe structure lies atop a poorly or strongly developed raised wing and is connected to the interior of the cell by means of tubular canals, the openings of which appear as small circular marks along the wing.

Cells may be seen in both valve and girdle views, but usually cannot be identified beyond the generic level in girdle view.

The valve surface is not flat or simply arched but is undulate. The undulations are always clearly visible and are usually accentuated by variations in the arrangements of the valve markings. The valve markings consist of costae and very fine striae that may in some cases be transverse but that often are arranged in angular systems related to imaginary transverse lines that correspond to the transverse midlines of the undulations. The whole effect of the markings and undulations is an impression that the valve is marked by broad transverse bands.

KEY TO THE TAXA OF Cymatopleura IN ILLINOIS

1. Valves transapically constricted near midvalve _ _ _ _ _ _ _ _ _ _ 3. *C. solea*
1. Valves with broadly convex margins _ 2
    2. Valve ends blunt _ 1. *C. elliptica*
    2. Valve ends broadly subacute _ _ _ _ _ _ _ _ 2. *C. elliptica* var. *nobilis*

1. Cymatopleura elliptica (Bréb.) W. Sm. var. **elliptica** *Fig. 99b.*

CR: Hustedt (1930).

DESCRIPTION: Valves broadly elliptic with blunt ends, undulations running from margin to margin; crests of undulations marked by diagonal striae; valleys marked by transverse striae.

DIMENSIONS: Length, 50–220 μm; width, 40–90 μm; striae, 15–20/10 μm; wing canals, 2.5–5/10 μm.

2. **Cymatopleura elliptica** var. **nobilis** (Hantz.) Hust. *Fig. 99c.*

CR: Hustedt (1930).
DESCRIPTION: Valves similar to var. *elliptica*, with the exception that the ends of var. *nobilis* are broadly subacute rather than blunt.
DIMENSIONS: See var. *elliptica*.

3. **Cymatopleura solea** (Bréb.) W. Sm. var. **solea** *(Fig. 99a.)*
CR: Hustedt (1930).
DESCRIPTION: Valves with concave margins and broadly cuneate ends; undulations of the valve surface apparently arising near one margin and ending as broad lobes near the other; wing canals prolonged onto the valve surface as thin costae that reach but do not cross the longitudinal midline; valve surface faintly and finely striate between the costae.
DIMENSIONS: Length, 30–300 μm; width, 12–40 μm; wing canals, 6–9/10 μm.

42. *Surirella* Turpin 1828

The genus *Surirella* is a member of the order Surirellales. Both valves of each cell have two keel-type raphes, one on each valve margin. The raphe structure consists of a narrow tubule lying atop a short or tall wing. The raphe proper is a slit in the tubule. The tubule is connected to the inside of the valve by means of a series of short transverse tubules or canals. Each of these canals appears as a pore at the crest of the wing (and usually also as a short strut on the wing side). In many species, costae on the valve surface appear to be continuations of the canals. Between the canals on the wing are openings sometimes called windows. These appear to be true perforations in the silica of the wing. In some species, the windows are more distinctive than the canals. Either can be counted to obtain the "wing canal" count.

In addition to the canals, costae, and windows, the valve surface may be completely or partially striate. The striae are often difficult to see and have not been shown in the drawings where this is the case. The valve markings are interrupted by a wide or narrow space along the longitudinal midline. In some cases this midline may be an actual ridge that may or may not bear spines.

Valves of *Surirella* are symmetric to the long axis but, with respect to the transverse axis, there are symmetric species (isopolar) and asymmetric ones (heteropolar). A few species appear to have variants also in which the valve plane is somewhat twisted around the long axis.

The valve surface in *Surirella* may be nearly flat or somewhat arched but does not have the transverse "ridge and valley" undulations characteristic of *Cymatopleura*.

KEY TO THE TAXA OF Surirella IN ILLINOIS

1. Valves symmetric to the transverse axis _____ 2
1. Valves asymmetric to the transverse axis _____ 4
   2. Valves with margins typically straight (rarely concave) at midvalve; ends typically short-cuneate; length under 70 μm; wing canals 6–8/10 μm _____ 1. *S. angusta*
   2. Valves with margins varying from straight (in valves longer than 70 μm) to convex and, more rarely, concave at midvalve; length often exceeding 70 μm; wing canals 1–2/10 μm _____ 3
3. Valve margins weakly concave at midvalve; valves often slightly twisted around the long axis, therefore not planar; costae arising from one margin not precisely meeting those arising from the other margin; costae often bearing small thorns _____ 4. *S. linearis* var. *constricta*
3. Valve margins convex in shorter valves, becoming straight in the longer valves; valves planar; costae not bearing thorns _____ _____ 3. *S. linearis* var. *linearis*
   4. Valves costate or striate or both almost all the way to the longitudinal centerline _____ 6
   4. Valves marked by canals, costae, or striae only in a narrow marginal band _____ 5
5. Very large and broad valves exceeding 100 μm in length _____ _____ 2. *S. guatimalensis*
5. Tiny valves rarely much longer than 10 μm _____ _____ 7. *S. ovata* var. *africana*
   6. Wing canals 4–10/10 μm; striae present as well, 16–20/10 μm, sometimes up to 36/10 μm (oblique light required) _____ 7

6. Wing canals and costae 0.7–3.0 in 10 μm; striae not visible between the costae _____ 12

7. Head end broadly rounded; foot end cuneate and valve margins weakly concave at midvalve; striae very fine, 30–36/10 μm __ 12. *S. suecica*

7. Valves broadly to narrowly ovate; striae usually not exceeding 20/10 μm _____ 8

8. Wing canals and costae 1.5–4.5/10 μm; costae usually not visible outside the border region; central part of valve elevated somewhat; striae around 16/10 μm; head end broadly rounded to broadly subacute; foot end narrower, very rarely twisted _____ 5. *S. ovalis*

8. Wing canals typically more than 4/10 μm; valves flat _____ 9

9. Valves almost circular _____ 8. *S. ovata* var. *crumena*

9. Valves ovate, linear-ovate, or narrowly ovate _____ 10

10. Valves narrowly ovate, the foot end very narrow _____
_____ 10. *S. ovata* var. *salina*

10. Valves ovate or linear-ovate _____ 11

11. Valves linear-ovate _____ 9. *S. ovata* var. *pinnata*

11. Valves ovate, the head part of the valve distinctly wider than the foot part _____ 6. *S. ovata* var. *ovata*

12. Valves typically over 150 μm long and over 50 μm wide; wing canals 0.7–1.5 in 10 μm _____ 11. *S. robusta*

12. Valves typically under 150 μm long and 40 μm wide or less; wing canals 2–3 in 10 μm _____ 13. *S. tenera*

1. Surirella angusta Kuetz. var. **angusta** *Fig. 101b.*

CR: Hustedt (1930).

DESCRIPTION: Valves isopolar; valve ends moderately cuneate; margins weakly convex to straight; wings rather narrow; wing canals small; transverse costae present; numerous fine transverse striae between the costae; longitudinal clear area narrow.

DIMENSIONS: Length, 18–70 μm; width, 6–15 μm; wing canals and costae, about 6–7.5/10 μm.

2. Surirella guatimalensis Ehr. var. *guatimalensis* *Fig. 103b*.

CR: Stoermer (1964).

DESCRIPTION: Valves heteropolar; head end broadly rounded; foot end broadly subacute; margins convex; wings fairly narrow; wing canals small, projected as short marginal costae onto the valve surface; numerous fine striae present between pairs of costae; most of the valve surface a broad unmarked plain.

DIMENSIONS: (my specimen): Length, 133 μm; width, 61 μm; wing canals, 3/10 μm; striae, 16/10 μm.

3. Surirella linearis W. Sm. var. **linearis** *Fig. 100a,b*.

CR: Hustedt (1930).

DESCRIPTION: Valves linear to linear-elliptic; valve margins convex in shorter valves and nearly straight in the longer valves; wing and wing canals well developed; wing canals prolonged onto the valve surface as costae that end short of the midline, the linear axial space marked by a longitudinal line in the larger specimens but not in the shorter (at least in the present materials). The longer specimens are almost identical in appearance to *S. biseriata* Bréb., but are narrower than the minimum width for that species and have more wing canals/10 μm.

DIMENSIONS: Length, 20–150 μm; width, 9–28 μm; wing canals, 2–3/10 μm.

4. Surirella linearis var. **constricta** Grun. *Fig. 101a*.

CR: Hustedt (1930).

DESCRIPTION: Valves linear with rounded-cuneate to subacute ends; valve margins weakly concave; wings and wing canals well developed; wing canals prolonged onto the valve surface as flat costae that reach the midline of the valve; costae on one side of the valve not precisely aligned with the costae on the other, appearing in some cases almost to alternate with them; valves sometimes planar or somewhat twisted about their long axis and sometimes beset with numerous small thorns. This latter character has been used to define var. *hel-*

vetica (Brun) Meister, but it is my feeling that the separation of this taxon should be rejected.

DIMENSIONS: See var. *linearis*.

5. **Surirella ovalis** Bréb. var. **ovalis** *Fig. 103c*.

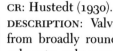

CR: Hustedt (1930).

DESCRIPTION: Valves heteropolar; head end varying from broadly rounded to broadly subacute; food end subacute; valves more or less flat near the margins, the central field of the valve typically arched. I have seen some specimens in which the valve plane is somewhat twisted about the long axis. Lowe (1972) has described a species, *S. ioensis* Lowe, which has this character, but I am almost convinced that my specimens are merely abnormal forms of *S. ovalis;* wing keels of *S. ovalis* narrow and the wing canals small; costae present but short; striae distinct, interrupted by narrow gaps along the longitudinal midline of the valve.

DIMENSIONS: Length, 20–100 μm; width, 10–40 μm; striae, around 16/10 μm; wing canals, 1.5–4.5/10 μm.

6. **Surirella ovata** Kuetz. var. **ovata** *Fig. 102f*.

CR: Hustedt (1930).

DESCRIPTION: Valves heteropolar with broadly rounded head ends and subacute foot ends; margins convex; wings narrow; wing canals narrow; costae and striae both present and reaching the midline of the valve, the longitudinal hyaline area therefore narrow.

DIMENSIONS: Length, 15–70 μm; width, 8–23 μm; wing canals and costae, 4–7/10 μm; striae, 16–20/10 μm.

7. Surirella ovata var. africana Choln. *Fig. 102d,e.*

CR: Cholnoky (1956).

DESCRIPTION: Valves heteropolar with broadly rounded head ends and acute foot ends; margins convex. These are very small valves with no visible markings with the exception of the wing canals.

DIMENSIONS: Length, 8.3–10.0 μm; width, 5 μm; wing canals, 10/10 μm.

8. Surirella ovata var. crumena (Bréb.) V.H. *Fig. 102b.*

CR: Hustedt (1930).

DESCRIPTION: Valve structure nearly identical to var. *ovata*, with the exception of shape, which is broadly elliptic and in some cases approaches circular.

DIMENSIONS: See var. *ovata*.

9. Surirella ovata var. pinnata (W. Sm.) Hust. *Fig. 102a.*

CR: Hustedt (1930).

DESCRIPTION: Valves weakly heteropolar with nearly parallel margins at midvalve; otherwise similar to var. *ovata*.

10. **Surirella ovata** var. **salina** (W. Sm.) Hust. *Fig. 102c.*

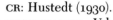

CR: Hustedt (1930).

DESCRIPTION: Valves strongly heteropolar with a broadly rounded head and margins that narrow strongly to form a narrow foot.

DIMENSIONS (my specimen): Length, 21 μm; width, 7 μm; costae, 9/10 μm.

11. **Surirella robusta** Ehr. var. **robusta** *Fig. 104a.*

CR: Hustedt (1930).

DESCRIPTION: Valves heteropolar with broadly rounded head and subacute foot; margins broadly convex; wing distinct; wing canals prolonged onto the valve surface as fairly broad ribs.

DIMENSIONS: Length, 150–400 μm; width, 50–150 μm; wing canals, 0.7–1.5/10 μm.

12. **Surirella suecica** Grun. var. **suecica** *Fig. 103a.*

CR: Hohn and Hellerman (1963).

DESCRIPTION: Valves heteropolar, the head end broadly rounded; valve margins slightly concave near midvalve; foot end cuneate; wing not strongly developed; wing canals prolonged as costae on the valve surface; valve finely striate between the costae; striation interrupted at midvalve by a narrow longitudinal clear space.

DIMENSIONS: Length, 18–25 μm; width, 6–9 μm; striae, 30–36/10 μm (oblique light required); wing canals, 8–9/10 μm.

13. Surirella tenera Greg. var. tenera *Fig. 104b.*

CR: Hustedt (1930).

DESCRIPTION: Valves heteropolar; head end rounded; foot end subacute; margins convex; wing prominent; wing canals prolonged onto the valve surface as broad costae; center line sometimes ornamented with short spines.

DIMENSIONS: Length, 40–170 μm; width, 13–40 μm; wing canals, 2–3/10 μm.

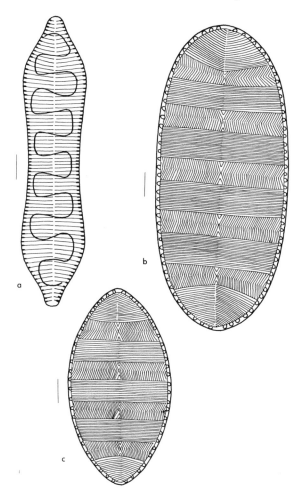

Fig. 99. a. *Cymatopleura solea* v. *solea*. **b.** *C. elliptica* v. *elliptica*. **c.** *C. elliptica* v. *nobilis*. (Scale lines equal 10 micrometers.)

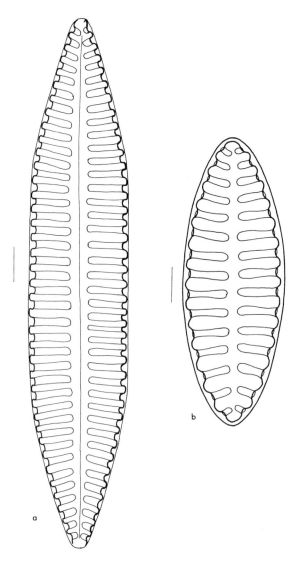

Fig. 100. a, b. *Surirella linearis* v. *linearis*. (Scale lines equal 10 micrometers.)

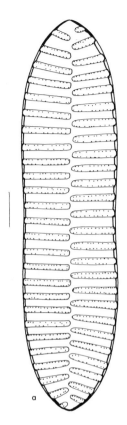

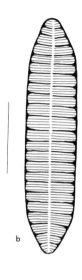

Fig. 101. a. *Surirella linearis* v. *constricta*. **b.** *S. angusta* v. *angusta*. (Scale lines equal 10 micrometers.)

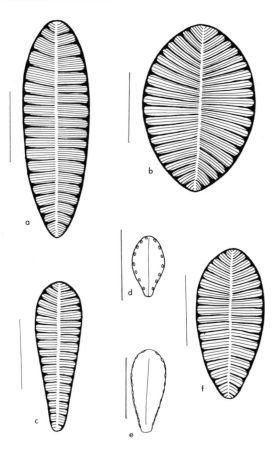

Fig. 102. *a. Surirella ovata* v. *pinnata*. *b. S. ovata* v. *crumena*. *c. S. ovata* v. *salina*. *d, e. S. ovata* v. *africana*. *f. S. ovata* v. *ovata*. (Scale lines equal 10 micrometers.)

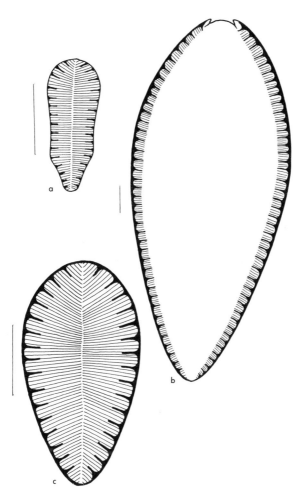

Fig. 103. *a. Surirella suecica* v. *suecica*. *b.* S. *guatimalensis* v. *guatimalensis*. *c.* S. *ovalis* v. *ovalis*. (Scale lines equal 10 micrometers.)

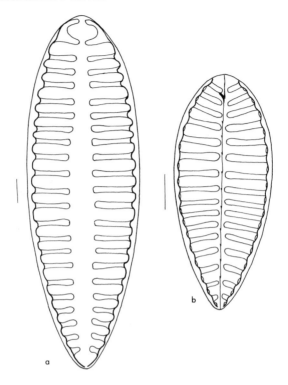

Fig. 104. a. *Surirella robusta* v. robusta. **b.** *S. tenera* v. *tenera*. (Scale lines equal 10 micrometers.)

Glossary

Many of the terms used in describing diatoms cannot be defined effectively with words alone. For this reason I have provided illustrations in place of, or in addition to, verbal definitions. Cross-references have been used in some cases, but most terms will be found in direct order only. Thus **Girdle bands** will be found, but **Bands, girdle** will not.

Active locomotion. Movements that require expenditure of energy by the diatom (as opposed to passive locomotion.)

Acuminate. *See Fig. 108ii.*

Acute. *See Fig. 108d–f,z.*

Alternating longer and shorter, regular (central striae). *See Fig. 109f.*

Alternating longer and shorter, irregular (central striae). *See Fig. 109g.*

Alveoli. Large chamberlike pores closed by thin, ultramicroscopically porous silica membranes. Under the light microscope alveoli appear to be large puncta.

Angular ends. *See Fig. 108c–f, ee.*

Apical axis. *See* **Long axis.**

Arched raphe branches. Curved raphe branches that would meet at an angle at midvalve instead of forming a continuous curve from pole to pole. *See also Fig. 112k.*

Arched valves. Valves with convex surfaces. *See also Fig. 112-o, q.*

Arc-shaped. *See* **Lunate.**

Area. Used by itself this term refers either to an axial area or to a central area.

Areolae. Modified pores found in some genera and similar in appearance to puncta.

Asymmetric (valve shape). To the long axis: *See Fig. 106i;* To the transverse axis: *See Fig. 106h.*

Asymmetric acute (valve end). *See Fig. 108z.*

Asymmetric capitate (valve end). *See Fig. 108dd.*

Asymmetric flat capitate (valve end). *See Fig. 108u.*

Asymmetric-lanceolate (valve shape). *See Fig. 107k, l.*

Asymmetric-rostrate (valve shape). *See Fig. 108aa.*

Asymmetric rounded-capitate (valve end). *See Fig. 108v.*

Attenuate (valve end). Elongate and tapered. *See also Fig. 108k, l.*

Atenuate-rostrate (valve ends). *See Fig. 108j.*

Axial. Pertaining to one of the main axes or lines of symmetry of a cell, but most often used in combinations referring to the long axis of a valve.

Axial area. An area free of markings along the long axis of a valve. *See also Fig. 109a.*

Axial band. A thickened band of

silica slightly wider than the raphe and running along the length of the long axis of the valve including the central area. This band is often faint outside the central area. *See also Fig. 112m.* (Do not confuse with the thickened ribs, *Fig. 112l,* around the raphe branches in some species.)

Axial illunination. *See* **Axial light.**

Axial light. The normal arrangement of the microscope light source in which the light is parallel to the optical axis of the microscope. *See also* **Oblique light.**

Axial locule. *See* **Central locule.**

Axis. Imaginary line of symmetry of a cell. *See also* **Long axis; Pervalvar axis; Transverse axis;** *Fig. 105b, e, f.*

Band. *See* **Axial band; Girdle band; Intercalary band; Longitudinal band.**

Bandlike colonies. Colonies of pennate diatoms in which the cells are joined together, valve to valve, with the long axes of the valves parallel. *See also Fig. 112x.*

Bandlike raphe (also called Lateral raphe). A raphe that appears wider than a line because the fissure is not perpendicular to the valve face. *See also Fig. 109k.*

Bayonet-shaped distal raphe ends. *See Fig. 112b.*

Beaklike (valve end). Narrow elongate-rostrate end. *See also Fig. 108n.*

Bifurcate. Y-shaped.

Bilateral symmetry. Symmetry to a line. *See also Fig. 106g.*

Bilobed wing. *See* **Wing;** *Fig. 111a, b.*

Biundulate (valve margin). *See Fig. 107q.*

Blunt (valve end). *See Fig. 108a.*

Bow-shaped raphe branches. *See* **Arched raphe branches.**

Bracket-shaped (valve shape). *See Fig. 107i.*

Branches, raphe. *See* **Raphe branches.**

Broad end with narrowly capitate tip. *See Fig. 108jj.*

Broadly elliptic (valve shape). *See Fig. 107a.*

Broadly rostrate (valve end). *See Fig. 108h, cc.*

Broadly rounded (valve end). *See Fig. 108b.*

Broadly subrostrate (valve end). *See Fig. 108g.*

Broad striae. Striae that are relatively thick in comparison with their length. *See also Fig. 110-0.*

Bubblelike "cells" (keel structure). *See Fig. 110d.*

Bulbous-capitate (valve end). *See Fig. 108l.*

Bundles of striae. *See Fig. 4a.*

Burned mount. Slides in which all organic matter has been removed by incineration, leaving only silica.

Butterfly-shaped fascia. *See* **Fascia;** *Fig. 109c.*

Canal. A narrow longitudinal depression accompanying the raphe in some genera. The term is used also for any of the numerous transverse members connecting the raphe structure on a wing keel to the inside of the valve. *See also* **Longitudinal furrows; Wing canal.**

Canal raphe structure. In the genera *Denticula, Epithemia,* and *Rhopalodia,* a raphe structure in which the raphe is a slit in a tubular member embedded in the valve surface rather than being a simple slit in the valve surface or a slit in a tubular member raised on a wing or keel.

Capitate (valve end). *See Fig. 108q–w, dd, jj.*

Cell. Properly, a complete diatom with its wall and protoplasm. I have occasionally used the word in a second sense (e.g., Bubblelike "cell") to refer to enclosures or apparent enclosures that are cell-like in appearance.

Central area. An area devoid of striae around the midpoint of the valve in pennate diatoms. *See also Fig. 109a.*

Central-axial area. The combined central and axial areas of a pennate diatom. Usually used when there is no sharp line of demarcation between the two areas. *See also Fig. 109b.*

Central field. In centric diatoms, a zone around the midpoint of the valve that has a different pattern of markings from the area outside it.

Central keel. *See Keel; Fig. 110c.*

Central locule. An elongate opening in the septum of *Mastogloia* species.

Central nodule. A thickening in the valve surface in the area between the proximal ends of raphe branches in pennate diatoms. In most cases, the nodule is confined to the midpoint of the valve and is not particularly prominent. Distinct central nodules can be dotlike, apically elongate, or, in the genera *Stauroneis* and *Capartogramma,* expanded transversely almost to the valve margins. Central nodules are most easily seen with low-power lenses, since these have a greater depth of focus.

Central pore (Bacillariales). A structure, different in appearance from the fibulae, at the midpoint of the keel in some species, and seemingly found only in those species that have been shown to have a central nodule. The nodule itself cannot be seen with the light microscope.

Central pore (Cymbella). A pore at or near the central nodule.

Central striae. The group of striae bordering the central area.

Central zone. *See* **Central field.**

Centric diatoms. Members of class Centrobacillariophyceae.

Chain. *See* **Filament.**

Chambers in septum (Mastogloia). A series of cell-like chambers bordering the central locule in the septum.

Clavate (valve shape). *See Fig. 107j.*

Cleaning. Various processes used to remove organic matter from diatoms prior to making permanent slides.

Club-shaped. *See* **Clavate.**

Coarse (costae, striae, puncta). Low in density and easily visible.

Colony. A group of cells mechanically joined together. As far as is known, diatoms in colonies are not connected protoplasmically. *See also* **Fig. 112s–y.**

Comma-shaped distal raphe ends. *See Fig. 112a.*

Complex raphe. A raphe in which the fissure changes structure and angle of penetration along its length, appearing, therefore, twisted as shown in *Fig. 109m.*

Concave (valve margin). *See Fig. 107h.*

Concentrically undulate (valve surface). *See Fig. 112n.*

Concentric zones. In centric diatoms the valve surface may be divided into inner and outer rings or zones with different patterns of markings. Since each ring lies entirely within the next larger, the rings are concentric.

Convergent striae. *See Fig. 110f.*

Convex (valve margin). *See Fig. 107h.*

Convex (valve surface). *See* Arched valves.

Costae. Thickened transverse, radial, or longitudinal structures that lie between, or that interrupt, the striae. Costae cannot be resolved into puncta, even with the aid of the electron microscope.

Cracticular plate. A septumlike plate in *Navicula cuspidata*. The plate is of the same shape as the valve but has neither raphe nor striae, having instead a series of large, irregularly shaped openings. In "cleaned" preparations, these plates may be found separated from the valves. *See also Fig. 32a.*

Cross-lineate striae. Striae marked by faint to distinct crosslines. *See also* **Lineae**; *Fig. 110l, q.*

Crosslines (on striae). *See* **Lineae.**

Cuneate (cell shape, girdle view). Wedge-shaped, but truncate at the narrow end.

Cuneate (valve end). *See Fig. 108hh.*

Dashlike mark. A structure found near the valve margin in *Thalassiosira fluviatilis*. *See also Fig. 3a.*

Dashlike puncta. *See* **Dashlike units or puncta.**

Dashlike units or puncta. In some species of pennate diatoms the striae are composed of transversely elongate "puncta." I have used the terms "dashlike unit" and "dashlike puncta" to describe their appearance. *See also Figs. 40b, 42g.*

Défaut régulier. A feature in many species of *Neidium* in which a stria near each end of the valve is shorter than the striae adjacent to it, giving the impression that there is a notch in the striation. Some species of *Caloneis* have a similar feature.

Density (of striae, etc.). The number per unit length.

Diagonal striae. *See Fig. 110h, j.*

Diameter. The maximum breadth of the valve of centric diatoms.

Distal raphe ends. The ends of raphe branches at the poles as shown in *Fig. 112a–d.*

Distal raphe fissures. *See* **Distal raphe ends.**

Distinctive ends. Valve ends that are set off from the adjacent margin by constriction or expansion or by an abrupt change in the rate of convergence of the margins. Rostrate and capitate ends are examples.

Distinctly punctate striae. Striae easily resolvable into puncta with the light microscope without the use of oblique light. This is relative, of course, and depends on the quality of the lens in use. *See also Fig. 110n.*

Distorted. Irregular in form, the degree of irregularity typically also variable from specimen to specimen.

Dorsal margin. In dorsiventral valves, the margin that has the greater convexity. *See also Fig. 107n.*

Dorsiventral. The condition of valves that have margins of different curvature. Curved cells that have parallel valve margins are not considered dorsiventral, but are described as lunate or as having convex and concave margins). *See also Fig. 107n.*

Double punctation. The condition of striae that are actually or apparently composed of two parallel rows of puncta. *See also* **Pseudostriae.**

Drawn out. *See* **Attenuate.**

Dry objective. A lens that does not employ oil immersion.

Elliptic (valve shape). *See Fig. 107a, b.*

Elliptic-lanceolate (valve shape). *See Fig. 107d.*

Ends (valves). The terminal regions of the valve. In some works, "ends" is used in the sense of "distinctive ends," the word "termini" being used to refer to other types.

Fan-shaped colonies. *See Fig. 112s.*

Fascia. A central area widened to one or both valve margins. A fascia is formed by the absence of striae and is not synonymous with "stauros," which is a widening of the thick central nodule. *See also Fig. 109c.*

Fibulae. Dotlike silica struts forming part of the keel. *See also* **Keel.**

Filament. A colony formed of cells united valve-to-valve. This term is used typically only for centric diatoms, the equivalent terms for pennate diatoms being bandlike colony or ribbonlike colony. *See also Fig. 112v.*

Filamentous colony. *See* **Filament.**

Filiform raphe. A raphe that appears as a thin line because the fissure is perpendicular to the valve surface. *See also Fig. 109j.*

Fine (striae, costae, etc.). Numerous, thin, difficult to see.

Flaring axial area. *See Fig. 109d.*

Flat capitate (valve end). *See Fig. 108t.*

Flat valve surface. *See Fig. 112p.*

Foot end (valve). In valves asymmetric to the transverse axis, the end of the narrower part of the valve.

Foot pole. *See* **Foot end.**

Foot portion of valve. The half of the valve containing the foot pole.

Foramina (singular: Foramen). Openings in the tubular striae of *Pinnularia* and *Caloneis* that lead to the inside of the valve. *See also Fig. 110k.*

Frustule. A complete cell wall of a diatom including valves, girdle bands, and intercalary bands (if present). *See also Fig. 106a.*

Gabled (valve cross section). *See Fig. 111e, i.*

Ghost striae. Faint transverse

marks in the central area of some species of *Synedra*. These marks are similar in spacing, but not in structure, to the true striae.

Girdle. The "side" of the diatom cell if the two valves are considered the "top" and "bottom."

Girdle band. One of two hooplike silica bands that serve to connect parts of the frustule. The two bands interlock with one another and with the adjacent valves or intercalary bands. *See also Fig. 105a, d, g.*

Girdle view. The side view of a cell. In pennate diatoms, "girdle view" refers to the long side of the cell. *See also Fig. 105d, g.*

Head end. In valves asymmetric to the transverse axis, the end of the wider half of the valve.

Head pole. *See* **Head end.**

Head portion of valve. The half of the valve that contains the head pole.

Height. In *Melosira*, one-half the length of the pervalvar axis of a cell. In most species, effectively the length of the valve mantle from valve face to girdle band.

Helical (shape). *See Fig. 107r.*

Helical (keels). *See Fig. 128d.*

Helical (striae). *See Fig. 106a.*

Heteropolar. Asymmetric to the transverse axis.

Hooked reflex valve end. *See Fig. 108ee.*

Horseshoe-shaped area, depression, or mark. *See Fig. 15b, d, h*

H-shaped area. A distinctive gap in the striation in some species of *Anomoeoneis* and *Navicula*. *See also Fig. 39a, b.*

Hyaline. Transparent almost to the point of invisibility.

Hyaline bands. *See* **Longitudinal clear areas.**

Imbricate (intercalary bands). *See Fig. 105h.*

Indented keel. Keel indented at midvalve. *See Fig. 93e.*

Indistinct central area. A central area that is little wider than the axial area or that is of indeterminate shape due to the elongation of a few striae.

Indistinctly punctate striae. Striae resolvable into puncta only by means of oblique light. *See also Fig. 110n.*

Inflated valve margin. The margin in an area of the valve that is distinctly wider than adjacent parts.

Inner field. *See* **Central field.**

Intercalary band. A silica band similar to a girdle band but interposed between the valve and the girdle band. *See also Fig. 105g, h.*

Intergrade. When cells intermediate in form between two named taxa are found they are called intergrading cells or intergrades. Their existence casts doubt on the separation of the named taxa.

Internal plates. Silica structures within a cell that are parallel to, and of the same shape as, the valves. Septa and craticular plates are examples. *See also* **Internal valves.**

Internal valves. In some species, notably in the genera *Meridion* and *Hantzschia*, for unknown reasons the protoplasm shrinks away from the valves followed by generation of a new pair of valves without cell division. The new valves are called internal valves and may or may not be different in structure from the outer valves. Unlike septa, which are formed in all cells of species that have them, internal valves are of sporadic occurrence.

Intersecting systems of striae. See Fig. 110i, j.

Interspaces. The spaces between the pairs of striae, costae, or tubules.

Isolated puncta. Puncta that are separated from the ends of the adjacent striae. Isolated puncta usually appear larger and rounder than puncta within the stria. See also Fig. 109h. The term is also sometimes used in the sense of scattered puncta when these are few in number.

Isopolar. Symmetric to the transverse axis.

Jelly pore. A visible pore at the ends of valves of some species and believed to be associated with the production of external gelatinous strands. The term "jelly pore" is an old one that is being replaced by other terms as the ultrastructure of the diatom cell becomes better known.

Keel. A raised longitudinal ridge found in marginal, or, less commonly, in central to subcentral, position on the valves of species in the Bacillariales and Surirellales. The raphe is a slit in a longitudinal tubular member located on top of the keel as shown in Figs. 110e; 111d, f, g, j. The tubular member is bridged by silica struts called fibulae. In some genera, the tubular member may be connected to the inside of the valve by narrow, tubular channels, sometimes called "canals." Keels have been given various names depending on their nature and position: **Central keel (Bacillariales).** Keel located near the midline of the valve. See also Fig. 110c. **Marginal keel (Bacillariales).** Keel located at the valve margin. See also Fig. 110a. **Submarginal keel (Bacillariales).** Keel located near the valve margin. See also Fig. 110b. **Wing keel (Surirellales).** A raised ridge, often very distinct, found along both valve margins of each valve in species of *Campylodiscus*, *Cymatopleura*, and *Surirella*. See also Figs. 110e; 111h, k.

Keel puncta. An older term equivalent to fibulae. The name was used because of the dotlike appearance of fibulae. See also **Keel.**

Keel-type raphe structure. See **Keel.**

Knoblike thickenings (Epithemia). As seen in girdle view, the thickened ends of the costae, apparently at the point of junction of the costae with the septum as shown in Fig. 106e.

Lanceolate (central-axial area). See Fig. 109b.

Lanceolate (valve shape). See Fig. 107c.

Lanceolate-elliptic(valve shape). See Fig. 107d.

Lateral raphe. See **Bandlike raphe.**

Length. The maximum length of the valve from end to end.

Lie. The way a valve or cell lies in a prepared mount. In species having valves of complex shape, this is often not a pure valve or girdle view. In these cases one refers to the "normal lie."

Lineae. Fine crosslines seen on the striae of some species. These lines are slitlike puncta. Striae marked in this way are called "lineate," "lineolate," or "cross-lineate," depending on the term favored by a particular author. *See also Fig. 110k, q.*

Linear (valve shape). *See Fig. 107e.*

Linear-elliptic (valve shape). *See Fig. 107b.*

Linear-lanceolate (valve shape). *See Fig. 107f.*

Lineate. *See Lineae.*

Linelike striae. Though all true striae can be shown to be punctate under the electron microscope as shown in *Fig. 110r*, in many species the puncta are so close together that they cannot be resolved with the light microscope even using a good lens and oblique light. I use the term "linelike" to describe striae of this type. The term "linelike" is therefore not synonymous with "lineate" or "lineolate." *See also Fig. 110m.*

Lineolate. *See Lineae.*

Locule. A large opening in a septum or plate. *See also* **Central locule.**

Long axis. An imaginary line running from pole to pole of valves or cells of pennate diatoms. Typically the long axis is straight, but in some genera such as *Gyrosigma*, the long axis is curved. *See also Figs. 105d, 106g.*

Longitudinal axial band. *See Fig. 112m.*

Longitudinal axis. *See* **Long axis.**

Longitudinal band. Any band that crosses the striae and runs roughly parallel to the valve margin.

Longitudinal band (Caloneis, Pinnularia). A band that crosses the striae. This band appears due to the alignment of small to large openings (foramina) in the inner surface of the tubular striae.

Longitudinal band (Neidium). A wide or narrow band of differently structured silica lying parallel to each valve margin. This band is not a clear area, nor, apparently, a row of openings. Each valve has at least these primary longitudinal bands and may have secondary bands as well. *See also Fig. 22.*

Longitudinal canal. *See* **Canal.**

Longitudinal clear areas. These longitudinal spaces in striation are formed by the alignment of gaps in each stria as shown in *Figs. 39g; 43b, c, g.*

Longitudinal clear spaces. *See* **Longitudinal clear areas.**

Longitudinal furrows. Longitudinal depressions accompanying the raphe branch structures in *Diploneis*. In some texts the furrows are called "canals."

Longitudinal lines. Any line that crosses all or most of the striae on one or both sides of the axial area. Such lines vary in structure with the genus. *See also* **Longitudinal band (Neidium); Longitudinal band (Caloneis, Pinnularia).**

Longitudinal midline. A real or, more typically, imaginary, line on the valve roughly corresponding to the longitudinal axis.

Longitudinal ribs (raphe structure). Silica ribs accompanying the raphe branches in some species.

Longitudinal striae. In species with visible puncta the puncta may be so arranged that they form longitudinal rows as well as transverse ones. *See also Fig. 110i.*

Lunate (valve shape). *See Fig. 107h.*

Lunate areas. Arc-shaped marks in the central area. *See also Fig. 109e.*

Mantle. *See* **Valve mantle.**

Marginal keel. *See* **Keel.**

Margins. In a strict sense the actual edge of a valve, but in practical terms, the apparent edge of the valve when it is seen in valve view or in normal lie.

Medial raphe branches. In dorsiventral valves, raphe branches that lie almost equidistant from both valve margins.

Medial striae. The striae forming the central area, but most often referring to the centermost striae on both sides of the central area.

Midline. *See* **Longitudinal midline.**

Midvalve. A point or region near the midpoint of the longitudinal axis.

Narrow lanceolate (valve shape). *See Fig. 107f.*

Narrow rostrate (valve ends). *See Fig. 108i.*

Narrow striae. *See Fig. 110-o.*

Needlelike (valve ends). *See Fig. 108m.*

Neidium-type distal raphe fissures. *See Fig. 112c.*

Nominate. *See* **Nominate variety.**

Nominate variety. The variety with the same name (epithet) as the species is the nominate variety or nominate.

Nondistinctive valve ends. Valve ends that are not set off by constriction or expansion of the margins.

Normal lie. *See* **Lie.**

Oblique light. An arrangement of the light source that allows only angled rays to enter the condenser. Oblique light increases contrast and effective resolution, though it cannot increase the absolute resolution of a lens system. Oblique light is generally achieved by inserting a sharp-edged, opaque card into the light path parallel to the structures (striae, etc.) under observation.

Oblique striae. *See* **Diagonal striae.**

Oblique-truncate (valve end). *See Fig. 108x, y.*

Obliquing. *See* **Oblique light.**

Oval (valve shape). *See* **Elliptic (valve shape).**

Parallel striae. Normally, striae that are parallel to each other and perpendicular to the long axis of the valve. *See also Fig. 110g.*

Parenthesis-shaped. *See* **Lunate.**

Passive motion. Motion by means of environmental forces and not requiring expenditure of energy by the cell.

Pennate diatoms. Members of class Pennatibacillariophyceae.

Perforate. With (usually numerous) holes.

Pervalvar axis. An imaginary line running from valve to valve and perpendicular to the valve surfaces at midvalve. *See also Fig. 105b.*

Pie-shaped wedges. A fanciful term for the appearance of the striated zones in species of *Coscinodiscus*. *See also Fig. 3b.*

Plane. Any flat surface, real or imaginary. *See also* **Valve plane.**

Plankton. Organisms that complete their life cycle, or at least the vegetative phase, free-floating in water.

Plates. *See* **Craticular plates; Internal plates; Septa.**

Polar nodule. A thickening at the end of the vale in which the raphe terminates or that may contain the entire raphe in the case of species with rudimentary raphes.

Polar pseudosepta. Pseudosepta expressed only at the ends of the valve. *See also* **Pseudosepta.**

Polar region. The region near the end of the valve that includes the polar nodule.

Pole. The extreme end of a valve.

Polymorphism. The exhibition by a species of a range of intergrading shapes or patterns in cells of similar size rather than a single more or less fixed shape or pattern.

Pore. A hole of some type, often with a complex ultrastructure that is not visible with the light microscope.

Primary longitudinal band. *See* **Longitudinal band (Neidium).**

Protoplasm. The living matter of a cell.

Protracted (valve ends). Semidistinctive ends produced by a slight retardation of the rate of convergence of the margins near the ends of the valve. Protracted ends may be almost rostrate, or, in narrow valves, may approach needlelike.

Protuberances. Various bumps, horns, etc., other than spines.

Proximal fissures. *See* **Proximal raphe fissures.**

Proximal raphe ends. *See* **Proximal raphe fissures.**

Proximal raphe fissures. The ends of the raphe branches that are nearest each other. *See also Fig. 112e–h.*

Pseudoraphe. A narrow, clear, unstriated space running from pole to pole and usually corresponding to an axial area, but without a raphe.

Pseudoraphe valve. In *Achnanthes* and *cocconeis* the valve with a pseudoraphe only.

Pseudoseptum. In some species the valve mantle is ingrown at the ends to form a narrow strip of silica parallel to the valve face. Unlike true septa, pseudosepta are not found separate from the valve in "cleaned" mounts. *See also Fig. 106c.*

Pseudostriae. I have coined this term to refer to an apparent structure consisting of a costa and parts of the two adjacent striae in some species of *Stephanodiscus*. Pseudostriae are an artifact of observation with the light microscope. *See also* **Tubular striae (Caloneis, Pinnularia).**

Pseudosulcus. In *Melosira*, if the valve mantle does not form a right angle with the valve face, and if cells are seen joined together in filaments, there will appear an indention at the point of junction of valve faces that is called a pseudosulcus. *See also Fig. 106a.*

Puncta. Under the light microscope, any dotlike structures presumed to be pores rather than thickenings or projections. The term is loosely used for a wide variety of pores and processes that are not identical ultrastructurally. I continue to use it because the distinctions between types of "puncta" cannot usually be seen with the light microscope.

Punctate striae. Striae resolvable into puncta (*Fig. 110n*). *See also* **Puncta.**

Punctate-striate. Marked by punctate striae.

Question-mark-shaped distal raphe ends. *See Fig. 112d.*

Radial costae. In centric diatoms, costae aligned with radii of the valve.

Radial symmetry. Symmetry with respect to a point and characteristic of the valve faces of centric diatoms.

Radiate striae. *See Fig. 110f.*

Raphe. A simple or complex longitudinal slit in the valve surface of many pennate diatoms. The presence of a raphe is associated with active locomotion. *See also* **Bandlike raphe** (also called **Lateral**); **Canal raphe structure; Complex raphe; Filiform raphe** (also called **Linelike**); **Keel; Rudimentary raphe; Simple raphe; Undulate raphe.**

Raphe branches. In naviculoid diatoms the raphe is divided into two halves. Each half is a raphe branch.

Raphe branches enclosed in silica ribs. *See Fig. 112l.*

Raphe ends. *See* **Distal raphe ends; Proximal raphe ends.**

Raphe rudiments. *See* **Rudimentary raphe.**

Raphe valve. In *Achnanthes, Cocconeis,* and *Rhoicosphenia,* the valve that has a fully developed raphe.

Recurved valve margins. *See* **Reflexed valve margins.**

Reflexed valve margins. Margins with reversal of curvature near the ends as shown in *Figs. 107-o; 108bb, ee.*

Resting spore. A hardened silica-walled structure that serves as a survival stage when growth conditions are unfavorable; also called Statospore.

Rhombic (valve shape). *See Fig. 107m.*

Ribbonlike colonies. Bandlike colonies of considerable length. *See also* **Bandlike colonies.**

Ribs. *See* **Costae.**

Rostrate (valve end). *See Fig. 108g–j, aa, cc.*

Rostrate-bulbous (valve end). *See Fig. 108l.*

Rostrate-capitate (valve end). *See Fig. 108-o.*

Rostrate-reflexed (valve end). *See Fig. 108bb.*

Rounded-subcapitate (valve end). A subcapitate end with a rounded terminus.

Rudimentary raphe. A raphe with two branches that are restricted to the region of the expanded polar nodule in *Eunotia,* (*See Fig. 112i*), or to a short space near the end of the valve in *Rhoicosphenia* (*See Fig. 112j*). In neither case are the branches connected by a central nodule. The short raphe branches of *Amphipleura* are not considered rudimentary because they are connected by a highly elongate central nodule.

Rudimentary raphe valve. In *Rhoicosphenia*, the valve bearing the rudimentary raphe.

Scalpeliform (valve end). *See Fig. 108ff, gg.*

Septum. A silica plate of simple or complex structure parallel to each valve and lying to the inside of the cell from it. Septa are often associated with intercalary bands. Septa are primary characters of some genera, such as *Tabellaria* and *Mastogloia*, and are secondary characters in others, for example, *Epithemia*. Certain other genera have species that produce similar structures in some individuals. Septa are illustrated in *Fig. 106d, e*. *See also* **Craticular plates; Internal valves.**

Set-off ends. *See* **Distinctive ends.**

Sickle-shaped distal raphe ends. *See Fig. 112d.*

Sigmoid (valve shape). *See Fig. 107g.*

Sigmoid raphe. *See Fig. 29c.*

Sigmoid wing base. *See Fig. 111c.*

Silicified. Made of or reinforced by silica. Weakly silicified structures are thin and usually barely visible. A heavily silicified structure will appear thick and dark.

Simple raphe. A two-branched raphe that is filiform, bandlike, or undulate, but not complex. In another sense, a simple raphe is any raphe that is not part of a canal-type or keel-type raphe structure.

Solitary (growth habit). Refers to species that typically are not united into colonies.

Spines. Sharp-ended projections of silica. *See also Figs. 105h, 106a.*

Star-shaped colony. *See Fig. 112t.*

Statospore. *See* **Resting spore.**

Stauros. A central nodule that is transversely widened and that usually reaches both valve margins. Though the stauros is usually accompanied by a striae-free fascia, it is the thickening that makes the stauros. This feature is characteristic of *Capartogramma* and *Stauroneis*.

Stigma. A complex pore different in structure and appearance from ordinary puncta. *See also Fig. 109i.*

Striae (singular: Stria). Striae are the linelike structures that form, together with the raphe and various bands, costae and puncta, the pattern of markings on the diatom valve. Striae vary in nature with the genus and indeed with the species. The electron microscope has shown that most striae are formed of rows of simple or complex pores. Other striae have been proven to be tubular structures that are ultramicroscopically perforate. Even though the term "stria" does not have a single absolute meaning, I have retained it since it is a useful general term and can be easily qualified to bring out differences. *See also Fig. 110f–r.*

Striate. Marked by striae.

Subacute (valve end). *See Fig. 108c.*

Subaerial (habitat). Any of the habitats existing in moist films on surfaces of living and nonliving matter.

Subcapitate (valve end). *See Fig. 108p.*

Subfossil. A term referring to organisms that exist in the living condition in some places but that do not form part of the living flora or fauna in the place where collected.

Submarginal. Close to, but not at, the valve margin or apparent valve margin.

Submarginal keel. *See* **Keel.**

Subrostrate (valve end). *See Fig. 108g.*

Subterminal. Close to, but not at, the extreme end of the valve.

Sulcus. A constriction of the valve mantle in some species of *Melosira*. *See also Fig. 106a.*

Symmetry. Types of symmetry are illustrated in *Fig. 106f–i.*

System of striae. All the striae that have a particular orientation: transverse, longitudinal, or diagonal. *See also Fig. 110g, i, j.*

Tangential undulations of valve surface (centric diatoms). Undulations parallel to a diameter of the valve.

Terminus (plural: termini). The end of a valve, either distinctive or nondistinctive.

Thick (costae, striae). Wide with respect to the spaces between. This width can be actual or can be apparent due to differences in the thickness of the silica.

Tip. In distinctive valve ends, the extreme end of the valve.

Transapical. Parallel to the transverse axis of a valve.

Transapical constrictions. Narrowings of the valve other than the gradual narrowing of the valve toward the ends.

Transverse axis. The short axis of the valve plane of pennate diatoms, perpendicular to the longitudinal axis at midvalve. *See also Fig. 105f.*

Transverse constrictions. *See Fig. 107p, q.*

Transverse striae. Striae more or less perpendicular to the longitudinal axis of the valve. Transverse striae may be convergent, parallel, or radial. Diagonal or oblique striae are so arranged that the striae on both sides of the axis form diagonal lines across the valve interrupted only by the axial area. Most diagonal striae are classed as transverse, though punctate diagonal striae may form complex systems.

Transverse undulation of valve surface (pennate diatoms). *See Fig. 112r.*

Triundulate (valve margin). *See Fig. 107p.*

Truncate (valve end). *See Fig. 108a.*

Truncate-capitate (valve end). *See Fig. 108t.*

Truncate-cuneate (valve end). A cuneate end with a blunt, rather than acute, tip.

T-shpaed proximal raphe ends. *See Fig. 112f.*

Tubular colony. *See Fig. 112y.*

Tubular raphe channel (Bacillariales, Epithemiales, Surirellales). The raphe in the genera of these orders is a slit in the tubular member located on a keel or in a canal.

Tubular striae (Caloneis, Pinnularia). The "striae" in these genera are actually finely perforate tubular structures that may be open to the inside of the valve over all or part of their length. The smaller openings are called foramina. *See also Fig. 110k, p.*

Undulate. Wavy in appearance.
Undulate raphe. A type of two-branch raphe intermediate in form between bandlike and complex. *See also Fig. 109l.*
Undulate valve margins. *See Fig. 107p, q.*
Undulate valve surfaces. *See Fig. 112n, r.*
Undulate wing base. *See Fig. 111b.*

Valvar plane. *See* **Valve plane.**
Valve. One of a pair of silica plates which, with the girdle bands and intercalary bands (if any), form the cell wall of a diatom. Using the analogy of a box, the valves form the bottom and top, and the girdle (and intercalary) bands make up the sides. The structure and form of the valves is the basis for most of diatom taxonomy. One valve is usually slightly larger than the other. The larger valve is the epivalve, the smaller is the hypovalve.
Valve body. The valve except the valve ends.
Valve face. *See* **Valve plane.** I have used the term "valve face" most often in connection with the valves of Melosira.
Valve mantle. At the edges of the valve the surface changes direction sharply, curving toward the girdle bands. Assuming we are talking about the upper valve, the mantle is a downturned rim. In some genera (such as *Melosira*) the man-

tle extends for a considerable distance down the side of the cell. *See also Fig. 106a, b.*
Valve margin. The edge or apparent edge of the valve when the valve is observed in valve view or normal lie.
Valve outline. The outline of a valve in valve view.
Valve plane. Strictly speaking, an imaginary plane that includes the valve margins and that is perpendicular to the pervalvar (valve-to-valve) axis of the cell. Used loosely, the "valve plane" is that portion of the valve inside the valve margins. *See also Fig. 106b.*
Valve surface. Roughly the same as "valve plane" used in the loose sense, that is, the valve exclusive of valve mantle.
Valve view. The valve when so observed that the entire circumference of the valve is visible. Some valves are so shaped (for example valves of *Entomoneis* and *Rhopalodia*) that they are not usually seen in a pure valve view. In such cases we refer to the valve in "normal lie" rather than in valve view. *See also Fig. 105c, e, f.*
Ventral margin. In dorsiventral valves, the margin with the lesser degree of curvature. *See also Fig. 107n.*

Wedge-capitate (valve end). *See Fig. 108w.*
Wedge-shaped (valve end). *See* **Cuneate** (valve end).
Wedge-shaped groups of striae. In some species of *Coscinodiscus*, the valve face is divided into wedge-shaped zones within which the striae are parallel to each other

and not strictly radial. Such groups are wedge-shaped groups. *See also Fig. 3b*.

Width. The width of a valve measured at midvalve (in the case of valves that are symmetric to the transverse axis) or at the point of maximum breadth otherwise. In *Rhopalodia* the width is measured from edge to edge with the valve in normal lie.

Wing (Entomoneis, Plagiotropis). A raised longitudinal ridge in the valve surface that is so tall with respect to the valve width that the valve typically lies on its side.

Wing base. *See Fig. 111b, c*.

Wing canal. A tubular member connecting the tubular raphe structure with the inside of the valve in many species of *Surirella*. *See also Fig. 110e*.

Wing crest. *See Fig. 111a, b*.

Wing keel. *See* **Keel**.

Wing lobes. *See Fig. 111a, b*.

Zigzag (colony form). *See Fig. 112w*.

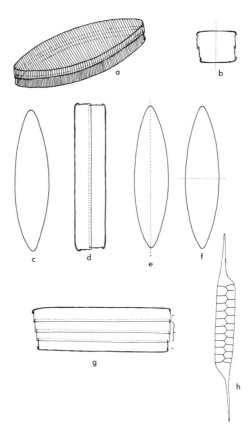

Fig. 105. a. Diatom cell showing valves and overlapping girdle bands. **b.** Cross section of diatom cell showing valves and overlapping girdle bands. Dotted line is pervalvar axis. **c.** Cell from **a** seen in valve view. **d.** Cell from **a** seen in girdle view. **e.** Cell in valve view. Dotted line is long or apical axis. **f.** Cell in valve view. Dotted line is transverse axis. **g.** Cell in girdle view showing girdle bands and intercalary bands. The girdle bands are indicated by a bracket. Each intercalary band is marked with a dash. **h.** Cell of *Rhizosolenia* showing the long spines and the numerous imbricate intercalary bands.

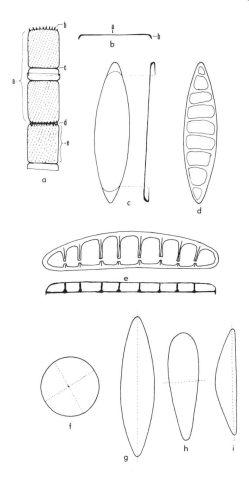

Fig. 106. a. Part of a filament of *Melosira* in girdle view. Note these features: a) a complete cell wall or frustule, b) short spines, c) sulcus, d) pseudosulcus, e) valve mantle. The circular valve face cannot be seen in this view. **b.** Cross section of valve of a pennate diatom. Letter "a" shows the position of the valve plane; letter "b" shows the curvature of the valve mantle. **c.** This pair of illustrations shows a valve with a pseudoseptum. The longitudinal section shows how the pseudoseptum is formed. **d.** A septum divided into locules. **e.** *Epithemia:* Above, a septum; below, a valve in girdle view showing knoblike thickenings at the ends of the costae. Striae omitted for clarity. **f.** Valve view of centric diatom showing radial symmetry. **g.** Pennate diatom in valve view showing bilateral symmetry (symmetry to a line). **h.** Pennate diatom in valve view showing asymmetry to the transverse axis. **i.** Pennate diatom in valve view showing asymmetry to the long or apical axis.

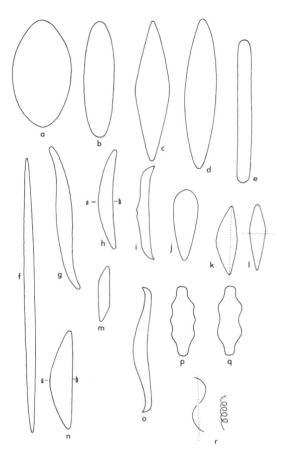

Fig. 107: Shapes. a. Broadly elliptic. **b.** Linear-elliptic. **c.** Lanceolate. **d.** Elliptic-lanceolate (lanceolate-elliptic). **e.** Linear. **f.** Linear-lanceolate (narrowly lanceolate). **g.** Sigmoid. **h.** Lunate. Letter "a" shows convex margin, letter "b" the concave margin. **i.** Bracket-shaped. **j.** Clavate. **k.** Asymmetric-lanceolate (asymmetry to long axis). **l.** Asymmetric-lanceolate (asymmetry to transverse axis). **m.** Rhombic. **n.** Dorsiventral: letter "a" indicates dorsal margin, letter "b" ventral margin. **o.** Cell showing margins reflexed at the ends. **p.** Valve showing triundulate margins. **q.** Valve showing biundulate margins. **r.** Helical shape.

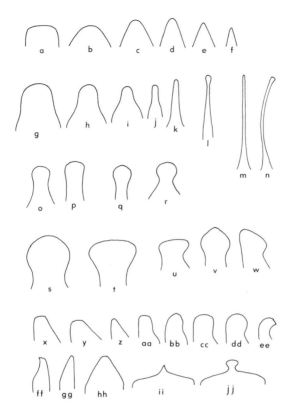

Fig. 108: Shapes of ends. a. Blunt (truncate). **b.** Broadly rounded. **c.** Subacute. **d–f.** Acute. **g.** Subrostrate. **h.** Broadly rostrate. **i.** Narrowly rostrate. **j.** Attenuate-rostrate. **k.** Attenuate. **l.** Attenuate with bulbous tip. **m.** Needlelike. **n.** Narrowly beaklike. **o.** Rostrate-capitate. **p.** Subcapitate. **q–s.** Capitate. **t.** Flat-capitate (truncate-capitate). **u.** Asymmetric flat-capitate. **v.** Asymmetric rounded-capitate. **w.** Wedge-capitate. **x, y.** Oblique-truncate. **z.** Asymmetric acute. **aa.** Asymmetric rostrate. **bb.** Rostrate-reflexed. **cc.** Broadly rostrate. **dd.** Asymmetric capitate. **ee.** Hooked reflexed. **ff, gg.** Scalpeliform. **hh.** Cuneate. **ii.** Acuminate tip. **jj.** Broad with narrowly capitate tip.

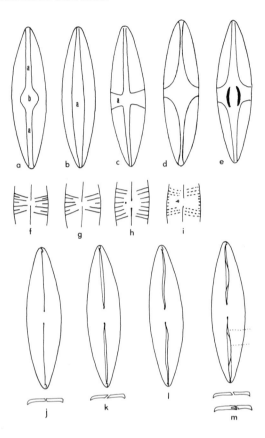

Fig. 109. a. Axial area (marked "a"); central area (marked "b"). **b.** Valve with combined central-axial area (marked "a"). **c.** Fascia-type central area (marked "a"). Type shown is butterfly-shaped but others may be rectangular or irregular. **d.** Valve showing flaring axial area. **e.** Valve showing pair of lunate areas or marks in central area. **f.** Striae regularly alternating longer and shorter at central area. **g.** Striae irregularly alternating longer and shorter. **h.** Central area with isolated punctum. **i.** The isolated mark in this central area is a stigma. **j.** Filiform raphe shown in valve view and cross section. **k.** Lateral (bandlike) raphe in valve view and cross section. **l.** Undulate raphe. **m.** Valve with complex raphe. The two valve cross sections show the structure of the raphe at the points marked with dotted lines.

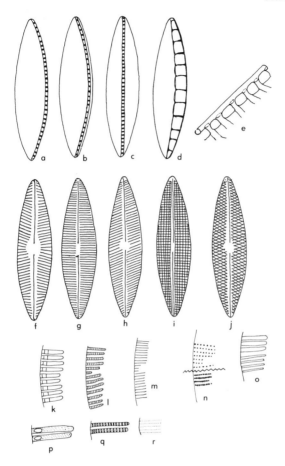

Fig. 110. a. Valve with marginal keel. **b.** Valve with submarginal keel. **c.** Valve with central keel. **d.** Valve with keel structure that appears like a row of bubblelike "cells". **e.** Portion of a wing or wing-keel showing the longitudinal tube that bears the raphe and the narrow tubules that connect this tube to the interior of the cell. Sketch based on an electron micrograph. **f.** Transverse striae convergent at the ends and radiate elsewhere. **g.** Parallel transverse striae. **h.** Oblique transverse striae. **i.** Longitudinal and transverse striae intersecting more or less at right angles. **j.** Intersecting systems of oblique striae. **k.** Tubular striae of *Pinnularia* showing alignment of pores to form a longitudinal band as seen with light microscope. **l.** Lineate (lineolate or crosslineate) striae as seen with·light microscope. **m.** Linelike striae as seen with light microscope. **n.** Punctate striae as seen with light microscope. The upper group are distinctly punctate, the lower group indistinctly punctate. **o.** Broad striae (upper group); narrow striae (lower group). **p.** Tubular striae of *Pinnularia* as they might appear under the electron microscope. Note the numerous fine pores. These cannot be seen with the light microscope. **q.** Lineate striae as they appear under the electron microscope. **r.** Linelike striae are resolvable into puncta when seen with the electron microscope.

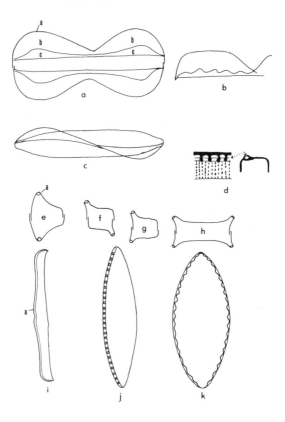

Fig. 111. a. Cell of *Entomoneis* showing bilobed wings of the valves. The small letters show: a) rounded wing crest, b) wing, c) wing base. **b.** One wing lobe of a species of *Entomoneis* showing a flattened wing crest and an undulate wing base. **c.** Valve view of *Entomoneis* showing sigmoid appearance of wing. You will, however, almost never see *Entomoneis* in valve view. **d.** Keel. Left-hand sketch shows a portion of a valve in valve view looking from inside the valve. The keel is at the top of the sketch. Note the thickened fibulae bridging the longitudinal tubular member and the thin line of the raphe. Right-hand sketch shows valve in cross section with keel at upper left. **e.** Cross section of a cell. The section has a "gabled" appearance. The position of the raphe is marked by "a". **f.** Cross section of *Nitzschia* cell showing diagonally opposed position of keels. **g.** Cross section of *Hantzschia* cell showing superimposed position of keels. **h.** Cross section of Surirella showing the four wing keels. **i.** Valve view of a "gabled" valve. The crest of the gable is marked with letter "a". **j.** Valve of *Nitzschia* (see cross section **f.**). **k.** Valve of *Surirella* (see cross section **h.**).

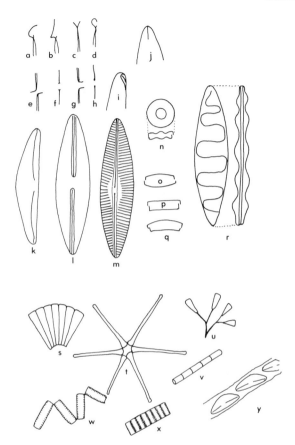

Fig. 112. a. Comma-shaped distal raphe end. *b.* Bayonet-shaped distal raphe end. *c.* *Neidium*-type distal raphe end. *d.* Sickle- or question-mark-shaped distal raphe end. *e.* Proximal raphe ends hooked in opposite directions. *f.* T-shaped proximal raphe ends. *g.* Proximal raphe ends hooked to same side of valve. *h.* Simple proximal raphe ends. *i.* Rudimentary-raphe branch of *Eunotia* located on polar nodule. *j.* Rudimentary-raphe branch of *Rhoicosphenia*. *k.* Valve showing bow-shaped (or arched) raphe branches. *l.* Raphe branches lying in thickened ribs. *m.* Raphe lying in a longitudinal axial band. *n.* Concentric undulation of valve surface in some centric diatoms. *o.* Arched valve surface (cross section). *p.* Flat or plane valve surface (cross section). *q.* Arched valve surface (cross section). *r.* Valve with transverse undulations (seen in valve and girdle views). *s.* Fan-shaped colony. *t.* Star-shaped colony. *u.* Dichotomously branched colony. *v.* Filamentous colony. *w.* Zigzag colony. *x.* Bandlike colony. *y.* Tubular Colony.

Valid Taxa Reported by Other Authors

The following valid taxa have been reported for one or more locations in Illinois or its border waters. I have reported border waters locations only when these were the only locations. For each taxon there is a brief description, often restricted to dimensions alone. When I have provided an illustration, it is not drawn from nature but is a stylized representation that shows the main features cited in the critical reference (CR). In addition, I have given the name(s) under which the taxon has been reported in Illinois if different from the name used here.

The purpose of the descriptions and illustrations is to alert you to possible identities of species that do not fit the descriptions in the earlier parts of the Systematic Section. I do not recommend using the material below to confirm any identification. Rather, you should refer to the named critical reference, if possible.

Achnanthes exilis (Kütz.) Cl. var. **exilis** *Fig. 115f, g.*
> CR: Hustedt (1930).
> REMARKS: Note widely spaced proximal raphe fissures.
> DIMENSIONS: Length, 13–30 μm; width, 4–6 μm; striae, around 20/10 μm at midvalve, becoming 27/10 μm at the ends.
> REPORTED LOCATIONS: Lake Michigan.

Achnanthes flexella (Kütz.) Brun var. **flexella** *Fig. 115h, i.*
> CR: Patrick and Reimer (1966).
> DIMENSIONS: Length, 20–50 μm; width, 10–18 μm; striae (raphe valve), 17–20/10 μm at midvalve, 28–30/10 μm at the ends; striae (pseudoraphe valve), 20–23/10 μm at midvalve, becoming 24–26/10μm at the ends.
> REPORTED IN ILLINOIS AS: *Achnanthidium flexellum* (Bréb.) Kütz., *Cocconeis flexella* (Kütz.) Cl., and *Eucocconeis flexella* (Kütz.) Cl.
> REPORTED LOCATIONS: Lake Michigan.

Achnanthes microcephala (Kütz.) Grun. var. **microcephala** *Fig. 115j, k.*
> CR: Patrick and Reimer (1966).
> REMARKS: I feel that this taxon may intergrade with *A. minutissima* Kütz.

DIMENSIONS: Length, 8–26 μm; width, 2–3 μm; striae (raphe valve), 28/10 μm at midvalve, becoming 34/10 μm at the ends; striae (pseudoraphe valve), 26/10 μm at midvalve, becoming 32/10 μm at the ends.
REPORTED LOCATIONS: Lake Michigan.

Actinocyclus niagarae H. L. Smith var. **niagarae.**

REMARKS: *Actinocyclus* is a marine genus and I doubt that any members of it belong to the living freshwater flora of Illinois.
REPORTED LOCATIONS: Lake Michigan.

Amphora calumetica (Thomas ex Wolle) M. Peragallo var. **calumetica** *Fig. 122a.*

CR: Patrick and Reimer (1975).
DIMENSIONS: Length, 47–87 μm; width (valve, normal lie) 11–13 μm; striae, 12–14/10 μm.
REPORTED IN ILLINOIS AS: *Amphiprora calumetica* Thomas.
REPORTED LOCATIONS: Lake Michigan.

Amphora normanii Rabh. var. **normanii** *Fig. 122b.*

CR: Patrick and Reimer (1975).
DIMENSIONS: Length, 15–40 μm; width (valve, normal lie), 4–6 μm; dorsal striae, 16–18/10 μm; ventral striae, 24–27/10 μm.
REPORTED LOCATIONS: Coles Co.

Amphora ovalis var. **pediculus** (Kütz.) V. H. ex De T. *Fig. 122c.*

CR: Patrick and Reimer (1975).
DIMENSIONS: length, 15–30 μm; width (valve, normal lie), 3.5–6.0 μm; striae, 15/10 μm at midvalve, becoming 18/10 μm at the ends.
REPORTED LOCATIONS: Kane Co.

Anomoeoneis follis (Ehr.) Cl. var. **follis** *Fig. 118a.*

CR: Patrick and Reimer (1966).
DIMENSIONS: Length, 12–54 μm; width, 12–19 μm; striae, 22–26/10 μm.
REPORTED IN ILLINOIS AS: *Navicula trochus* Ehr.
REPORTED LOCATIONS: Lake Michigan.

Anomoeoneis serians (Bréb. ex Kütz.) Cl. var. **serians** *Fig. 118c*.

CR: Patrick and Reimer (1966).

DIMENSIONS: Length, 50–100 μm; width, 12–18 μm; striae, 19–21/10 μm.

REPORTED IN ILLINOIS AS: *Navicula serians* Bréb.

REPORTED LOCATIONS: Lake Michigan.

Caloneis alpestris (Grun.) Cl. var. **alpestris** *Fig. 120a*.

CR: Patrick and Reimer (1966).

DIMENSIONS: Length, 45–92 μm; width, 6–15 μm; striae, 20–24/10 μm.

REPORTED LOCATIONS: Lake Michigan.

Caloneis bacillaris (Greg.) Cl. var. **bacillaris**.

REMARKS: Application of this name to Illinois specimens is uncertain. The reported specimens possibly belong to var. *thermalis* (Grun.) A. Cl.

REPORTED IN ILLINOIS AS: *Navicula bacillaris* Greg.

REPORTED LOCATIONS: Lake Michigan.

Caloneis ventricosa var. **alpina** (Cl.) Patr.

CR: Patrick and Reimer (1966).

REMARKS: Similar in shape to *Caloneis limosa*, but lacking the lunate marks in the central area.

DIMENSIONS: Length, 20–45 μm; width, 7–8 μm; striae, 18–22/10 μm.

REPORTED IN ILLINOIS AS: *Caloneis silicula* var. *alpina* Cl.

REPORTED LOCATIONS: Coles Co.

Campylodiscus noricus Ehr. var. **noricus** *Fig. 126d, e*.

CR: Hustedt (1930).

REMARKS: Valve almost circular, but bent sharply around one axis so that it has a saddle-shape.

DIMENSIONS: Diameter, 60–150 μm; wing canals, 2–3/10 μm.

REPORTED LOCATIONS: Lake Michigan.

Cyclotella antiqua W. Sm. var. **antiqua** *Fig. 113a*.

CR: Hustedt (1930).

DIMENSIONS: Diameter, 10–30 μm; striae, around 17/

10 μm; shadow lines (marks seen in the striated zone as you change to a different level of focus), 6/10 μm.
REPORTED LOCATIONS: Lake Michigan.

Cyclotella bodanica Eulenst. var. bodanica *Fig. 113d*.

CR: Hustedt (1930).
REMARKS: Similar in appearance to *C. comta*, but the isolated puncta at the ends of shortened striae in *C. bodanica* are very distinct, while those of *C. comta* are indistinct, apparently few in number, and sometimes not seen at all. Valves of *C. bodanica* are strongly concentrically undulate.
DIMENSIONS: Diameter, 20–80 μm, striae, around 13/10 μm; shadow lines (see description of *C. comta* for explanation), around 5/10 μm; isolated puncta (as noted above), 1–5, typically 2–4.
REPORTED LOCATIONS: Lake Michigan.

Cyclotella bodanica var. michiganensis Skv. *Fig. 113b*.

CR: Skvortzow (1937).
REMARKS: Differs from var. *bodanica* in the appearance of the central field.
REPORTED LOCATIONS: Lake Michigan.

Cyclotella bodanica var. stellata Skv. *Fig. 113c*.

CR: Skvortzow (1937).
REMARKS: Differs from var. *bodanica* in the appearance of the central field.
REPORTED LOCATIONS: Lake Michigan.

Cyclotella comta var. glabriuscula Grun.

CR: Hustedt (1930).
REMARKS: Differs from var. *comta* in having a central field that is densely and regularly punctate-striate.
REPORTED LOCATIONS: Lake Michigan.

Cyclotella glomerata Bachmann var. glomerata *Fig. 113e*.

CR: Hustedt (1930).
REMARKS: Cells united into filamentous colonies (though not as tightly as in *Melosira*); central field either devoid of markings or with a rosette of dashlike marks.

DIMENSIONS: Diameter, 4–10 μm; striae thin, around 13–15/10 μm.
REPORTED LOCATIONS: Lake Michigan.

Cyclotella kützingiana Thwaites var. **kützingiana** *Fig. 113f.*
CR: Hustedt (1930).
REMARKS: Valve surface weakly tangentially undulate; central field very broad and either devoid of marks or with a few scattered puncta.
DIMENSIONS: Diameter, 10–45 μm; striae fine.
REPORTED LOCATIONS: Grundy Co., Mason Co.

Cyclotella melosiroides (Kirch.) Lemm. var. **melosiroides** *Fig. 113h.*
CR: Hustedt (1930).
REMARKS: Cells united into filamentous colonies; valves tangentially undulate.
DIMENSIONS: Diameter, 4–13 μm; striae 16–18/10 μm.
REPORTED LOCATIONS: Lake Michigan.

Cyclotella ocellata Pant. var. **ocellata** *Fig. 113g.*
CR: Hustedt (1930).
REMARKS: Central field with several large circular marks and a few fine small puncta.
DIMENSIONS: Diameter, 6–20 μm; striae, around 15/10 μm.
REPORTED LOCATIONS: Lake Michigan.

Cyclotella operculata (Ag.) Kütz. var. **operculata** *Fig. 113i.*
CR: Hustedt (1930).
REMARKS: Valves tangentially undulate; central field very finely punctate and separated from the striated outer zone by a ring of marks that appear broad and bright from the outside of the valve, but sharp, dark, and dotlike when seen from the inside of the valve or, by adjusting focus, from the outside of the valve.
DIMENSIONS: Diameter, 6–30 μm; striae, 13–15/10 μm.
REPORTED LOCATIONS: Lake Michigan.

Cyclotella pseudostelligera Hust. var. **pseudostelligera** *Fig. 113j.*

> CR: Huber-Pestalozzi (1942).
> REMARKS: Note presence of a ring of stout spines near the margin.
> DIMENSIONS: Diameter, 7–8 μm; striae, around 18/10 μm.
> REPORTED LOCATIONS: Kane Co.

Cymatopleura elliptica var. **hibernica** (W. Sm.) Hust. *Fig. 125a.*

> CR: Hustedt (1930).
> REMARKS: Differs from var. *elliptica* (Bréb.) W. Sm. in shape.
> REPORTED IN ILLINOIS AS: *Cymatopleura hibernica* W. Sm.
> REPORTED LOCATIONS: Lake Michigan.

Cymatopleura elliptica f. **spiralis** Boyer.

> CR: Tiffany and Britton (1952).
> REMARKS: Differs from var. *elliptica* (Bréb.) W. Sm. in having the valves slightly twisted around the long axis instead of being more or less flat.
> REPORTED IN ILLINOIS AS: *Cymatopleura spiralis* Chase.
> REPORTED LOCATIONS: Lake Michigan.

Cymatopleura solea var. **apiculata** (W. Sm.) Ralfs.

> CR: Hustedt (1930).
> REMARKS: Said to differ from var. *solea* (Bréb.) W. Sm. in having a more apiculate tip and more deeply constricted margins. Var. *solea* is highly variable in shape and I think it likely that this is only an extreme form.
> REPORTED IN ILLINOIS AS: *Cymatopleura apiculata* W. Sm.
> REPORTED LOCATIONS: Lake Michigan.

Cymatopleura solea var. **subconstricta** Muell.

> CR: A.S.A. (1874–1959), Taf. 245 as var. *constricta* f. *minor* Muell.

REMARKS: Likely to be only a shape variant of var. *solea* (Bréb.) W. Sm.

REPORTED LOCATIONS: Coles Co.

Cymbella aequalis W. Sm. var. **aequalis** *Fig. 122d.*

CR: Hustedt (1930).

DIMENSIONS: Length, 15–54 μm; width, 4–12 μm; striae, 11–14/10 μm.

REPORTED LOCATIONS: Coles Co.

Cymbella amphicephala Naeg. ex. Kutz. var. **amphicephala** *Fig. 122e.*

CR: Patrick and Reimer (1975).

REMARKS: Very similar in appearance to *C. naviculiformis* Auersw. ex Heib., but with a central area no wider than the axial area and with striae that are very difficult to resolve into puncta with the light microscope.

DIMENSIONS: Length, 20–40 μm; width, 6–8 μm; striae at midvalve, 12–14/10 μm (dorsal), 15–16/10 μm (ventral); striae (end of valve) 18–20/10 μm.

REPORTED LOCATIONS: Coles Co.

Cymbella inaequalis (Ehr.) Rabh. var. **inaequalis** *Fig. 122j.*

CR: Patrick and Reimer (1975).

DIMENSIONS: Length, 50–220 μm; width, 19–50 μm; striae, 6–9/10 μm at midvalve, becoming 10–12/10 μm at the ends; puncta, 16–20/10 μm.

REPORTED IN ILLINOIS AS: *Cymbella ehrenbergii* Kütz.

REPORTED LOCATIONS: Lake Michigan.

Cymbella lacustris (Ag.) Cl. var. **lacustris** *Fig. 122i.*

CR: Hustedt (1930).

DIMENSIONS: Length, 25–60 μm; width, 7–12 μm; striae, 9–12/10 μm, coarsely lineate.

REPORTED IN ILLINOIS AS: *Colletonema lacrustre* V.H.

REPORTED LOCATIONS: Coles Co.

Cymbella laevis Naeg. ex Kutz. var. **laevis** *Fig. 122f.*

CR: Patrick and Reimer (1975).

REMARKS: Note the finely lineate striae.

DIMENSIONS: Length, 20–35 μm; width, 6–10 μm; dorsal striae, 12/10 μm at midvalve, becoming 17–18/10 μm at the ends; ventral striae, 13–16/10 μm at midvalve, becoming 18–20/10 μm at the ends.
REPORTED LOCATIONS: Lake Michigan.

Cymbella lanceolata (Ag.) Ag. var. **lanceolata.**
 CR: Patrick and Reimer (1975).
 REMARKS: Almost identical in appearance to *C. aspera* (Ehr.) H. Perag., but having strongly radiate terminal striae and distal raphe ends that are deflected almost at right angles to the raphe.
 DIMENSIONS: Length, 100–210 μm; width, 20–28 μm; striae, 9–10/10 μm at midvalve; crosslines on striae, 12–15/10 μm.
 REPORTED LOCATIONS: Lake Michigan.

Cymbella leptoceros (Ehr.?) Grun. var. **leptoceros** *Fig. 122g.*
 CR: Hustedt (1930).
 REMARKS: Note the coarsely punctate or lineate striae.
 DIMENSIONS: Length, 17–60 μm; width, 7–13 μm; striae, 9–12/10 μm.
 REPORTED LOCATIONS: Lake Michigan.

Cymbella turgida (Greg.) Cl. var. **turgida.**
 CR: Hustedt (1930).
 REMARKS: Similar in shape to *C. minuta* var. *pseudogracilis* (Choln.) Reim., but typically much larger.
 DIMENSIONS: Length: 30–100 μm; width, 9–25 μm; striae, 7–9/10 μm near midvalve, somewhat finer at the ends.
 REPORTED LOCATIONS: Clark Co.

Denticula tenuis Kutz. var. **tenuis** *Fig. 124i.*
 CR: Patrick and Reimer (1975).
 DIMENSIONS: Length, 6–60 μm; width, 3–7 μm; costae, 5–7/10 μm; striae, 22–30/10 μm.
 REPORTED LOCATIONS: Coles Co.

Diatoma anceps (Ehr.) Kirch. var. **anceps** *Fig. 114e.*
 CR: Patrick and Reimer (1966).

DIMENSIONS: Length, 12–50 μm; width, 4–8 μm; striae, 17–22/10 μm; costae, 3–6/10 μm.

REPORTED LOCATIONS: Lake Michigan.

Diatoma hiemale (Roth) Heib. var. **hiemale** *Fig. 114f.*

CR: Patrick and Reimer (1966).

DIMENSIONS: Length, 30–100 μm; width, 7–13 μm; striae, 18–20/10 μm; costae, 2–4/10 μm.

REPORTED IN ILLINOIS AS: *Diatoma hiemale* var. *genuinum* Grun.

REPORTED LOCATIONS: Lawrence Co.

Diatoma hiemale var. **mesodon** (Ehr). Grun.

CR: Patrick and Reimer (1966).

REMARKS: Said to be more elliptic than var. *hiemale* and without distinctive ends. Colonies zigzag in form.

DIMENSIONS: Length, 12–40 μm; width, 6–15 μm; costae, 2–4/10 μm; striae, 18–24/10 μm.

REPORTED LOCATIONS: Lawrence Co.

Diatoma tenue Ag. var. **tenue.**

CR: Patrick and Reimer (1966).

REMARKS: Similar in shape to var. *elongatum* Lyngb., but margins not constricted to form subcapitate ends.

DIMENSIONS: Length, 20–55 μm; width, 3–5 μm; striae, 16–20/10 μm; costae, 6–10/10 μm.

REPORTED IN ILLINOIS AS: *Diatoma elongatum* var. *tenuis* (Ag.) V.H.

REPORTED LOCATIONS: Mason Co.

Didymosphenia geminata (Lyngb.) M. Schmidt var. **geminata** *Fig. 122h.*

CR: Patrick and Reimer (1975).

DIMENSIONS: Length, 100–140 μm; width, 25–43 μm; striae, 8–10/10 μm.

REPORTED IN ILLINOIS AS: *Gomphonema geminatum* (Lyngb.) Ag.

REPORTED LOCATIONS: Lake Michigan.

Diploneis elliptica (Kütz.) Cl. var. **elliptica** *Fig. 119a.*

CR: Patrick and Reimer (1966).

DIMENSIONS: Length 20–130 μm; width, 10–60 μm; costae 8–14/10 μm; alveoli, 12–14/10 μm at midvalve.
REPORTED IN ILLINOIS AS: *Navicula elliptica* Kütz.
REPORTED LOCATIONS: Lake Michigan.

Diploneis puella (Schum.) Cl. var. **puella** *Fig. 119b.*
CR: Patrick and Reimer (1966).
DIMENSIONS: Length, 13–27 μm; width, 6–14 μm; costae, 14–18/10 μm.
REPORTED IN ILLINOIS AS: *Navicula elliptica* var. *minutissima* V. H.
REPORTED LOCATIONS: Lake Michigan.

Epithemia adnata (Kütz.) Bréb. var. **adnata** *Fig. 124a.*
CR: Patrick and Reimer (1975).
DIMENSIONS: Length, 15–150 μm; width, 7–14 μm; costae typically 3–5/10 μm, rarely more; striae, 12–14/10 μm.
REPORTED IN ILLINOIS AS: *Epithemia zebra* (Ehr.) Kütz.
REPORTED LOCATIONS: Lake Michigan.

Epithemia argus var. **alpestris** Grun. *Fig. 124b, c.*
CR: Patrick and Reimer (1975).
DIMENSIONS: Length, 55–62 μm; width, 7–10 μm; costae, around 3/10 μm; striae, 11–14/10 μm.
REPORTED IN ILLINOIS AS: *Epithemia alpestris* W. Sm., and *Epithemia argus* var. *amphilepta* Grun.
REPORTED LOCATIONS: Lake Michigan.

Epithemia muelleri Fricke var. **muelleri** *Fig. 124d.*
CR: Patrick and Reimer (1975).
DIMENSIONS: Length, 40–100 μm; width, 12–18 μm; costae, 1–2/10 μm; striae, 10–12/10 μm.
REPORTED LOCATIONS: Lake Michigan.

Epithemia ocellata (Ehr.) Kutz. var. **ocellata** *Fig. 124e.*
CR: Patrick and Reimer (1975).
DIMENSIONS: Length, 25–45 μm; width, 5–11 μm; costae, 2–3/10 μm; striae, 10–14/10 μm.
REPORTED LOCATIONS: Lake Michigan.

Epithemia turgida var. westermannii (Ehr.) Grun. *Fig. 124j.*
CR: Patrick and Reimer (1975).
DIMENSIONS: Length, 40–120 μm; width, 16–34 μm; costae, 3–5/10 μm; striae, around 8/10 μm.
REPORTED IN ILLINOIS AS: *Eunotia westermannii* Ehr.
REPORTED LOCATIONS: Mississippi River.

Eunotia formica Ehr. var. **formica** *Fig. 115l.*
CR: Patrick and Reimer (1966).
DIMENSIONS: Length: 40–160 μm; width, 7–13 μm; striae, 8–12/10 μm.
REPORTED LOCATIONS: Lake Michigan.

Fragilaria brevistriata Grun. var. **brevistriata** *Fig. 114h.*
CR: Patrick and Reimer (1966).
DIMENSIONS: Length, 12–28 μm; width, 3–5 μm; striae, 13–17/10 μm.
REPORTED LOCATIONS: Kane Co.

Fragilaria capucina Desmaz. var. **capucina** *Fig. 114k.*
CR: Patrick and Reimer (1966).
DIMENSIONS: Length, typically 40–100 μm; width, 2–5 μm; striae, 14–18/10 μm.
REPORTED IN ILLINOIS ALSO AS: *Fragilaria rhabdosoma* Ehr.
REPORTED LOCATIONS: Champaign Co., Coles Co.

Fragilaria constricta Ehr. var. **constricta** *Fig. 114i.*
CR: Patrick and Reimer (1966).
DIMENSIONS: Length, 20–70 μm; width, 6–16 μm; striae, 13–18/10 μm.
REPORTED LOCATIONS: Mississippi River.

Fragilaria crotonensis Kitton var. **crotonensis** *Fig. 114g.*
CR: Patrick and Reimer (1966).
DIMENSIONS: Length, 40–170 μm; width, 2–4 μm; striae, 15–18/10 μm.
REPORTED IN ILLINOIS ALSO AS: *Synedra crotonensis* Grun.
REPORTED LOCATIONS: Lake Co., Mason Co.

Fragilaria leptostauron var. **rhomboides** (Grun.) Hust. *Fig. 114j.*

> CR: Hustedt (1930), reported as *F. harrissonii* var. *rhomboides* Grun.; Patrick and Reimer (1966).
> REMARKS: Differs from var. *leptostauron* (Ehr.) Hust. in shape only and possibly also may intergrade with *F. leptostauron* var. *dubia* (Grun.) Hust.
> REPORTED LOCATIONS: Lake Michigan.

Fragilaria virescens Ralfs var. **virescens** *Fig. 114l.*

> CR: Patrick and Reimer (1966).
> DIMENSIONS: Length, 12–120 μm; width, 5–10 μm; striae, 15–19/10 μm.
> REPORTED LOCATIONS: Grundy Co., Mason Co., Morgan Co.

Frustulia rhomboides (Ehr.) De T. var. **rhomboides.**

> CR: Patrick and Reimer (1966).
> REMARKS: Similar in shape to var. *amphipleuroides*, but has proximal raphe fissures close together.
> DIMENSIONS: Length, 70–160 μm; width, 15–30 μm; striae (both transverse and longitudinal), 20–30/10 μm.
> REPORTED IN ILLINOIS ALSO AS: *Navicula rhomboides* Ehr., and *Vanheurckia rhomboides* (Ehr.) Bréb.
> REPORTED LOCATIONS: Macon Co.

Frustulia rhomboides var. **viridula** (Bréb.) Cl. *Fig. 116e.*

> CR: Patrick and Reimer (1966).
> REMARKS: Separated from var. *rhomboides* (Ehr.) De T. by the appearance of the valve and the shape of the central area.
> REPORTED IN ILLINOIS AS: *Frustulia viridula* (Bréb.) De T., and *Schizonema viridulum* Bréb.
> REPORTED LOCATIONS: Lake Michigan.

Gomphoneis eriense (Grun.) Skv. & Meyer var. **eriense** *Fig. 123a.*

> CR: Patrick and Reimer (1975).
> DIMENSIONS: Length, 32–56 μm; width, 13–14 μm; striae, 13–16/10 μm.

REPORTED IN ILLINOIS AS: *Gomphonema eriense* Grun.

REPORTED LOCATIONS: Lake Michigan.

Gomphoneis herculeana (Ehr.) Cl. var. herculeana *Fig. 123b.*

CR: Patrick and Reimer (1975).

DIMENSIONS: Length, 60–100 μm; width, 20–22 μm; striae, 10–12/10 μm.

REPORTED IN ILLINOIS AS: *Gomphonema herculeanum* Ehr.

REPORTED LOCATIONS: Lake Michigan.

Gomphoneis herculeana var. robusta (Grun.) Cl. *Fig. 123c.*

CR: Patrick and Reimer (1975).

DIMENSIONS: Length, 55–65 μm; width, 17–21 μm; striae, 11–12/10 μm.

REPORTED IN ILLINOIS AS: *Gomphonema robustum* Grun.

REPORTED LOCATIONS: Lake Michigan.

Gomphonema acuminatum var. elongatum (W. Sm.) Carr. *Fig. 123i.*

CR: Patrick and Reimer (1975).

REMARKS: Differs from var. *acuminatum* Ehr. in shape and usually in size.

DIMENSIONS: Length, 70–110 μm; width, 12–17 μm; striae, 9–12/10 μm.

REPORTED IN ILLINOIS AS: *Gomphonema acuminatum* var. *intermedia* Grun.

REPORTED LOCATIONS: Lake Michigan.

Gomphonema acuminatum var. trigonocephalum (Ehr.) Grun. *Fig. 123g.*

CR: Hustedt (1930).

REMARKS: Differs from var. *acuminatum* Ehr. in shape. Some specimens that nearly fit the description of var. *trigonocephalum* may in fact be extreme variants of the nominate variety. The name *trigonocephalum* should probably not be applied to specimens with undulate margins, however slight the undulations.

REPORTED IN ILLINOIS AS: *Gomphonema trigonocephalum* Ehr.
REPORTED LOCATIONS: Lake Michigan.

Gomphonema bohemicum Reich. & Fricke var. **bohemicum** *Fig. 123j*.

> CR: Hustedt (1930).
> DIMENSIONS: Length, 10–50 μm; width, 2.5–7.0 μm; striae, 10–14/10 μm.
> REPORTED LOCATIONS: Kane Co.

Gomphonema dichotomum Kütz. var. **dichotomum** *Fig. 123h*.

> CR: Patrick and Reimer (1975).
> DIMENSIONS: Length, 23–63 μm; width, 4–8 μm; striae, 8–13/10 μm.
> REPORTED IN ILLINOIS AS: *gomphonema gracile* var. *dichotoma* (Kütz.) Grun.
> REPORTED LOCATIONS: Lake Michigan.

Gomphonema intricatum Kütz var. **intricatum** Fig. 123e.

> CR: Patrick and Reimer (1975).
> DIMENSIONS: Length, 30–70 μm; width, 5–9 μm; striae, 9–11/10 μm.
> REPORTED LOCATIONS: Coles Co.

Gomphonema intricatum var. **vibrio** (Ehr.) Cl. *Fig. 123f*.

> CR: Patrick and Reimer (1975).
> DIMENSIONS: Length, 75–110 μm; width, 10–12 μm; striae, 8–10/10 μm.
> REPORTED IN ILLINOIS AS: *Gomphonema vibrio* Ehr.
> REPORTED LOCATIONS: Lake Michigan.

Gomphonema semiapertum Grun. var. **semiapertum** *Fig. 123d*.

> CR: Patrick and Reimer (1975).
> DIMENSIONS: Length, 60–106 μm; width, 11–13 μm; striae, 10–12/10 μm.
> REPORTED LOCATIONS: Lake Michigan.

Gyrosigma eximium (Thwaites) Boyer var. **eximium** *Fig. 116f*.

> CR: Patrick and Reimer (1966).

DIMENSIONS: Length, 60–100 μm; width, 9–13 μm; transverse striae, 23–24/10 μm; longitudinal striae, 28–30/10 μm.

REPORTED IN ILLINOIS AS: *Pleurosigma eximium* V.H.

REPORTED LOCATIONS: Lake Michigan.

Gyrosigma spencerii (Quek.) Griff. & Henfr. var. **spencerii** *Fig. 116g.*

CR: Patrick and Reimer (1966).

DIMENSIONS: Length, 95–140 μm; width, 13–15 μm; transverse striae, 18–20/10 μm; longitudinal striae, 22–24/10 μm.

REPORTED IN ILLINOIS ALSO AS: *Gyrosigma kützingii* (Grun.) Cl., and *Pleurosigma kützingii* Grun.

REPORTED LOCATIONS: Lake Michigan, Mississippi River.

Gyrosigma wormleyi (Sulliv.) Boyer var. **wormleyi** *Fig. 116h.*

CR: Patrick and Reimer (1966).

DIMENSIONS: Length, 75–115 μm; width, 13–17 μm; transverse striae, 20–22/10 μm; longitudinal striae, 23–25/10 μm.

REPORTED IN ILLINOIS AS: *Pleurosigma wormleyi* Sulliv.

REPORTED LOCATIONS: Lake Michigan.

Hantzschia amphioxys f. **capitata** Muell.

CR: Hustedt (1930).

REMARKS: Said to differ from var. *amphioxys* (Ehr.) Grun. in the shape of the ends, which are distinctly capitate. It is my experience, however, that *H. amphioxys* is highly variable in shape and that many specimens have rostrate-capitate ends. I have therefore not separated var. *capitata* from the nominate variety. Specimens that could be assigned to var. *capitata* (if one chooses to do so) can be found almost everywhere I have found var. *amphioxys*.

Mastogloia smithii Thwaites ex W. Sm. var. **smithii** *Fig. 116a, b.*

CR: Patrick and Reimer (1966).

DIMENSIONS: Length, 20–65 μm; width, 8–16 μm;

striae, 18–19/10 μm; chambers in septum, 6–8/10 μm.
REPORTED LOCATIONS: Lake Michigan.

Mastogloia smithii var. amphicephala Grun. *Fig. 116c, d.*
CR: Patrick and Reimer (1966).
REMARKS: Differs from var. *smithii* in shape.
REPORTED LOCATIONS: Lake Michigan.

Melosira distans (Ehr.) Kütz. var. **distans** *Fig. 114a.*
CR: Hustedt (1930).
DIMENSIONS: Diameter, 4–20 μm; height, 4.0–8.5 μm; striae on mantle, 12–15/10 μm.
REPORTED IN ILLINOIS AS: *Gallionella distans* Ehr.
REPORTED LOCATIONS: Mississippi River.

Melosira juergensii C. A. Ag. var. **juergensii** *Fig. 114b.*
CR: Hustedt (1930).
DIMENSIONS: Diameter, 6–38 μm; height, 13–22 μm; striae on the valve mantle very fine, around 28/10 μm.
REPORTED IN ILLINOIS AS: *Melosira subflexilis* Kütz.
REPORTED LOCATIONS: Mississippi River.

Navicula anglica Ralfs var. **anglica.**
CR: Hustedt (1930).
REMARKS: Similar in shape and size to *N. gastrum* (Ehr.) Kütz. I have followed Patrick and Reimer (1966) on the matter of *N. gastrum*, particularly in accepting as part of that species those specimens that have striae that do not alternate longer and shorter in the central area. With this reading of *N. gastrum*, there is little left in the description to distinguish it from the more recently described species, *N. anglica*. I have not attempted to separate the two. For those who may wish to do so, I recommend the critical reference, in which these dimensions are found for *N. anglica*: Length, 20–40 μm; width, 8–14 μm; striae, 9–12/10 μm. In Illinois *N. anglica* has been reported from Lake Michigan as *Navicula tumida* W. Sm.

Navicula bacillum Ehr. var. **bacillum** *Fig. 119j.*
CR: Patrick and Reimer (1966).
DIMENSIONS: Length, 30–89 μm; width, 10–20 μm;

striae, 12–14/10 μm at midvalve, becoming 22/10 μm at the ends.
REPORTED LOCATIONS: Lake Michigan.

Navicula cryptocephala var. veneta (Kütz.) Rabh.
CR: Patrick and Reimer (1966).
REMARKS: Patrick and Reimer differentiate this taxon on the basis of ends that are somewhat protracted-subrostrate rather than capitate to rostrate-capitate, and on the basis of dimensions. I have not made the separation of the two taxa since it is my belief that they intergrade. For those who wish to maintain var. *veneta* as separate, the reported dimensions are: Length, 13–26 μm; width, 5–6 μm; sriae, 14–16/10 μm. Specimens fitting the description of var. *veneta* can be found in many of the locations reported for the nominate variety.

Navicula exigua Greg. ex Grun. var. **exigua** *Fig. 119c*.
CR: Patrick and Reimer (1966).
DIMENSIONS: Length, 17–30 μm; width, 9–12 μm; striae, 12–14/10 μm.
REPORTED LOCATIONS: Lake Michigan.

Navicula graciloides A. Mayer var. **graciloides** *Fig. 119e*.
CR: Patrick and Reimer (1966).
DIMENSIONS: Length, 30–40 μm; width, 6–8 μm; striae, 10–13/10 μm.
REPORTED LOCATIONS: Kane Co.

Navicula oblonga var. **subcapitata** Pant. *Fig. 119h, i*.
CR: Hustedt (1930).
REMARKS: Differs from the nominate variety in the shape of the ends.
REPORTED LOCATIONS: Lake Michigan.

Navicula platystoma Ehr. var. **platystoma** *Fig. 119f*.
CR: Hustedt (1930).
DIMENSIONS: Length, 37–50 μm; width, 15–20 μm; striae, 16–18/10 μm, becoming 25/10 μm near the ends.
REPORTED LOCATIONS: Lake Michigan.

Navicula reinhardtii (Grun.) Grun. var. **reinhardtii** *Fig. 119d.*

CR: Patrick and Reimer (1966).

DIMENSIONS: Length, 35–70 μm; width, 11–18 μm; striae, 7–9/10 μm.

REPORTED LOCATIONS: Lake Michigan.

Navicula rhynchocephala Kütz. var. **rhynchocephala** *Fig. 119g.*

CR: Patrick and Reimer (1966).

DIMENSIONS: Length, 35–60 μm; width, 10–13 μm; striae, 8/10 μm at midvalve, becoming 12/10 μm at the ends.

REPORTED LOCATIONS: Coles Co., Macon Co.

Neidium hitchcockii (Ehr.) Cl. var. **hitchcockii** *Fig. 118b.*

CR: Patrick and Reimer (1966).

REMARKS: Note that in addition to the primary longitudinal band there are longitudinal bands along both sides of the axial area.

DIMENSIONS: Length, 35–100 μm; width, 8–15 μm; striae, around 20/10 μm.

REPORTED IN ILLINOIS AS: *Navicula hitchcockii* Ehr.

REPORTED LOCATIONS: Lake Michigan.

Neidium iridis (Ehr.) Cl. var. **iridis** *Fig. 118d.*

CR: Patrick and Reimer (1966).

REMARKS: Similar in shape to *N. decens* (Pant.) Stoermer, but in *N. iridis* the primary longitudinal band is strongly expressed.

DIMENSIONS: Length, 50–190 μm; width, 16–40 μm; striae, 14–18/10 μm.

REPORTED IN ILLINOIS AS: *Navicula iridis* Ehr.

REPORTED LOCATIONS: Mason Co.

Neidium iridis var. **amphigomphus** (Ehr.) A. Mayer *Fig. 118e.*

CR: Patrick and Reimer (1966).

REMARKS: differs from the nominate variety primarily in shape.

REPORTED IN ILLINOIS AS: *Neidium amphigomphus* (Ehr.) Pfitz., and *Navicula iridis* var. *amphigomphus* (Ehr.) V.H.

REPORTED LOCATIONS: Lake Michigan.

Neidium productum (W. Sm.) Cl. var. **productum.**
CR: Patrick and Reimer (1966).
REMARKS: Similar in shape to *Fig. 19a.*
DIMENSIONS: Length, 60–100 μm; width, 20–30 μm; striae, 16–18/10 μm.
REPORTED IN ILLINOIS AS: *Navicula iridis* var. *affinis* (Ehr.) V.H., and *Navicula iridis* var. *producta* (W. Sm.) V.H.
REPORTED LOCATIONS: Lake Michigan.

Nitzschia clausii Hantzsch var. **clausii** *Fig. 124g.*
CR: Hustedt (1930).
REMARKS: Sigmoid in both valve and girdle view.
DIMENSIONS: Length, 20–55 μm; width, 3–5 μm; striae 32–38/10 μm; fibulae, 10–12/10 μm.
REPORTED LOCATIONS: Coles Co.

Nitzschia closterium (Ehr.) W. Sm. var. **closterium** *Fig. 124f.*
CR: Hustedt (1930).
DIMENSIONS: Length, 32–260 μm; width, 2–6 μm; fibulae, 12–16/10 μm; striae not visible with the light microscope.
REPORTED LOCATIONS: Lake Michigan.

Nitzschia sinuata (W. Sm.) Grun. var. **sinuata** *Fig. 124h.*
CR: Hustedt (1930).
DIMENSIONS: Length, 20–50 μm; width, 5–8 μm; striae, 18/10 μm; fibulae, 5–6/10 μm.
REPORTED LOCATIONS: Lake Michigan.

Pinnularia appendiculata (Ag.) Cl. var. **appendiculata** *Fig. 120b.*
CR: Patrick and Reimer (1966).
DIMENSIONS: Length, 18–36 μm; width, 4–6 μm; striae, 16–18/10 μm.
REPORTED IN ILLINOIS AS: *Navicula appendiculata* Kütz.
REPORTED LOCATIONS: Lake Michigan.

Pinnularia boyeri Patr. var. **boyeri** *Fig. 120f.*
CR: Patrick and Reimer (1966).

DIMENSIONS: Length, 100–200 μm; width, 15–22 μm; striae, 10–14/10 μm.
REPORTED IN ILLINOIS AS: *Pinnularia tabellaria* Ehr., and *Navicula tabellaria* Ehr.
REPORTED LOCATIONS: Lake Michigan.

Pinnularia braunii (Grun.) Cl. var. braunii *Fig. 120c.*

CR: Patrick and Reimer (1966).
DIMENSIONS: Length, 30–60 μm; width, 8–12 μm; striae, 10–12/10 μm.
REPORTED LOCATIONS: Cumberland Co.

Pinnularia brevicostata Cl. var. brevicostata *Fig. 120e.*

CR: Patrick and Reimer (1966).
DIMENSIONS: Length, 70–135 μm; width, 12–20 μm; striae, 7–10 μm.
REPORTED IN ILLINOIS AS: *Pinnularia brevicostata* var. *leptostauron* Cl.
REPORTED LOCATIONS: Coles Co.

Pinnularia cardinalis (Ehr.) W. Sm. var. cardinalis *Fig. 121e.*

CR: Patrick and Reimer (1966).
DIMENSIONS: Length, 150–320 μm; width, 30–45 μm; striae, 4–5/10 μm.
REPORTED IN ILLINOIS AS: *Navicula cardinalis* Ehr.
REPORTED LOCATIONS: Lake Michigan.

Pinnularia divergens W. Sm. var. divergens *Fig. 120d.*

CR: Patrick and Reimer (1966).
DIMENSIONS: Length, 50–140 μm; width, 13–20 μm; striae, 9–12/10 μm.
REPORTED IN ILLINOIS AS: *Navicula divergens* Ralfs.
REPORTED LOCATIONS: Lake Michigan.

Pinnularia globiceps Greg. var. globiceps *Fig. 121b.*

CR: Hustedt (1930).
DIMENSIONS: Length, 30–40 μm; width, 8–10 μm; striae, 15–18/10 μm.
REPORTED IN ILLINOIS AS: *Navicula globiceps* Greg.
REPORTED LOCATIONS: Lake Michigan.

Pinnularia maior (Kütz.) Rabh. var. **maior** *Fig. 121c.*
CR: Patrick and Reimer (1966).
DIMENSIONS: Length, 140–200 μm; width, 25–40 μm;
striae, 5–7/10 μm.
REPORTED IN ILLINOIS AS: *Navicula major* Kütz.
REPORTED LOCATIONS: Lake Michigan.

Pinnularia nobilis (Ehr.) Ehr. var. **nobilis** *Fig. 121d.*
CR: Patrick and Reimer (1966).
DIMENSIONS: Length, 200–350 μm; width, 34–50 μm;
striae, 4–5/10 μm.
REPORTED IN ILLINOIS AS: *Navicula nobilis* (Ehr.)
Kütz.
REPORTED LOCATIONS: Lake Michigan.

Pinnularia parvula (Ralfs) Cleve-Euler var. **parvula** *Fig. 121a.*
CR: Patrick and Reimer (1966).
DIMENSIONS: Length, 40–65 μm; width, 7–8 μm;
striae, 8–11/10 μm.
REPORTED IN ILLINOIS AS: *Pinnularia parva* Greg.,
and *Navicula parva* Greg.
REPORTED LOCATIONS: Lake Michigan.

Pleurosigma angulatum (Quek.) W. Sm. var. **angulatum** *fig. 117a, b.*
CR: Patrick and Reimer (1966).
DIMENSIONS: Length, 130–360 μm; width, 30–60 μm;
striae, 17–19 μm.
REPORTED LOCATIONS: Mason Co.

Rhizosolenia eriensis H. L. Smith var. **eriensis** *Fig. 114d.*
CR: Hustedt (1930).
DIMENSIONS: Length of cell in girdle view, 40–150 μm;
width, 5–15 μm; intercalary bands, 2–4/10 μm.
REPORTED LOCATIONS: Mason Co.

Rhizosolenia longiseta Zach. var. **longiseta** *Fig. 114c.*
CR: Hustedt (1930).
DIMENSIONS: Length of cell in girdle view, 70–200 μm;
width, 4–10 μm.

REPORTED IN ILLINOIS AS: *Rhizosolenia gracilis* H. L. Smith.
REPORTED LOCATIONS: Lake Michigan.

Rhopalodia gibba var. ventricosa (Ehr.) Grun.

CR: Patrick and Reimer (1975).
REMARKS: Said to differ from the nominate variety in size and in being more elliptic in girdle view. It is my experience that the nominate variety is highly variable in size and shape. I do not recognize var. *ventricosa*. For those who wish to make the separation, the CR lists dimensions as: Length, 25–100 μm; width of cell in girdle view, 18–28 μm; costae, 5–8/10 μm; striae, 1–2 rows between each pair of costae. Specimens fitting this description can be found in almost all locations I have reported for var. *gibba*. Reported in Illinois by other authors as: *Rhopalodia ventricosa* (Kütz.) Muell., and *Epithemia ventricosa* Kütz.

Stauroneis acuta W. Sm. var. acuta *Fig. 117c*.

CR: Patrick and Reimer (1966).
REMARKS: Note the presence of pseudosepta at the ends.
DIMENSIONS: Length, 80–170 μm; width, 15–27 μm; striae, 12–16/10 μm.
REPORTED LOCATIONS: Lake Michigan.

Stauroneis obtusa Lagerst. var. obtusa *Fig. 117e*.

CR: Patrick and Reimer (1966).
REMARKS: Note presence of pseudosepta at the ends.
DIMENSIONS: Length, 32–76 μm; width, 5–10 μm; striae, 18–20/10 μm at midvalve, becoming 24–26/10 μm at the ends.
REPORTED LOCATIONS: Border waters.

Stauroneis producta Grun. var. producta *Fig. 117d*.

CR: Hustedt (1959).
REMARKS: Note the presence of pseudosepta at the ends.

DIMENSIONS: Length, 30–50 μm; width, 8–11 μm; striae, 22–28/10 μm, finely punctate.
REPORTED LOCATIONS: Lake Michigan.

Stephanodiscus astraea (Ehr.) Grun. var. **astraea** *Fig. 113k.*
CR: Hustedt (1930).
DIMENSIONS: Diameter, 30–70 μm; striae, around 9/10 μm, each composed of 2–4 rows of puncta.
REPORTED LOCATIONS: Lake Michigan.

Stephanodiscus astraea var. **intermedia** Fricke.
CR: Hustedt (1930).
REMARKS: Similar to var. *astraea* but diameter 20–25 μm and striae composed of 3–5 rows of puncta.
REPORTED LOCATIONS: Lake Michigan.

Stephanodiscus niagarae var. **magnifica** Fricke.
CR: A.S.A. Taf. 227.
REMARKS: Valves appear similar to the nominate variety but are strongly concentrically undulate and are heavily silicified.
REPORTED LOCATIONS: Lake Michigan.

Surirella biseriata Bréb. var. **biseriata** *Fig. 125c.*
CR: Hustedt (1930).
REMARKS: Similar in appearance to *S. linearis* W. Sm., but with more pronounced wing keels and different dimensions.
DIMENSIONS: Length, 80–350 μm; width, 30–80 μm; wing canals, 1–2/10 μm.
REPORTED LOCATIONS: Lake Michigan.

Surirella biseriata var. **bifrons** (Ehr.) Hust. *Fig. 125b.*
CR: Hustedt (1930).
REMARKS: Differs from the nominate variety in shape.
REPORTED LOCATIONS: Lake Michigan.

Surirella didyma Kutz. var. **didyma** *Fig. 126a.*
CR: Hustedt (1930).
DIMENSIONS: Length, 65–90 μm; width, 17–19 μm; wing canals, 3–3.5/10 μm.
REPORTED LOCATIONS: Lake Michigan.

Surirella elegans Ehr. var. **elegans** *Fig. 126b.*
CR: Hustedt (1930).
DIMENSIONS: Length 130–425 μm; width, 40–90 μm;
wing canals, 1.2–2.0/10 μm.
REPORTED LOCATIONS: Lake Michigan.

Surirella oregonica Ehr. var. **oregonica** *Fig. 127b.*
CR: Tiffany and Britton (1952).
DIMENSIONS: Length, 180–220 μm; width, 90–110
μm; wing canals, 0.8–1.5/10 μm.
REPORTED LOCATIONS: Lake Michigan.

Surirella patella Kütz. var. **patella** *Fig. 126c.*
CR: Hustedt (1930).
DIMENSIONS: Length, 40–90 μm; width, 19–30 μm;
wing canals, 2.4–3.0/10 μm; striae, 14–16/10 μm.
REPORTED LOCATIONS: Coles Co.

Surirella robusta var. **splendida** (Ehr.) V.H. *Fig. 127c.*
CR: Hustedt (1930).
DIMENSIONS: Length, 75–250 μm; width, 40–60 μm;
wing canals, 1.2–2.5/10 μm.
REPORTED IN ILLINOIS AS: *Surirella splendida* (Ehr.)
Kütz.
REPORTED LOCATIONS: Fulton Co., Mason Co.

Surirella spiralis Kütz. var. **spiralis** *Fig. 127a.*
CR: Hustedt (1930).
DIMENSIONS: Length, 100–260 μm; width, 60–160
μm; wing canals, 0.6–1.2/10 μm.
REPORTED LOCATIONS: Coles Co., Fulton Co., Mason
Co.

Synedra capitata Ehr. var. **capitata** *Fig. 115a.*
CR: Patrick and Reimer (1966).
DIMENSIONS: Length, 125–300 μm; width, 7–10 μm;
striae, 8–11/10 μm.
REPORTED LOCATIONS: Mason Co.

Synedra dorsiventralis Muell. var. **dorsiventralis** *Fig. 115e.*
CR: Mueller (1911).

DIMENSIONS: Length, 61–153 μm; width, 6.5–11.0 μm; striae, 12–13/10 μm, finely punctate.

REPORTED LOCATIONS: Coles Co.

Synedra rumpens Kütz. var. **rumpens** *Fig. 115d*.

CR: Patrick and Reimer (1966).

DIMENSIONS: Length, 27–70 μm; width, 2–4 μm; striae, 18–20/10 μm.

REPORTED LOCATIONS: Lake Michigan.

Synedra ulna var. **obtusa** V.H. *Fig. 115b*.

CR: Patrick and Reimer (1966).

DIMENSIONS: Length, 150–200 μm; width, 6–8 μm; striae, 8–11/10 μm.

REPORTED IN ILLINOIS AS: *Synedra ulna* var. *aequalis* (Kütz.) Hust., and *Synedra obtusa* W. Sm.

REPORTED LOCATIONS: Coles Co.

Genera Expected in Illinois but not Yet Reported

Attheya T. West 1860

Attheya belongs to family Biddulphiaceae, order Biddulphiales, class Centrobacillariophyceae. Each valve has two long spines on opposite ends of the long axis. Numerous imbricate intercalary bands are present in the girdle. Only one species is likely to be found:

Attheya zachariasi Brun var. **zachariasi** *Fig. 128a.*

CR: Hustedt (1930).
REMARKS: Cells and valves always seen in girdle view. This is a planktonic species with cells that do not survive chemical cleaning or boiling. Burned mounts should be used whenever it is suspected. It seems to be found frequently in the company of *Rhizosolenia* species.
DIMENSIONS: Length of spines, 40–60 μm; length of cell in girdle view, often over 200 μm including the spines; width of cell, 12–40 μm; striae not evident.

Cylindrotheca Rabenhorst 1848–60; 1864.

Cylindrotheca belongs to family Bacillariaceae, order Bacillariales, class Pennatibacillariophyceae. Each valve of a cell has a single keel-type raphe structure. The only species likely to be encountered is:

Cylindrotheca gracilis (Bréb.) Grun. var. **gracilis** *Fig. 128d.*

CR: Hustedt (1930).
REMARKS: Valves narrowly linear-lanceolate; cells twisted around the long axis so the keels appear to be helically arranged and to intersect at several points. The frustules contain little silica and are completely destroyed by chemical "cleaning" and boiling. Only con-

ventional "burned mounts" can be used and even in these almost all that can be seen are the keels and the ghostly valve outline. The illustration shows the structure far more distinctly than it usually appears! This species has been found occasionally in Iowa in waters of high nitrogen content.

DIMENSIONS: Length, 60–230 μm; width, 4–5 μm; fibulae, around 20/10 μm; striae not visible.

Opephora Petit 1888

Opephora belongs to family Fragilariaceae, order Fragilariales, class Pennatibacillariophyceae. Cells are asymmetric to the transverse axis in valve view and are wedge-shaped in girdle view. Only one species is likely to be encountered in this area:

Opephora martyi Herib. var. **martyi** *Fig. 128b, c.*

CR: Patrick and Reimer (1966).

REMARKS: I have found some specimens (*Fig. 6e–g*) that seem to fit the generic characteristics of *Opephora*. However, I am convinced that they are simply atypical valves of the common species *Fragilaria pinnata*. If *F. pinnata* is present in the sample, be careful about assigning any asymmetric specimens to *Opephora* before making a careful check. *O. martyi* has very broad striae that are almost always clearly poroid or cross-lineate.

DIMENSIONS: Length, 5–60 μm; width, 4–8 μm; striae, 4.5–8.0/10 μm.

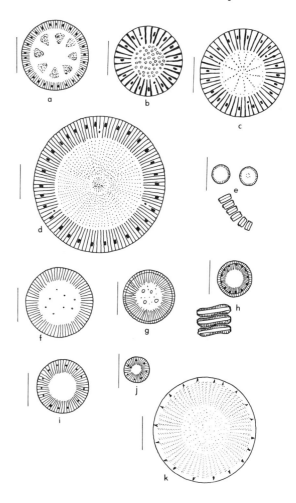

Fig. 113. a. Cyclotella antiqua v. *antiqua*. *b. Cyclotella bodanica* v. *michiganensis*. *c. Cyclotella bodanica* v. *stellata*. *d. Cyclotella bodanica* v. *bodanica*. *e. Cyclotella glomerata* v. *glomerata*. *f. Cyclotella kützingiana* v. *kützingiana*. *g. Cyclotella ocellata* v. *ocellata*. *h. Cyclotella melosiroides* v. *melosiroides*. *i. Cyclotella operculata* v. *operculata*. *j. Cyclotella pseudostelligera* v. *pseudostelligera*. *k. Stephanodiscus astraea* v. *astraea*. (Scale lines equal 10 micrometers. All illustrations stylized.)

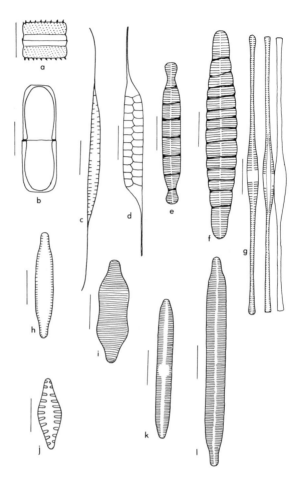

Fig. 114. a. *Melosira distans* v. *distans*. **b.** *Melosira juergensii* v. *juergensii*. **c.** *Rhizosolenia longiseta* v. *longiseta*. **d.** *Rhizosolenia eriensis* v. *eriensis*. **e.** *Diatoma anceps* v. *anceps*. **f.** *Diatoma hiemale* v. *hiemale*. **g.** *Fragilaria crotonensis* v. *crotonensis*. **h.** *Fragilaria brevistriata* v. *brevistriata*. **i.** *Fragilaria constricta* v. *constricta*. **j.** *Fragilaria leptostauron* v. *rhomboides*. **k.** *Fragilaria capucina* v. *capucina*. **l.** *Fragilaria virescens* v. *virescens*. (Scale lines equal 10 micrometers. All illustrations stylized.)

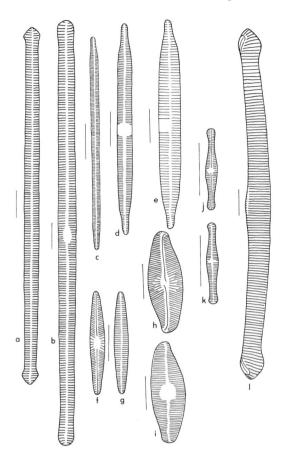

Fig. 115. a. *Synedra capitata* v. *capitata*. **b.** *Synedra ulna* v. *obtusa*. **c.** *Synedra tenera* v. *tenera*. **d.** *Synedra rumpens* v. *rumpens*. **e.** *Synedra dorsiventralis* v. *dorsiventralis*. **f.** *Achnanthes exilis* v. *exilis* (raphe valve). **g.** *Achnanthes exilis* v. *exilis* (pseudoraphe valve). **h.** *Achnanthes flexella* v. *flexella* (raphe valve). **i.** *Achnanthes flexella* v. *flexella* (pseudoraphe valve). **j.** *Achnanthes microcrephala* v. *microcephala* (raphe valve). **k.** *Achnanthes microcephala* v. *microcephala* (pseudoraphe valve) **l.** *Eunotia formica* v. *formica*. (Scale lines equal 10 micrometers. All illustrations stylized.)

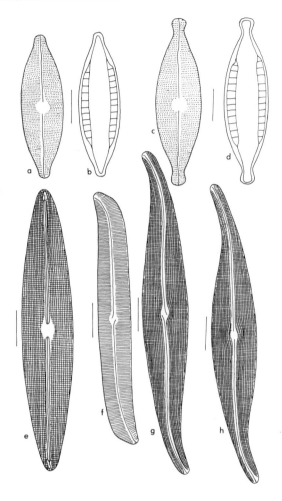

Fig. 116. a. *Mastogloia smithii* v. *smithii* (valve). **b.** *Mastogloia smithii* v. *smithii* (septum). **c.** *Mastogloia smithii* v. *amphicephala* (valve). **d.** *Mastogloia smithii* v. *amphicephala* (septum). **e.** *Frustulia rhomboides* v. *viridula*. **f.** *Gyrosigma eximium* v. *eximium*. **g.** *Gyrosigma spencerii* v. *spencerii*. **h.** *Gyrosigma wormleyi* v. *wormleyi*. (Scale lines equal 10 micrometers. All illustrations stylized.)

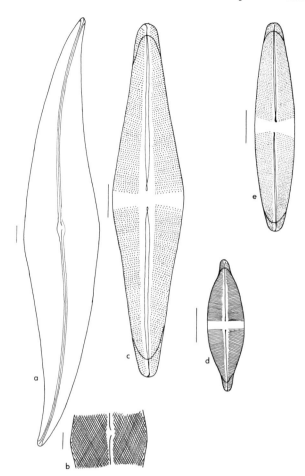

Fig. 117. a. *Pleurosigma angulatum* v. *angulatum*. **b.** *Pleurosigma angulatum* v. *angulatum* (detail showing striae). **c.** *Stauroneis acuta* v. *acuta*. **d.** *Stauroneis producta* v. *producta*. **e.** *Stauroneis obtusa* v. *obtusa*. (Scale lines equal 10 micrometers. All illustrations stylized.)

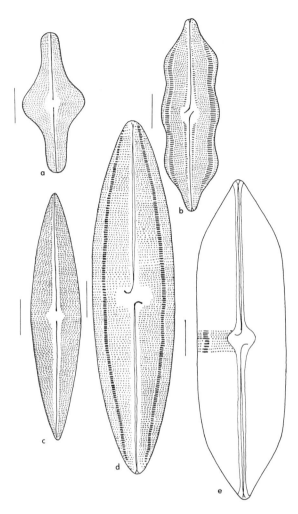

Fig. 118. a. Anomoeoneis follis v. follis. **b.** Neidium hitchcockii v. hitchcockii. **c.** Anomoeoneis serians v. serians. **d.** Neidium iridis v. iridis. **e.** Neidium iridis v. amphigomphus. (Scale lines equal 10 micrometers. All illustrations stylized.)

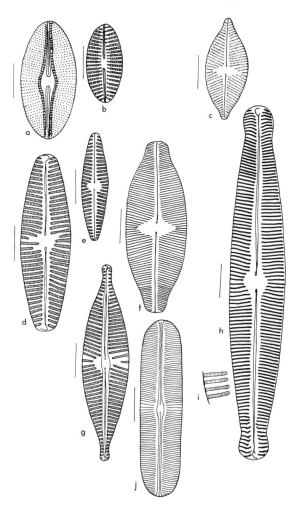

Fig. 119. a. *Diploneis elliptica* v. *elliptica*. **b.** *Diploneis puella* v. *puella*. **c.** *Navicula exigua* v. *exigua*. **d.** *Navicula reinhardtii* v. *reinhardtii*. **e.** *Navicula graciloides* v. *graciloides*. **f.** *Navicula platystoma* v. *platystoma*. **g.** *Navicula rhynchocephala* v. *rhynchocephala*. **h.** *Navicula oblonga* v. *subcapitata*. **i.** *Navicula oblonga* v. *subcapitata* (showing detail of striae). **j.** *Navicula bacillum* v. *bacillum*. (Scale lines equal 10 micrometers. All illustrations stylized.)

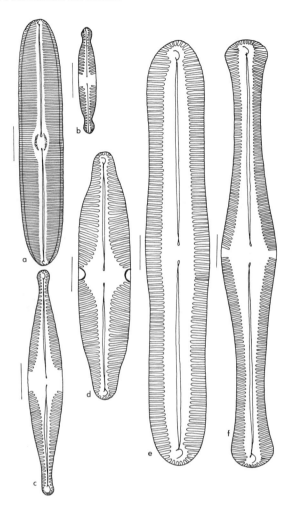

Fig. 120. a. *Caloneis alpestris* v. *alpestris.* **b.** *Pinnularia appendiculata* v. *appendiculata.* **c.** *Pinnularia braunii* v. *braunii.* **d.** *Pinnularia divergens* v. *divergens.* **e.** *Pinnularia brevicostata* v. *brevicostata.* **f.** *Pinnularia boyeri* v. *boyeri.* (Scale lines equal 10 micrometers. All illustrations stylized.)

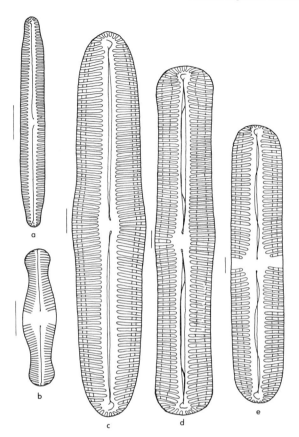

Fig. 121. *a. Pinnularia parvula* v. *parvula*. *b. P. globiceps* v. *globiceps*. *c. P. maior* v. *maior*. *d. P. nobilis* v. *nobilis*. *e. P. cardinalis* v. *cardinalis*. (Scale lines equal 10 micrometers. All illustrations stylized.)

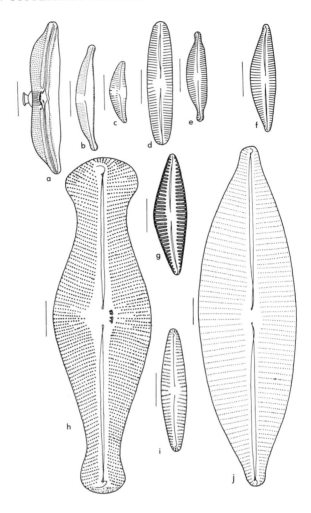

Fig. 122. a. *Amphora calumetica* v. *calumetica*. **b.** *Amphora normanii* v. *normanii*. **c.** *Amphora ovalis* v. *pediculus*. **d.** *Cymbella aequalis* v. *aequalis*. **e.** *Cymbella amphicephala* v. *amphicephala*. **f.** *Cymbella laevis* v. *laevis*. **g.** *Cymbella leptoceros* v. *leptoceros*. **h.** *Didymosphenia geminata* v. *geminata*. **i.** *Cymbella lacustris* v. *lacustris*. **j.** *Cymbella inaequalis* v. *inaequalis*. (Scale lines equal 10 micrometers. All illustrations stylized.)

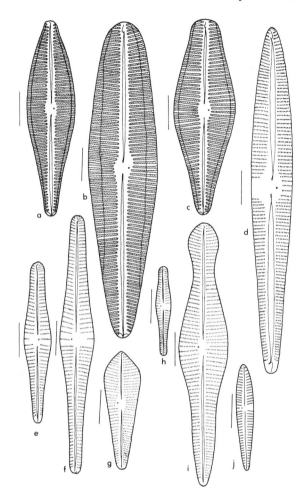

Fig. 123. *a. Gomphoneis eriense* v. *eriense. b. Gomphoneis herculeana* v. *herculeana. c. Gomphoneis herculeana* v. *robusta. d. Gomphonema semiapertum* v. *semiapertum. e. Gomphonema intricatum* v. *intricatum. f. Gomphonema intricatum* v. *vibrio. g. Gomphonema acuminatum* v. *trigonocephalum. h. Gomphonema dichotomum* v. *dichotomum. i. Gomphonema acuminatum* v. *elongatum. j. Gomphonema bohemicum* v. *bohemicum.* (Scale lines equal 10 micrometers. All illustrations stylized.)

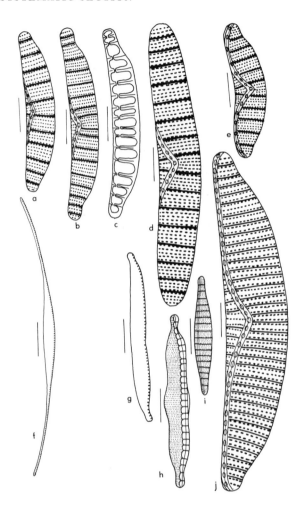

Fig. 124. a. *Epithemia adnata* v. adnata. **b.** *Epithemia argus* v. *alpestris* (valve). **c.** *Epithemia argus* v. *alpestris* (septum). **d.** *Epithemia muelleri* v. *muelleri*. **e.** *Epithemia ocellata* v. *ocellata*. **f.** *Nitzschia closterium* v. *closterium*. **g.** *Nitzschia clausii* v. *clausii*. **h.** *Nitzschia sinuata* v. *sinuata*. **i.** *Denticula tenuis* v. *tenuis*. **j.** *Epithemia turgida* v. *westermannii*. (Scale lines equal 10 micrometers. All illustrations stylized.)

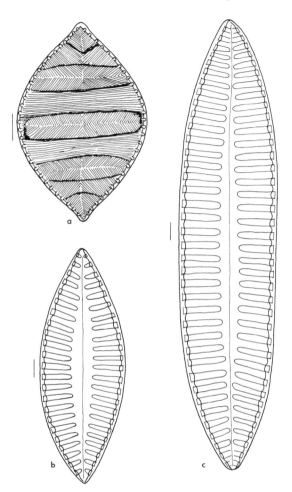

Fig. 125. a. *Cymatopleura elliptica* v. *hibernica*. **b.** *Surirella biseriata* v. *bifrons*. **c.** *Surirella biseriata* v. *biseriata*. (Scale lines equal 10 micrometers. All illustrations stylized.)

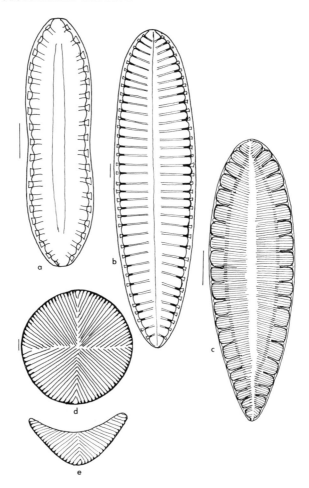

Fig. 126. a. *Surirella didyma* v. *didyma*. **b.** *Surirella elegans* v. *elegans*. **c.** *Surirella patella* v. *patella*. **d.** *Campylodiscus noricus* v. *noricus* (valve view). **e.** *Campylodiscus noricus* v. *noricus* (as typically seen). (Scale lines equal 10 micrometers. All illustrations stylized.)

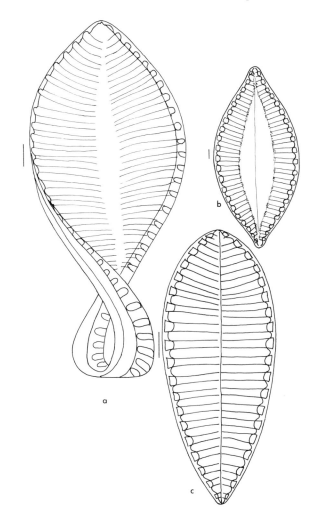

Fig. 127. a. *Surirella spiralis* v. *spiralis*. **b.** *S. oregonica* v. *oregonica*. **c.** *S. robusta* v. *splendida*. (Scale lines equal 10 micrometers. All illustrations stylized.)

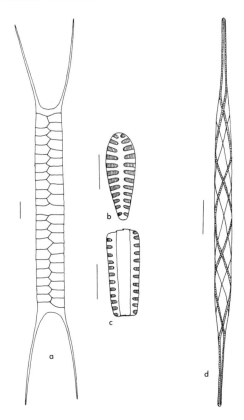

Fig. 128. a. *Attheya zachariasi* v. *zachariasi*. **b.** *Opephora martyi* v. *martyi* (valve view). **c.** *Opephora martyi* v. *martyi* (girdle view). **d.** *Cylindrotheca gracilis* v. *gracilis*. (Scale lines equal 10 micrometers. All illustrations stylized.)

Synonyms and Names of Uncertain Application

The names in the following list are recorded using the spellings and authorities listed in Britton (1944) and in some cases those in later reports on Illinois diatoms. Since it was my intention to give the names as reported, the spellings and authorities may not be the same as those given for these names in standard references. In a few cases, it has been necessary to make spelling cross-references to guide the user to the name as spelled in the main part of the Systematic Section or the list Valid Taxa Reported by Other Authors.

The names in the list below fall into two categories: synonyms of valid taxa and names of uncertain application. The first category contains both synonyms that have arisen through multiple descriptions of the same taxon and ones that have resulted from taxonomic decisions involving a change of rank (from species to variety, for example).

Names of uncertain application include names that cannot be applied with certainty to any specimen or that have been indiscriminately applied to specimens belonging to several taxa. In addition, some names that are valid have fallen into this category with respect to Illinois either because they have been listed without an authority or because there is doubt in my mind whether the names have been correctly applied.

Conventions used in the list are as follows:

1. A species name given without a variety is assumed to be the nominate variety.

2. If the genus is represented by a letter, it is assumed to be the same as the genus that begins the line.

3. Spelling cross-references are given as "See" references.

4. The "equal" sign should be read "is the same as."

5. The question mark following the name of a taxon indicates uncertainty about the synonymy. It does not imply that the name is invalid.

6. In cases of synonymy, the synonym is listed with its reported authority, but the valid taxon is not.

Achanthes lanceolata var. *rostrata* Hust. = *A. lanceolata* var. *dubia*.
Achnanthidium flexellum (Breb.) Kütz. = *Achnanthes flexella*.

Achnanthidium lanceolatum Bréb. = *Achnanthes lanceolata*.

Achnanthidium lanceolatum var. *dubia* Grun. = *Achnanthes lanceolata* var. *dubia*.

Amphiprora calumetica Thomas = *Amphora calumetica*.

Amphiprora ornata Bail. = *Entomoneis ornata*.

Anomoeoneis exilis (Kütz.) Cl. = *Anomoeoneis vitrea*.

Asterionella gracillima (Hantz.) Heib. = *Asterionella formosa*.

Caloneis fasciata (Lagerst.) Cl. = *C. bacillum*.

Caloneis silicula (Ehr.) Cl.: Name improperly applied as synonymous with *Navicula limosa* Kütz.

Caloneis silicula var. *alpina* Cl. = *C. ventricosa* v. *alpina*.

Caloneis silicula var. *genuina* Cl. = *C. ventricosa* var.(?)

Caloneis silicula var. *gibberula* (Kütz.) Cl.: Name of uncertain application in Illinois.

Caloneis silicula var. *inflata* (Grun.) Cl.: Name of uncertain application in Illinois.

Caloneis silicula v. *undulata* (Grun.) Cl.: Name of uncertain application in Illinois.

Caloneis trinodis (Lewis) Meist.: Name of uncertain application. In Illinois probably = *C. lewisii*.

Cocconeis communis Habirsh. = *C. pediculus*.

Cocconeis flexella (Kütz.) Cl. = *Achnanthes flexella*.

Cocconeis lineata Ehr. = *C. placentula* var. *lineata*.

Cocconema cymbiforme Rabh. = *Cymbella tumida*(?)

Colletonema lacustre V.H. = *Cymbella lacustris*.

Coscinodiscus asteromphalus Ehr.: Name applied in error as synonymous with *C. asteromorphus*.

Coscinodiscus asteromorphus: Name of uncertain application.

Cymatopleura apiculata W. Sm. = *C. solea* var. *apiculata*.

Cymatopleura hibernica W. Sm. = *C. elliptica* v. *hibernica*.

Cymatopleura solea var. *vulgaris* Meist. = *C. solea* var. *apiculata*.

Cymatopleura spiralis Chase = *C. elliptica* f. *spiralis*.

Cymbella anglica Lagerst. = *C. naviculiformis*.

Cymbella cistula var. *maculata* (Kütz.) Grun. = *C. cistula* var. *cistula*.

Cymbella ehrenbergii Kütz. = *C. inaequalis*.

Cymbella gastroides Kütz. = *C. aspera*.

Cymbella gracilis (Rabh.) Cl.: Name of uncertain application.

Cymbella maculata Kütz. = *C. cistula*.

Cymbella parva (W. Sm.) Cl.: Name of uncertain application.

Cymbella rotundata Chase = *C. triangulum*.

Cymbella stomatophora Grun. = *C. tumida*.

Cymbella stromatophora: See *C. stomatophora*.

Cymbella ventricosa Kütz. = *C. minuta*.

Diatoma elongatum var. *tenuis* (Ag.) V.H. = *D. tenue* var. *tenue*.

Diatoma hiemale var. *genuina* Grun. = *D. hiemale* var. *hiemale*.

Encyonema caespitosum Kütz. = *Cymbella minuta*.

Encyonema prostratum Ralfs = *Cymbella prostrata*.

Encyonema triangulum Kütz. = *Cymbella triangulum*.

Encyonema ventricosum Kütz. = *Cymbella minuta*.

Epithemia alpestris W. Sm. = *E. argus* var. *alpestris*.

Epithemia argus var. *amphicephala* Grun. = *E. argus* var. *alpestris*.

Epithemia gibba Kütz. = *Rhopalodia gibba*.

Epithemia ventricosa Kütz. = *Rhopalodia gibba* var. *ventricosa*.

Epithemia zebra (Ehr.) Kütz. = *E. adnata*.

Eucocconeis flexella (Kütz.) Cl. = *Achnanthes flexella*.

Eunotia lunaris (Ehr.) Grun. = *E. curvata*.

Eunotia major: See *E. maior*.

Eunotia monodon var. *major* (W. Sm.) Hust. = *E. maior*.

Eunotia westermannii Ehr. = *Epithemia turgida* var. *westermannii*.

Fragilaria crotonensis var. *prolongata* Grun. = *F. crotonensis* var. *crotonensis*.

Fragilaria harrisonii (W. Sm.) Grun. var. *harrisonii* = *F. leptostauron* var. *leptostauron*.

Fragilaria harrisonii var. *rhomboides* Grun. = *F. leptostauron* var. *rhomboides*.

Fragilaria mutabilis (W. Sm.) Grun. = *F. pinnata*.

Fragilaria parasitica (W. Sm.) Grun. = *Synedra parasitica*.

Fragilaria rhabdosoma Ehr. = *F. capucina*.

Frustulia viridula (Bréb.) De T. = *F. rhomboides* var. *viridula*.

Gallionella distans Ehr. = *Melosira distans*.

Gomphonema acuminatum var. *coronatum* (Ehr.) Rabh. = *G. acuminatum* var. *acuminatum*.

Gomphonema acuminatum var. *intermedia* Grun. = *G. acuminatum* var. *elongatum*.

Gomphonema acuminatum var. *laticeps* (Ehr.) A.S. = *G. acuminatum* var. *acuminatum*.

Gomphonema acuminatum var. *turris* (Ehr.) Cl. = *G. turris*.

Gomphonema capitatum Ehr. = *G. truncatum* var. *capitatum*.

Gomphonema constrictum var. *capitatum* (Ehr.) Grun. = *G. truncatum* var. *capitatum*.

Gomphonema constrictum var. *capitatum* f. *turgidum* (Ehr.) A. Mayer = *G. truncatum* var. *turgidum*.

Gomphonema constrictum Ehr. var. *constrictum* = *G. truncatum* var. *truncatum*.

Gomphonema coronatum Ehr. = *G. acuminatum*.

Gomphonema curvatum Kütz. = *Rhoicosphenia curvata*.

Gomphonema eriense Grun. = *Gomphoneis eriense*.

Gomphonema geminatum (Lyngb.) Ag. = *Didymosphenia geminata*.

Gomphonema gracile var. *dichotomum* (Kütz.) V.H. = *G. dichotomum*.

Gomphonema herculeanum Ehr. = *Gomphoneis herculeana*.

Gomphonema laticeps Ehr. = *G. acuminatum*.

Gomphonema montanum var. *subclavatum* Grun. = *G. subclavatum*.

Gomphonema robustum Grun. = *Gomphoneis herculeana* var. *robusta*.

Gomphonema semiapterum = *G. semiapertum*.

Gomphonema trigonocephalum Ehr. = *G. acuminatum* var. *trigono-cephalum*.

Gomphonema turgidum Ehr. = *G. truncatum* var. *turgidum*.

Gomphonema vibrio Ehr. = *G. intricatum* var. *vibrio*.

Gyrosigma kuetzingii (Grun.) Cl. = *G. spencerii*.

Lysigonium granulatum (Ehr.) Kuntze = *Melosira granulata*.

Lysigonium varians (Ag.) De T. = *Melosira varians*.

Melosira crenulata (Ehr.) Kütz. = *M. italica*.

Melosira granulata var. *spinosa* Balach. = *M. granulata* var. *granulata*.

Melosria lacustris: Name of uncertain application.

Melosira laevis (Ehr.) Ralfs = *M. italica*.

Melosira subflexilis Kütz. = *M. juergensii*.

Meridion circulare var. *zinkenii* (Kütz.) Grun. = *M. circulare* var. *circu-lare*.

Meridion constrictum Ralfs = *M. circulare* var. *constrictum*.

Microneis minutissima (Grun.) Cl. = *Achnanthes minutissima*.

Navicula ambigua Ehr. = *N. cuspidata* var. *ambigua*.

Navicula ambigua f. *craticularis* = *N. cuspidata* var. *ambigua*(?)

Navicula amphirhyncus Ehr. = *Neidium iridis* var. *amphirhynchus*.

Navicula appendiculata Kütz. = *Pinnularia appendiculata*.

Navicula bacillaris Greg. = *Caloneis bacillaris*.

Navicula bicapitata Lag. = *Pinnularia biceps*.

Navicula brebissonii Kütz. = *Pinnularia brebissonii*.

Navicula cardinalis Ehr. = *Pinnularia cardinalis*.

Navicula dicephala (Ehr.) W. Sm. var. *dicephala* = *N. elginensis*.

Navicula dicephala var. *elginensis* (Greg.) Cl. = *N. elginensis*.

Navicula dilatata Ehr.: Name of uncertain application.

Navicula divergens Ralfs = *Pinnularia divergens*.

Navicula dubia Ehr. = *Neidium dubium*.

Navicula elliptica Kütz. = *Diploneis elliptica*.

Navicula elliptica var. *minutissima* V.H. = *Diploneis puella*.

Navicula exilis Grun. = *Anomoeoneis vitrea*.

Navicula exilis Kütz.: Should have been reported as *N. exilis* Grun.

Navicula gibba Kütz.: Name of uncertain application.

Navicula globiceps Greg. = *Pinnularia globiceps*.

Navicula gracilis Ehr. = *N. tripunctata*.

Navicula hemiptera Kütz. = *Pinnularia hemiptera*.

Navicula hitchcockii Ehr. = *Neidium hitchcockii*.

Navicula iridis Ehr. = *Neidium iridis*.

Navicula iridis var. *affinis* (Ehr.) V.H. = *Neidium productum*.

Navicula iridis var. *amphigomphus* (Ehr.) V.H. = *Neidium iridis* var. *am-phigomphus*.

Navicula iridis var. *amphirhyncus* (Ehr.) V.H. = *Neidium affine* var. *amphirhynchus*.

Navicula iridis var. *firma* (Grun.) Kütz. = *Neidium iridis* var. *iridis*.

Navicula iridis var. *producta* (W. Sm.) V.H. *Neidium productum*.

Navicula lacunarum Grun. = *Caloneis bacillum*.

Navicula limosa Kütz. = *Caloneis limosa*.

Navicula limosa var. *subinflata* Grun.: Name of uncertain application in Illinois.

Navicula limosa var. *undulata* Grun.: Name of uncertain application in Illinois.

Navicula major Kütz. = *Pinnularia maior*.

Navicula mesolepta Ehr. = *Pinnularia mesolepta*.

Navicula nobilis (Ehr.) Kütz. = *Pinnularia nobilis*.

Navicula parva Ralfs: Should have been reported as *N. parvula* Ralfs.

Navicula parvula Ralfs = *Pinnularia parvula*.

Navicula punctata (Kütz.) Donk. = *N. tuscula*.

Navicula rhomboides Ehr. = *Frustulia rhomboides*.

Navicula rhyncocephala: See *N. rhynchocephala*.

Navicula saugerri Desmaz. = *N. seminulum*.

Navicula serians Bréb. = *Anomoeoneis serians*.

Navicula sigma Ehr.: Name of uncertain application.

Navicula silicula var. *gibberula* (Kütz.) Grun.: Name of uncertain application.

Navicula tabellaria Ehr. = *Pinnularia boyeri*(?)

Navicula tenella Bréb. = *N. radiosa* var. *tenella*.

Navicula termes Ehr. = *Pinnularia termes*.

Navicula trinodis Lewis = *Caloneis lewisii*(?)

Navicula trochus Ehr. = *Anomoeoneis follis*.

Navicula tumida W. Sm. = *N. anglica*.

Navicula varians Greg. = *N. gastrum*.

Navicula viridis Kütz. = *Pinnularia viridis*.

Neidium amphigomphus (Ehr.) Pfitz. = *N. iridis* var. *amphigomphus*.

Neidium amphirhyncus (Ehr.) Pfitz. = *N. affine* var. *amphirhynchus*.

Neidium dilatatum (Ehr.) Cl.: Name of uncertain application.

Neidium iridis var. *amphirhyncus* (Ehr.) V.H. = *N. affine* var. *amphirhynchus*.

Neidium iridis var. *firma* (Kütz.) V.H. = *N. iridis* var. *iridis*.

Nitzschia amphibia var. *intermedia* Mayer = *N. amphibia* var. *amphibia*.

Nitzschia amphioxys (Ehr.) Kütz. = *Hantzschia amphioxys*.

Nitzschia linearis var. *tenuis* (W. Sm.) Grun. = *N. linearis* var. *linearis*.

Nitzschia palea var. *tenuirostris* Grun. = *N. palea* var. *palea*.

Pinnularia brevicostata var. *leptostauron* Cl. = *P. brevicostata* var. *brevicostata*.

Pinnularia gibba (V.H.) Boyer: Name of uncertain application.

Pinnularia hemiptera (Kütz.) W. Sm. = *P. acuminata* var. *acuminata*.

Pinnularia interrupta f. *bicapitata* (Lag.) Fritsch = *P. biceps*.

Pinnularia major: See *P. maior.*

Pinnularia parva Greg. = *P. parvula.*

Pinnularia radiosa W. Sm. = *Navicula radiosa.*

Pinnularia tabellaria Ehr. = *P. boyeri*(?)

Pinnularia termes Ehr. = *P. biceps.*

Pinnularia viridis var. *elliptica* Meist. = *P. viridis* var. *viridis.*

Pleurosigma attenuatum W. Sm. = *Gyrosigma attenuatum.*

Pleurosigma eximium V.H. = *Gyrosigma eximium.*

Pleurosigma kuetzingii Grun. = *Gyrosigma spencerii.*

Pleurosigma spencerii (Quek.) W. Sm. = *Gyrosigma spencerii.*

Pleurosigma wormleyi Sulliv. = *Gyrosigma wormleyi.*

Rhizosolenia gracilis H.L. Smith = *R. longiseta.*

Rhopalodia ventricosa (Kütz.) Muell. = *R. gibba* var. *ventricosa.*

Schizonema viridulum Bréb. = *Frustulia rhomboides* var. *viridulum.*

Schizonema vulgare Thwaites = *Frustulia vulgaris.*

Sphinctocystis elliptica (Kütz.) Kuntze = *Cymatopleura elliptica.*

Sphinctocystis librilis (Ehr.) Hass. = *Cymatopleura solea.*

Stauroneis anceps var. *birostris* (Ehr.) Cl. = *S. anceps* var. *anceps.*

Stauroneis baileyi Ehr. = *S. phoenicenteron.*

Stauroneis gracilis Ehr. = *S. phoenicenteron* f. *gracilis.*

Stauroneis linearis Ehr. = *S. anceps* var. *linearis.*

Stauroneis phoenicenteron var. *amphilepta* (Ehr.) Cl. = *S. phoenicenteron* f. *gracilis.*

Stauroneis phoenicenteron var. *baileyi* (Ehr.) Cl. = *S. phoenicenteron* var. *phoenicenteron.*

Stephanodiscus astraea var. *minutula* (Kütz.) Grun. = *S. minutus.*

Striatella fenestrata (Lyngb.) Kuntze = *Tabellaria fenestrata.*

Striatella flocculosa (Roth) Kuntze = *Tabellaria flocculosa.*

Surirella apiculata W. Sm. = *S. angusta.*

Surirella cardinalis Kitton = *S. guatimalensis.*

Surirella elegans var. *norvegica* (Eulenst. in A.S.A.) Brun. = *S. elegans* var. *elegans.*

Surirella minuta Bréb. = *Surirella ovata.*

Surirella norvegica Eulenst. = *S. elegans.*

Surirella ovalis var. *minuta* (Bréb.) Kirch. = *S. ovata*(?)

Surirella ovalis var. *pinnata* V.H. = *S. ovata* var. *pinnata*(?)

Surirella saxonica Auers. = *S. robusta.*

Surirella splendida (Ehr.) Kütz. = *S. robusta* var. *splendida.*

Synedra acus var. *delicatissima* W. Sm. = *S. delicatissima.*

Synedra acus var. *radians* (Kuetz.) Hust. = *S. radians.*

Synedra biceps W. Sm. = *Eunotia flexuosa.*

Synedra chasei Thomas = *S. ulna* var. *danica.*

Synedra crotonensis Grun in V.H. = *Fragilaria crotonensis.*

Synedra crotonensis var. *prolongata* Grun. = *Fragilaria crotonensis* var. *crotonensis.*

Synedra danica Kütz. = *S. ulna* var. *danica*.

Synedra hyalina Prov.: This name is of uncertain application in Illinois.

Synedra jourascensis Herib. = *S. ulna* var. *spathulifera*.

Synedra lanceolata Kütz. = *S. ulna* var. *ulna*.

Synedra longissima W. Sm. = *S. ulna* var. *longissima*.

Synedra obtusa W. Sm. = *S. ulna* var. *obtusa*.

Synedra spathulifera Grun. = *S. ulna* var. *spathulifera*.

Synedra splendens Kütz. = *S. ulna* var. *ulna*.

Synedra tenuissima Kütz. = *S. radians*(?)

Synedra ulna var. *aequalis* (Kütz.) Hust. = *S. ulna* var. *obtusa*.

Synedra ulna var. *biceps* (Kütz.) Schönf. = *S. ulna* var. *longissima*.

Synedra vaucheriae (Kütz.) Kütz. = *Fragilaria vaucheriae*.

Synedra vitrea Kütz. = *S. ulna* var. *ulna*.

Vanheurckia rhomboides (Ehr.) Bréb. = *Frustulia rhomboides* var. *rhomboides*.

Authorities

This is a list of the names used as authorities for the diatom taxa referred to in this book. The names are entered in direct order, alphabetized letter by letter. Letters with umlauts are alphabetized as if the umlaut were not present. Where more than one form of abbreviation is used for a single author, all are listed. Multiple authors connected by "and," "et" or "&" are alphabetized as single authorities, other combinations must be accessed by checking the individual abbreviations. In cases where the form of name begins with initials, a cross-reference is made from the last name only. Names listed without any expansion are last names for which I do not know the initials.

A. Cl.: Astrid Cleve (*married name:* Astrid Cleve-Euler)
Ag.: C. A. Agardh
Agardh: C. A. Agardh
A. Mayer: Anton Mayer
Andrews
Arch.: R. E. M. Archibald
Archibald: R. E. M. Archibald
Arnott: G. A. W. Arnott
A.S.: Adolf Schmidt
A.S.A.: Adolf Schmidt et al., Atlas der Diatomaceenkunde (1874–1959)
Auers.: B. Auerswald
Bachm.: H. Bachmann
Bachmann: H. Bachmann
Bail.: J. W. Bailey
Bailey: J. W. Bailey
Balach.: E. N. Balachontzev (E. N. Bolochoncew)
Berk.: M. J. Berkeley
Biswas: S. Biswas
Bleisch: Max Bleisch
Bory: J. B. M. Bory de Saint-Vincent
Boyer: C. S. Boyer
Bréb.: Alphonse de Brébisson
Brébisson: Alphonse de Brébisson

Brun: J. Brun
Brutschy: A. Brutschy
C. A. Ag.: C. A. Agardh
C. A. Agardh
Carlson: G. W. F. Carlson
Carr.: W. Carruthers
Chase: H. H. Chase
Choln.: B. J. Cholnoky
Cholnoky: B. J. Cholnoky
Cl.: P. T. Cleve
Cl.-Eul.: Astrid Cleve-Euler
Cleve: P. T. Cleve
Cleve, A.: Astrid Cleve
Cleve-Euler: Astrid Cleve-Euler (*born:* Astrid Cleve)
De Notaris: G. De Notaris
Desmaz.: J. B. H. T. Desmazières
De T.: G. B. De Toni
De Toni: G. B. De Toni
Donk.: Arthur S. Donkin
Donkin: Arthur S. Donkin
Ehr.: C. G. Ehrenberg
Eulenst.: Theodor Eulenstein
Foged: Niels Foged
Fricke: Friedrich Fricke
Fritsch: F. E. Fritsch
F. W. Mills
Gmelin: J. F. Gmelin

Gray: S. F. Gray
Greg.: William Gregory
Grev.: Robert K. Greville
Greville: Robert K. Greville
Griff. & Henfr.: J. W. Griffith & A.
  Henfrey
Grun.: Albert Grunow
Grunow: Albert Grunow
Habirsh.: Frederick Habirshaw
Hantz.: C. A. Hantzsch
Hantzsch: C. A. Hantzsch
Hass.: Arthur H. Hassall
Hassall: Arthur H. Hassall
Heib.: P. A. C. Heiberg
Herib: Joseph Heribaud
Hilse: W. Hilse
H. L. Smith
Hohn & Hellerman: Matthew
  Hohn & Joan Hellerman
H. Peragallo
Hust.: Friedrich Hustedt
J. W. Bailey
Kirch.: O. Kirchner
Kirchn.: O. Kirchner
Kirchner: O. Kirchner
Kitton: F. Kitton
Krasske: Georg Krasske
Krieger: W. Krieger
Kuetz.: F. T. Kützing
Kuetzing: F. T. Kützing
Kufferath: H. Kufferath
Kuntze: O. Kuntze
Kütz: F. T. Kutzing
Kützing: F. T. Kutzing
Lag.: N. G. W. Lagerstedt
Lagerst.: N. G. W. Lagerstedt
Lemm.: E. Lemmermann
Lewis: F. Lewis
Lowe: Rex L. Lowe
Lund: J. W. G. Lund
Lyngb.: Hans Christian Lyngbye
Lyngbye: Hans Christian Lyngbye
Mang.: E. Manguin
Manguin: E. Manguin
Mayer: Anton Mayer

Meist.: F. Meister
Meister: F. Meister
Mills: F. W. Mills
M. Peragallo
M. Schmidt
Muell.: Otto Mueller
Muell., O.: Otto Mueller
Muell., O. F.: O. F. Mueller
Naeg.: Naegeli
Näge.: Naegeli
Nitz.: Christian Ludwig Nitzsch
Norman: George Norman
Oestr.: E. Oestrup
O. F. Mueller
O'Meara: E. O'Meara
O. Muell.: Otto Mueller
O. Müll.: Otto Mueller
Pant.: Joseph Pantocsek
Patr.: Ruth Patrick
Peragallo, H.
Peragallo, M.
Peragallo & Herib.: Peragallo &
  Heribaud
Peragallo & Peragallo: H. Peragallo
  & M. Peragallo
Peters.: J. Boye Petersen
Petersen: J. Boye Petersen
Petit: P. Petit
Pfitz.: Ernst Pfitzer
Pfitzer: Ernst Pfitzer
Prov.: S. Provažek
Quek.: J. T. Quekett
Rabenhorst: Ludwig Rabenhorst
Rabh.: Ludwig Rabenhorst
Ralfs: J. Ralfs
Reich. & Fricke: H. Reichelt &
  Friedrich Fricke
Reim.: Charles W. Reimer
Ross: Robert Ross
Roth: A. W. Roth
Schmidt, A.: Adolf Schmidt
Schmidt, M.
Schoeman: F. R. Schoeman
Schönf: H. von Schönfeldt
Schultze: Max Schultze

Schum.: Julius Schumann
Schumann: Julius Schumann
S. F. Gray
Skv.: B. W. Skvortzow
Skv. & Meyer: B. W. Skvortzow &
  K. I. Meyer
Smith, William
Stod.: C. Stodder
Stoermer: Eugene F. Stoermer
Sulliv.: W. S. Sullivant
Sulliv. & Wormley: W. S. Sullivant
  & T. G. Wormley
Thomas: B. W. Thomas
Thw.: George Henry Thwaites

Thwaites: George Henry Thwaites
Turpin: P. J. F. Turpin
V.H.: Henri van Heurck
Wallace: John Wallace
Weinhold
West: T. West
William Smith
Wisl. et Poretz.: S. M. Wislouch &
  V. S. Poretzky
Wolle: Francis Wolle
W. Sm.: William Smith
Yerm.: Nicholas Yermoloff
Zach.: O. Zacharias
Zanon: D. Vito Zanon

Supplementary Section

Introduction to the Diatoms

The Supplementary Section is written for the use of the person beginning the study of diatoms. Much of the following material will be of little use to the experienced algologist, save perhaps the methods for preparing permanent mounts. Some of the remarks on collecting methods will be of interest only to the true beginner who has no previous experience with the tools of the trade. I will not apologize for this, since even the most experienced researcher can remember a time when he or she had no idea of what a plankton net was.

Position of the Diatoms in the Plant Kingdom

Diatoms are single-celled plants that together form a division of the plant kingdom called the Bacillariophyta (in some classifications called Bacillariophycophyta). While this is the view I hold, there is also an opinion that the diatoms constitute a class within the division Chrysophyta (or Chrysophycophyta). In such a classification the diatoms are assigned to class Bacillariophyceae.

General Characteristics of Diatoms

Habit

Diatoms live as isolated cells or in colonies of various types. Isolated cells may be free-floating or attached to various substrates. Diatom colonies appear to be formed by mechanical connection only. There do not seem to be protoplasmic connections between cells of the colony nor any functional distinction or division of "labor" between cells. The size of colonies appears to be limited mainly by the strength of the physical connections and the number of cell divisions occurring after initiation of the colony. Cells of colonial species may be held together by gelatinous material in the form of strands, pads or amorphous matrices, by interlocking spines, or by remnants of old girdle bands. I have illustrated several of the more common colonial forms in connection with the Glossary under **Colony**.

The Cell

Diatoms are eukaryotic organisms—each cell has a true membrane-bound nucleus. Chromosomes, though small, are probably similar in structure to those of higher organisms. Nucleoli have been observed in some species.

The protoplast of a diatom cell is bound by a plasma membrane similar in appearance to that observed in other eukaryotic organisms. Within the protoplast one finds most of the cellular structures observed in other plant cells: membrane-bound vacuoles, chloroplasts (usually called chromatophores), mitochondria, Golgi bodies (dictyosomes), ribosomes, and an endoplasmic reticulum.

Besides the more common organelles, diatom cells also contain some special structures. So-called vesicular complexes and oil-storage structures are known. At least one species of *Rhopalodia* appears to have an inclusion that is nearly identical in structure to a blue-green algal cell and that indeed may be a parasite or symbiote (Drum and Pankratz 1965). Of particular importance are the silica-deposition vesicles that are concerned with formation of the silica cell wall.

The Cell Wall

The most characteristic feature of the diatom cell is its complex silica cell wall. The taxonomy of diatoms is based almost entirely on the features of this wall. Each cell wall consists of two plates, called valves, held together by a series of silica bands that include two girdle bands and often several to many intercalary bands similar in form to the girdle bands. The complete wall structure is called a frustule. If you are unfamiliar with the structure of diatoms, you should at this time go to the Glossary and study the definitions of **Frustule, Valve, Girdle band, Intercalary band, Raphe, Keel, Canal, Striae,** and **Puncta.** The Glossary definitions will refer you to the appropriate illustrations. The cell wall lies outside the plasma membrane of the cell.

Duke and Reimann (in Werner, ed., 1977) have outlined one possible mechanism for the formation of the wall. In this scenario, silica is first deposited in membrane-bound vesicles within the cytoplasm, after which the vesicles coalesce. Completed silica structures are then discharged through the plasma membrane of the cell. The pattern of silica deposition is generally believed to be under genetic control, though at the time of this writing the mechanism is unknown.

Physiology

With few exceptions diatoms are photoautotrophs. As with higher plants the primary photosynthetic pigment is chlorophyll-a The secondary chlorophyll is chlorophyll-c. Accessory pigments including b-carotene, fucoxanthin, diadinoxanthin, diatoxanthin, and neofucoxanthin (Darley, in Werner, ed., 1977), plus some colored oils, which give the diatoms their characteristic brown or golden-brown color.

Some diatoms appear to be auxotrophic, that is, photosynthetic but requiring organic supplements (particularly vitamins) for growth. A fairly small number of species (Lewin and Hellebust in Werner, ed., 1977) have been shown to be facultatively or obligately heterotrophic.

As with other photosynthetic organisms, diatoms are capable of storing energy by means of primary food reserves. The principal reserves are various lipids and the water-soluble carbohydrate chrysolaminarin (Darley in Werner, ed., 1977). In addition to intracellular storage, diatoms produce extracellular polysaccharides of various compositions that form the stalks, tubes, and gelatinous matrices involved in colony formation.

Movement

Vegetative cells of diatoms may move passively or actively in their environment if not attached to a substrate. Passive motion depends on changes in cell buoyancy and on the activity of winds and currents. Centric diatoms and pennate diatoms that lack raphes exhibit only passive motion. Pennate species that have one or more raphes per cell exhibit active locomotion as well. The mechanism of this movement has not been explained to everyone's satisfaction, but Harper (in Werner, ed., 1977) has summarized a number of proposed theories that involve the secretion of mucilaginous materials along the raphe. Diatoms in any case do not move by the undulation of the protoplast or by means of pseudopodia, and while flagella have been demonstrated in the gametes of certain centric diatoms, no such structures appear to be present in any vegetative cells.

Vegetative Reproduction and the Macdonald-Pfitzer Hypothesis

Diatoms increase their numbers by asexual cell division following mitosis. One valve of the parent cell is retained by each daughter cell and two new valves are formed, one for each daughter cell. This leads to some characteristic peculiarities in the cell-size structure of clones in a number of species. Often, perhaps typically, one valve of a cell (the epivalve) is slightly larger than the other (the hypovalve). In the course of division, the two new valves produced both function as hypovalves. Thus at the end of a division the daughter cell that receives the original epivalve of the parent will be the same size as the parent, while the daughter cell that receives the original hypovalve of the parent (which becomes the epivalve of this daughter) will be somewhat smaller. This size reduction is most noticeable in length, but width decreases also.

Under this pattern, proposed in the Macdonald-Pfitzer hypothesis, only one cell in a clone will retain the initial maximum dimensions, all other cells being smaller. Since the same rule applies to all divisions, the average cell size of a clone will decrease with every division.

If a clone develops according to the Macdonald-Pfitzer pattern, and many divisions take place, it may exhibit a rather large size range. Consider, however, that only one cell in a clone can have the maximum size and only one the current minimum length. Near-maximum and near-minimum sizes are also rare, so when you see a large clonal population most of the cells will be near the midpoint of the current range of size, thus giving you the impression of a *narrow* size range. True maximum size ranges for the species are established by long periods of observation of many clones at various stages of development including sexually reproducing populations.

Sexual Reproduction

The foregoing discussion of the Macdonald-Pfitzer hypothesis leaves one very important question unanswered: Where do the maximum-size cells come from? Obviously, if the effect of repeated cell division is a progressive diminution in average cell size, there must come a stopping point and nothing in the hypothesis as given so far explains how the shrinking process is reversed. As it happens, this is where the sexual cycle fits in.

Unlike most algae, vegetative diatom cells are diploid. The haploid condition is encountered only in the gametes. Prior to meiosis a vegetative cell is changed physiologically into a gametogenic cell, but this transformation is not accompanied by any pronounced morphological changes, so potentially gametogenic cells cannot be identified as such with certainty. It does appear that for species following the Macdonald-Pfitzer pattern the gametogenic cells tend to be approximately one half the maximum length for the species. Little is known about the factors that induce the onset of sexual reproduction.

Sexual reproduction of diatoms has been observed only rarely in nature and almost never in laboratory culture. Those who have been most successful in this aspect of diatom study have achieved their knowledge in part through patient and frequent observation of naturally occurring populations. If you have an interest in diatoms as living organisms (and have the requisite patience) I can think of no type of diatom study that offers greater potential for producing new knowledge than this. Even if the study of sexual processes is not your principal goal, you should make extensive observations any time you enounter a sexually reproducing population since so little is known about this phase of the life cycle of most species.

Sexual Reproduction in the Centric Diatoms

Those centric diatoms that have been studied typically have an oogamous reproductive pattern with large, nonmotile eggs fertilized by small motile sperm. Drebes (in Werner, ed., 1977) notes that the sperm are uniflagellate with the flagellum forward-directed and of the flimmer type. The egg is formed within the original cell wall after meiotic division of the nucleus in the gametogenic cell and degeneration of three of the resulting haploid nuclei. Sperm, at least in some cases, appear to be formed in a more complicated way. Drebes (in Werner, ed., 1977) reports that several mitotic divisions take place in a parent cell resulting in formation of numerous spermatogonia. The parent cell wall is shed and each released spermatogonium undergoes meiosis to produced four motile sperm. Fertilization takes place within the cell wall surrounding the egg. The zygote increases considerably in size, forces the two valves apart, and continues to grow rapidly as a naked protoplast. The zygote is called an auxospore or growth spore and its enlargement restores the size lost during the preceding series of vegetative divisions.

A thin wall or perizonium eventually forms around the auxospore. The auxospore nucleus undergoes mitosis. The cell does not divide, but one valve is formed. One daughter nucleus aborts and the second undergoes mitosis. The cell again does not divide, but a second valve is formed and one daughter nucleus aborts, leaving a single large diploid cell with two valves. This cell sheds the perizonium and begins to divide according to the vegetative pattern.

Sexual Reproduction in the Pennate Diatoms

Pennate diatoms typically have isogamous sexual processes, though Drebes (in Werner, ed., 1977) has reported the existence of some oogamous species. The gametes that have been observed are non-flagellate and move, if at all, by amoeboid motion.

A few species appear to be autogamous, but in most two parent cells are involved. Each of the parent cells undergoes meiosis to form gamete nuclei. Although four gametes per cell are possible, usually some of the nuclei degenerate and each parent cell produces only one or two gametes. Pairs of gametes fuse to produce diploid zygotes. The zygotes enlarge to about twice the length of the parent cells (thus they are auxospores) and form a thin cell wall or perizonium. As with the centric diatoms, the first mitotic division of the zygote produces one valve, one viable nucleus, and one aborted nucleus. The second mitotic division produces a second valve and again one viable and one aborted nucleus, leaving a diploid vegetative cell with one nucleus and two valves. Early postauxospore cells have valves that are often highly atypical of the species and that may be grossly deformed, but subsequent divisions produce normal valves. (Incidentally, this feature of reproduction is useful to the taxonomist since it allows verification of suspicions about maximum cell sizes for a species.)

While this pattern is but one of several that have been reported for pennate diatoms, it shares with the others the end result of restoring size lost during vegetative divisions.

Principles of Diatom Classification

For a long time diatoms have been recognized as a well-defined group in the plant kingdom. Most often in earlier classifications the diatoms were considered a class (Bacillariophyceae) within the division Chrysophyta (or Chrysophycophyta as it is sometimes written). In view of the peculiar cell-wall structure found in all diatoms, however, it seems reasonable to consider the diatoms to be a division in their own right. The proper name of the division would be Bacillariophyta (or Bacillariophycophyta). I have taken this position, but the matter is by no means settled.

If diatoms are considered a division, then the group contains two classes, the centric diatoms or Centrobacillariophyceae, and the pennate diatoms or Pennatibacillariophyceae. Systems regarding the diatoms as a class recognize these two groups as orders.

Centric diatoms take their name from the radial symmetry of markings on the valve surface. The valves range in outline from elliptical to more typically circular or subcircular. No true centric diatoms have any sort of raphe, though the absence of a raphe is not in itself sufficient to assign a genus to this class. The centric diatoms are the older of the two classes, with fossil specimens well known in Cretaceous sediments and claimed for some Jurassic materials. Sexual reproduction, as far as is known, is oogamous and most that have been studied appear to produce uniflagellate sperm.

The pennate diatoms are characterized by bilateral symmetry of markings and most have valves that are distinctly longer than broad. Most pennate diatom species have at least one fully developed raphe per cell. Some do not. Presence of a true raphe is sufficient for assignment of a species to this class. Sexual reproduction in the pennates appears to be isogamous in most of the few cases where it is known at all. Flagellate gametes have not been observed. The pennate diatoms are believed to have descended from centric ancestors. Indeed the fossil genus *Raphidodiscus* appears to be intermediate in characteristics between the two classes. Pennate diatoms have been observed in late Cretaceous sediments but became widespread during the middle Tertiary.

Both centric and pennate diatoms are subdivided further into orders (suborders in schemes that recognize the diatoms as a class and the pennate and centric groups as orders), families, genera, species, varieties, and forms.

The centric diatoms are primarily marine and few species are found in Illinois (although those that exist here often occur as large populations). The characters of orders and families are of little practical use in identifying the freshwater species of centrics. Those who are interested may find them described in Hustedt (1930*a*).

On the other hand, the orders of pennate diatoms have some utility in the practical taxonomy of freshwater diatoms, so I will say something abut them. The orders are separated on the basis of the raphe. (See the Glossary for definition of raphe types). Those species that lack any sort of raphe on either valve are assigned to genera in the order Fragilariales. Species with rudimentary raphes on both valves belong to the Eunotiales. Those with a fully developed two-branch raphe on one valve and no raphe or a rudimentary raphe on the other are placed in the Achnanthales. The remaining orders are characterized by the presence of one or more true raphes on each valve. In three orders the raphe is a slit in a narrow longitudinal tube. When the tube is embedded in the valve surface and generally away from the valve margins, the species belong to the Epithemiales. When there is a tubular raphe structure located atop a single raised wing or keel on each valve, one is dealing with the Bacillariales, and when each valve has two such wings or keels (one along each valve margin), each with a tubular raphe structure, the species belong to the Surirellales. The last order of pennate diatoms is the Naviculales. In these diatoms there is one two-branch raphe on each valve, but the raphe is not associated with a tubular structure in the valve surface.

Orders of pennate diatoms are divided into families, but only the Naviculales are represented by more than one family in Illinois. Indeed some of the orders have but a single family, so the familial and ordinal characters are the same. While the orders of pennate diatoms seem to make biological sense, it is quite possible that the families are largely artificial, so beyond recognizing that they have been described I will say no more about them.

For practical purposes you will be dealing with genera and species in your identifications. Most (but not all) of the genera appear to be true biological units composed of closely related species. I have provided a brief description of each of the genera found in Illinois at the beginning of the Systematic Section dealing with its component species.

Each genus includes the characteristics of the order to which it belongs and other characteristics that seem to be common to some groups of species in the order but not to all. Formerly, colony form and valve-shape symmetry were the main criteria. More recently, colony form has been rejected as significant in some cases (thus the genus *Cymbella* contains the species formerly assigned to three genera: *Cymbella Encyonema*, and *Cocconema*, based on colony form) but retained in others (*Fragilaria* and *Synedra*, for example). Sometimes the position of the raphe structure on the valve surface is considered significant (*Hantzschia* and *Nitzschia*), sometimes the nature of the striae (*Pinnularia* versus *Navicula*). The list of generic characters observable with the light microscope is quite long. In recent years, several ultrastructural characters have been observed that may come to be used as the true basis for separating genera.

Practically speaking, you will quickly come to recognize a species as belonging to one traditional genus or another on the basis of its overall appearance.

The Diatom Species

A truism in plant systematics is that the species is the only genuine unit in nature, all higher levels of classification being based on inferences about ancient relationships. A species consists of all individual organisms of a particular type. This is easy enough to say. The problems arise when one attempts to define the criteria for determining the "particular types."

The best natural definition is that members of a species are those individuals that can mate and produce fertile offspring. Closely related species can sometimes be induced to interbreed successfully by the artificial removal of reproductive barriers, but one can still argue that the natural condition should be the criterion.

Unfortunately, nature does not always cooperate in the work of the taxonomist. Diatoms reproduce chiefly by vegetative means, and the sexual cycle is observed only infrequently in nature and rarely in laboratory culture. With the loss of the reproductive criterion, diatomists have been forced to resort to inferences based on cell morphology to which may sometimes be added criteria based on chemical composition.

Traditionally, diatom systematics has been based on the appearance of the cell wall. Some attempts have been made to use protoplast characteristics or the form of colonies, but these have generally not proved to be practical in the circumscription of species. It is the cell wall, with its vast variety of form and marking, that has provided the diatomist with the means of classifying these organisms. The use of electron microscopes (both transmission and scanning) has added ultramicroscopic features to those already revealed by the light microscope.

The problem with using morphological characters has been and continues to be in deciding how much variation can be accepted in a single species. While it is almost certain that the basic cell shapes and patterns of markings are genetically controlled, it is not at all certain how strong this control is and even less certain how variations in marking and shape correlate with variations in physiological activity and capacity for interbreeding. We make the assumption that observed stability in morphological features is paralleled by other types of stability, but we can seldom prove it.

Still, there do appear to be morphological discontinuities separating many of the "species," which suggest that interbreeding is not taking place between them and that they would be shown to be biological species, could the more rigorous test of reproductive incompatibility be applied.

Besides those species that are obviously distinct there are groups of species that tend to intergrade with each other in form and are thus difficult to separate with certainty. You will find these especially in the genera *Nitzschia* and *Navicula*. While these groups of species are vexing, they offer the best hope for investigating speciation in a part of the plant kingdom that has heretofore resisted such analysis.

For practical purposes one must accept that at the present time attempts must be made to establish limits of variation and to define species based on these limits.

Characters Used in Describing Species

The characters used in separating species vary with the genus. The ones most often used are the shape and size of the cell, the density, form and angle of striae, the form of the raphe (if present in the genus), and the nature and number of specialized structures such as spines, protuberances, stigmata, isolated puncta, and costae. Those you find mentioned in the descriptions have been defined in the Glossary and illustrated.

Limits of Variation: Narrow versus Broad Definitions of Species

Some characteristics seem to be relatively invariable. The density of striae seems to fluctuate within narrow limits, especially in clonal populations. Many of the specialized structures, such as stigmata and isolated puncta, are either present or absent and rarely exhibit both states within the populations of any one species.

On the other hand, the operation of the cell division pattern outlined in the Macdonald-Pfitzer hypothesis leads us to expect a greater degree of variability in cell size, so within reason we expect most species to exhibit a fairly substantial range of values for length and width. The main question here is how large we should expect this range to be.

Cell shape is another matter. With each cell division the parent valves are distributed to the daughter cells. While the parent valve may not function as a template, I have never seen a case where one valve is radically different in shape from the other. We would thus expect that all the cells in a clone should have similar shapes, at least for cells within a narrow length range. (I emphasize "narrow length range" since it has been observed that the diminution of length proceeds at a greater rate than that of width in many pennate diatoms, which has the effect of producing a gradual change in shape as the clone develops.) On the other hand, it is not at all certain that the initial cell shape is fixed for all species. Environmental conditions at the time of auxospore formation may produce variations in the basic form that are preserved throughout the duration of the clone but that will disappear during the following auxospore stage. You will frequently encounter populations of cells of deformed shape that appear to have arisen from such a process and others with more subtle variations that may also have arisen in this way.

So far, we can see certain characters that seem to have narrow ranges of variability and certain others that appear to vary gradually but definitely over the life of a clone. A narrow definition of a species requires comparative stability in the first group and predictable, continuous variation in the second. A narrowly defined species would be one in which there is a fixed initial shape, a more or less fixed maximum cell size, and a characteristic rate of decrease in size with each division leading to a predictable cell shape after any number of divisions.

The broad view of species differs primarily in rejecting the notion that there is a fixed intial size and shape. The flexibility in these characteristics can be supposed to be a result of inherent genetic variation, influence of environmental factors, or both.

Both the narrow and broad views of species are means of accounting for the variation observed in nature. Those with a narrow view of permissable variation in a species will react to specimens outside this range by describing a new species. Those with a broader species concept will tend to "shoehorn" the odd cells into existing species by stretching their descriptions to fit. Neither tendency is without merit, though it is obvious that in any particular case only one can be right. I tend toward narrow definitions in species that I seldom see but broader ones for those that I have encountered often and in large populations. When I have included what you feel to be too much variation in a species you will find many who agree with you and have obliged by describing additional species.

Varieties and Forms

Among higher plants the variety (varietas) abbreviated var. or v., and the form (forma), abbreviated fo. or f., are used in the description of groups of individuals within a species that can be separated by minor but more or less stable characters. Diatom species have been so frequently subdivided into varieties that it is customary to report them as trinomials (*Gomphonema parvulum* v. *parvulum*, for example). In fewer cases the variety is subdivided into forms (*Stauroneis anceps* v. *anceps* f. *gracilis*, for example). Often a shorthand is used to record forms of the so-called nominate variety (the variety whose name (eptithet) is identical to that of the species) thus: *Stauroneis anceps* f. *gracilis*.

You will find that the varieties of diatoms show variation almost of the same magnitude as that found among species. The description of varieties has been one means of broadening the description of a species without glossing over the discontinuities of variation.

Final Remarks on Classification

Every described genus began with a single described species. Diatoms have been studied since the time that microscopes first became powerful enough to observe them. With early microscopes one could discern shapes and coarse detail but not the fine detail on which modern descriptions are based. Thus in 1822 Bory was able to distinguish boat-shaped cells, which he designated *Navicula*. The early descriptions of this genus referred mainly to the presence of symmetry to the long and transverse axes of the cell. As microscopes improved and more specimens were described, it became evident that consideration of additional characters would allow breaking up the assemblage into more natural genera. Thus began the process of splitting that continues to this day. The additional characters observable with the electron microscopes may well lead to rearrangements as profound as those that gave us our present understanding of diatom genera and species.

You will find that some genera are quite constant in their characteristics while others, like *Navicula*, often seem to have more differences between species than are found between some genera. *Navicula* will eventually be broken up into many genera, but for the present we must accept the fact that it remains the genus into which all naviculoid species fall when they do not fit any of the other genera.

Diatom Communities and Adaptations for Survival

There are few places on earth that are utterly devoid of diatoms. Of the hundreds of samples I took during the course of my study in Illinois, only a few appeared to lack any diatoms and most revealed 20 or more species. Still, not all species are found in all places. Diatoms, as is the case of other organisms, are best adapted for survival in particular types of habitat and each type of habitat will have a distinctive diatom flora or community. Let us discuss some of these habitats.

Diatoms are algae, and when one thinks of algae one thinks first of the open-water assemblages of the plankton. Which organisms can be considered "true" plankters is a matter of debate. Some purists insist that only those species that complete their life cycles in the free-floating condition qualify, but Patrick (in Werner, ed., 1977) holds the broader view that plankton consists of organisms that spend the greatest portion of their life cycles in the free-floating condition.

In Illinois, the centric diatoms and those pennate species lacking any sort of true raphe constitute the majority of the true plankters. Many of these have obvious modifications that retard sinking. *Melosira* and *Fragilaria,* for example, form filamentous and ribbonlike colonies respectively, while cells of many species of *Synedra* are very long by comparison with their width. Cells of some other planktonic diatoms have long spines. In each case an increase in effective surface area, which retards sinking, has been obtained without a notable increase in cell volume. Still other species achieve buoyancy by the storage of oils.

The near-shore diatom communities tend to have a greater diversity of species corresponding to a greater variety of microhabitats.

To begin with, there are the bottom-dwelling or benthic communities living on the surface of loose sediments. These diatoms are typically motile, as you would imagine, since the flocculent materials of the bottom are constantly being moved about by waves and currents. A nonmotile organism would tend to be quickly buried.

Not all near-shore communities are as tenuous as those of the bottom floc. Many diatom species have means to become attached to solid organic or inorganic substrates. The German term "Aufwuchs" and the term "periphyton" have been coined to refer to these communities. Depending on whether the substrate is plant, animal, or mineral, the communities may be called "epipiphytic," "epizoic," or "epilithic." The diatom communities of the Aufwuchs are complex and rich in species. The study of their development is one of the most interesting aspects of diatom ecology, and the study of diatom communities on artificial solid substrates has yielded a series of useful environmental indicators.

Associated with the attached communities, one finds a large group of species sometimes given the name "tychoplankton." Unlike the true plankters, these organisms maintain their position in the water column by active locomotion rather than by passive means such as changes in buoyancy.

The foamy spindrift found in lakes, while not a permanent habitat, is frequently rich in diatoms accumulated from various parts of the lake.

As one moves out of the lake, pond, or stream onto the wet shore or stream bank, one can find the communities of the psammon, located in wet sand, and subaerial communities on the damp surfaces of mud and other substrates.

In various places, but especially on the floodplains near rivers, there may be temporary pools or mud puddles that fill during periods of rain or high water. The same spot may go through wet and dry cycles repeatedly over a period of months and in some cases, years. There is frequently an algal flora that is different from that of adjacent permanent wet areas because it is composed of species that have the ability to survive radical changes in environment.

In low-lying areas, such as roadside ditches, one often encounters marshy patches with cattails and other marshland plants. The wet stems of these plants are often covered with epiphytes and unusual assemblages of species may inhabit the decaying vegetation around the rooted plants.

More extensive marshlands and overgrown ponds exhibit attached communities that are unlike those in roadside ditches since drought is almost never a factor. The true cypress swamps in the southern part of Illinois are different again due to the varying water levels and the low light intensity on the swamp floor.

Before leaving wetlands, one must consider certain rare types that are present or to be expected in Illinois. While normal marshland water may not differ radically in hardness from open waters in the same area, there may be occasional examples of acid-water marshes or bogs. These will exhibit unusual floras, rich in the species of *Eunotia* and *Pinnularia*. I have not sampled such a bog in Illinois, but have found marshy patches exhibiting some of the same characteristics, especially in the southern part of the state. On the opposite end of the hardness scale are the alkaline raised bogs or fens. I have not found any of these in Illinois but expect them. In Iowa these marshy spots arise around seeps or artesian springs that are fed by extremely hard (500 mg/ml $CaCO_3$) water from certain aquifers. Evaporation concentrates the dissolved matter to the point where precipitation occurs. In Iowa, fens are associated with the terminal moraines of the last stage of the Wisconsin glaciation. Should you see a marshy spot on a morainal hillslope, investigate it. The odor of hydrogen sulfide is often an indication, since sulfur bacteria are common components of fen flora. If you do find a fen, both the vascular plant flora and the diatom flora will be unusual. In Illinois we have a third rare type of wetland habitat: the salt spring. The most famous of these brine springs is called Saline Wells and is located near the town of Equality in southern Illinois. These springs were used at one time as commercial brine sources. I have to say, unfortunately, that the diatom flora did not appear exotic, at least on the occasion when I sampled them.

So far, we have been dealing with habitats that are fairly obviously wet. Diatoms, however, can thrive in habitats that are apparently dry or at least dry by comparison. I have already mentioned temporary pools. It is worth sampling these even when in their dry phase, since some of the more hardy species may thrive in thin films of water. Forest floors and the bark of trees are often covered with bryophyte/lichen communities. Some species of diatoms grow in the water films covering these plants. Damp stones and wooden posts may reveal a weak green color indicative of an algal flora and diatoms will be a part of it. Even ordinary soil in lawns or gardens can support a limited but interesting diatom flora. All of these habitat types may be characterized as subaerial. The drier ones are called xeric with respect to the algae, even if not so with respect to the vascular plants.

Collecting and Preserving Diatoms

The methods selected for diatom collecting depend first on the reason behind the collecting. It is beyond the scope of this work to discuss the formal collecting methods used in ecological work. Such methods should be sought in textbooks devoted to ecological methods and in the journal literature. I emphasize this because modern ecological studies demand rigorous statistical approaches. Failure to follow accepted formal methods will render the results of the study worthless in the eyes of most researchers.

Instead, I will discuss the methods of collection appropriate to a nonquantitative survey of an area, since even quantitative approaches require such initial surveys to find out which species are present.

Collecting Equipment and Its Use

Plankton Sampling

Sampling plankton is largely a matter of concentrating the organisms. Traditionally this has been done with a plankton net made of fine-meshed silk or nylon cloth. In recent years, however, researchers interested in quantitative measures have often preferred to use the so-called whole-water sample.

The plankton net is convenient, easily transportable, and capable of gathering large quantities of concentrated biological material even from fairly sparsely populated waters, but it does have limitations. To work efficiently, the net must allow free passage of water. This is facilitated by comparatively large mesh openings. To be effective, it must be able to trap even very small organisms—which can only be accomplished by making the mesh openings very small. The plankton net must therefore be a compromise design. This compromise is avoided by the whole-water sample.

A whole-water sample is collected by filling a container with a large and known volume of water from the sample site. This water is then treated with preservatives to kill the planktonic organisms. Concentration takes place by settling or can be effected by centrifugation. If done correctly, few if any of the organisms are lost. Neither, unfortunately, are the suspended clay and silt that are collected along with the plankton. Furthermore, sparsely populated

435

waters will produce a fairly small yield of organisms even after concentration. While this is not particularly important in quantitative
work, it is a disadvantage in survey collecting where the intent is to
discover the less abundant organisms as well as the dominant ones.
For this reason I recommend the plankton net for general-purpose
studies.

Plankton Nets

The net is a conical silk or nylon bag with a stiff metal rim at the
large end and a hole at the small end to receive a sample bottle. As
supplied, most nets have a pair of strings for attaching the bottle
and a harness of wires for attaching the towing cord.

Plankton nets come in various sizes. I prefer the six-inch diameter for general work. Larger sizes, as well as the extremely expensive Wisconsin nets, are intended for offshore towing in areas free
from snags.

There is also a choice of net material and mesh size. Silk bolting
cloth is traditional, but many nets are now made of nylon cloth,
which is considerably more durable. For diatom work it is best to
choose the finest weave available in either fabric. Silk nets are assigned numbers, the finest mesh being #25. While #20 is accept-
able for most purposes, the #12 cloth is simply too coarse to be of
any use. Nylon nets have a different numbering system, but equivalent weaves can be found.

I recommend that you purchase a bottle adapter for any net that
lacks one as supplied. It is almost impossible to untie wet strings,
and the bottle adapter (which allows you to unscrew the sample
bottle) is a decided improvement. Finally, you will need to have a
strong cord, twelve to fifteen feet long. One end is attached to the
wire harness, the other is formed into a loop to fit over your wrist.

Plankton nets are not cheap and are rather fragile. Take care to
avoid snags when using them. Never put a net away dirty or wet,
and *never* leave glass bottles attached during transport or storage.
Silk bolting cloth decays rapidly. Small holes will appear first near
the bottle-end of the net. Very small holes can be sealed with clear
nail polish. Larger holes or tears can be mended with scrap silk
from an old net, but the stitches must be sealed.

Use of the Plankton Net

Working from Shore: Attach the sample bottle. Slip the loop in the
free end of the cord over one wrist. Lower the net into the water to

wet the net (be sure to fill the bottle with water or it will float). While holding the free end of the cord with one hand, throw the net into a snag-free area with the other. Draw the net slowly back to shore, lifting it out before it touches bottom. Repeat until you have a sufficient sample. Remove and label the sample bottle (or transfer contents to a labeled bottle). Wash the net if possible before taking another sample.

Offshore Towing: Attach the free end of the cord to a fitting in the boat, *not* to your wrist. Reduce speed until you are just making headway. Lower the net into the water (be sure to fill the bottle or it will float). Keep the net away from hull and propellers. Inspect the net frequently. (It is easy to clog a net with plankton and there is no point in continuing to tow if this happens.) Remove and label the sample.

Working in Areas with Snags or Little Open Water: Take the actual water sample with a large bottle or bucket and pour it through the net. This method can be used to make semiquantitative samples as well from any body of water since you can measure the amount of water sampled even if you cannot be sure that all the organisms in it are captured.

Nonplanktonic Habitats

Materials

Sample bottles, labels, pipettes (small laboratory pipettes and one large kitchen baster), knife with a stout blade, grappling hook, plankton net, boots.

Procedures

Bottom Samples: Bottom-dwelling diatoms may exist as attached communities or as less stable communities on the surface of loose sediments. Small pebbles with attached diatoms can be removed intact. Larger rocks, if they can be lifted, may be scraped to remove the diatoms. Sampling of rocks in place can be attempted by scraping (larger mats may be loosened in this way), by pipette (weakly attached materials will yield to this approach), or by the simultaneous use of the knife to detach materials and a pipette or baster to capture them. The patchy communities on sand or mud can be sampled with small pipettes drawn gently over the surface as the bulb is released. Flocculent assemblages may be sampled with pipettes or the baster or by hand-gathering of materials.

Attached Communities on Organic Substrates: Small macrophytes can be collected whole. Larger ones should be scraped to remove the attached growths. A grappling hook (easily fabricated from stiff wires, a short piece of pipe to hold them together and provide weight, and a rope) may be useful in pulling in sunken branches and pondweeds. Attached communities can sometimes be found on animals as well as plants. While plankton and animal remains such as the shells of mollusks can be collected without much damage to the ecosystem, be very careful when dealing with living macrofauna. I would suggest working with a zoologist if you are interested in these communities.

"Tychoplankton": These assemblages occur in and around mats of vegetation and debris. Direct sampling by hand is useful. Alternatively, the material can be squeezed over a bottle or into a plankton net, thus capturing the diatoms without retaining the coarse matter.

Psammon: Put the sand into a sample bottle, add water and stir. Allow the sand to settle, pour off the water into another bottle and discard the sand. Alternatively, sample the wet sand for processing as is.

Damp Surfaces and Soil: Scrape the top layer off with a knife. Try to retain as little as possible of the substrate.

Bryophyte/Lichen Communities: Collect whole-plant samples.

Final Remarks on Collecting

Wading and Boots: I prefer to collect from shore whenever possible for reasons of safety. If you are used to wading, use whatever type of wading boots you wish. If not, stick to the shore. When you are working in marshy areas you *will* need some type of boots to protect your feet from punctures and your legs from those creatures that may regard them as a good meal. I do not advise lace-up boots for this purpose since you may have to get out of them in a hurry if you get stuck and laces will make this impossible. Boots are also necessary when sampling highly contaminated areas (rubber gloves would be appropriate here as matching protection for your hands).

Collecting Alone: If you must do so, let someone know exactly where you will be and when you plan to return. Better still, take a companion, even if that person is only interested in a walk in the country. Especially, do not go wading alone.

Preserving the Environment: Remember as you sample that your purpose is to study nature, not destroy it. You will never process more than a few grams of sample, so do not collect by the kilogram. If you must remove a rock for scraping, put it back where it was. Try to avoid uprooting aquatic vegetation when you can scrape it in place. Ideally no one should be able to detect that you have sampled an area. Familiarize yourself with lists of endangered species in your area and take care not to disturb them.

Sampling on Private Land, in Parks, or in Preserves: Obtain permission from landowners and abide by their wishes. If you must sample in parks or preserves be aware that permits are often, if not always, required. Check with the ranger or manager before you start.

Care of Living Samples

If you intend to examine the material in living condition you should leave plenty of air space above the sample. Do not allow the material to overheat (some researchers prefer to pack their samples in ice during transit). Once in the lab, open the bottles immediately and refrigerate (do not freeze) them if examination must be postponed.

Preservation of Samples

The fixation of samples for protoplasmic study is beyond the scope of this book. For present purposes, preservation methods are used only to halt growth and prevent decay.

In the past, preservation was accomplished through the use of such chemicals as formalin and formaldehyde. Recent studies have shown these agents to be more hazardous than previously believed. I do not advise their use in diatom study, the more so because they interfere with processing samples for making permanent slides.

There are safer means for preservation. The first is freezing, the second is boiling, and the third is storage in ethyl alcohol. If you wish only to halt growth (thus preserving the numerical proportions of the various species at the time of collection), all three methods will work. If you wish to preserve a part of the sample for later study and do not wish to keep a freezer for the purpose, storage in alcohol or boiling will do.

Heating the sample to the boiling point will kill all living algae. Use of a pressure cooker or autoclave will increase the level of protection by killing as well fungal and bacterial spores. A cookbook will provide instructions for safe use of pressure cookers. Alcohol preservation requires 95% ethyl alcohol (denatured alcohol is fine). Add alcohol in an amount equal to the volume of the sample.

Cautions: Keep in mind that if you use heat as a preservation method, you will have to re-treat the sample each time you open it if you wish to retain the protection against fungi and bacteria. If you use alcohol, you will have to reduce the alcohol content before treating the sample to make permanent slides, since high concentrations of alcohol present a risk of fire during operations involving heating.

Observation of Living Material and Preparation of Permanent Mounts

Observation of Living Material

I strongly recommend that a preliminary examination be made of the fresh material in each sample before fixation and certainly before chemical "cleaning" to remove protoplasm. There are several reasons for this. First, some diatoms have lightly silicified cell walls that are almost always destroyed by chemical cleaning methods. Second, colonial form is important in separating species of some genera from similar-appearing species of others. Vigorous chemical cleaning methods destroy colonial form. Third, when studying the ecology and biogeography of diatoms, it is important to know whether the organisms were alive or dead at the time of collection. It is almost impossible to make this judgment on the basis of the silica cell wall alone since the wall of a dead (sometimes long dead) cell may look little different under the light microscope from one that was living at the time of collection. An ideal technique would be one that allowed identification of diatom cells in the living condition, but unfortunately no such technique exists. Preliminary examination of the fresh sample will at least give you an impression of the proportion of living to nonliving cells.

Materials

Microscope, slides, cover glasses (#1 or #2 glass, not plastic. I prefer 18 or 22 mm square, but some use circular glasses of similar size), distilled water, scalpel, pipettes (disposable Pasteur pipettes are useful). To avoid contamination, pipettes should be thoroughly cleaned after use before using in another sample. For research purposes it is better not to reuse pipettes at all.

Procedure

For liquid samples: mix sample thoroughly, withdraw a subsample with a clean pipette, and transfer to slide. For coarse materials: scrape the substrate with a scalpel and suspend the scrapings in a drop of distilled water on the slide. Cover the drop of sample or suspension with a cover glass. The most useful objective lens for observing living material is the 20–25 × dry, with a high dry (43 ×) for closer study.

Plankton samples should be searched for lightly silicified cells. One way of spotting these is to look for isolated spots of color. Most often these spots prove to be bits of debris or isolated cells of other organisms, but occasionally you may discern around them the faint outlines of the cell walls of *Attheya* or *Rhizosolenia*. (It is well to familiarize yourself with the descriptions of these two genera.) A second thing to seek in plankton is the presence of colonial species. While most of these are not hard to identify in "cleaned" mounts, it is good to know in advance that you should expect them. With plankton it is almost always worth the time to make a burned mount (see below) even if your primary method of making permanent slides is one of the chemical methods. A burned mount will preserve both colonial form and thin-walled cells.

Most other types of sample do not present a problem with lightly silicified cells, but if you are collecting in an area highly contaminated with nitrogenous wastes, yet well aerated, be on the lookout for *Cylindrotheca*. (Almost all that can be seen is the helical keels.) More important with nonplanktonic samples is to form an impression of what diatoms appear to be alive at the time of collection. You cannot make identification to species in most cases, but it is interesting to note that more than one type of, say, *Gomphonema* was living at the time of collection. One means of correlating observations of living material with those on "cleaned" mounts is by use of the burned mount (described below). Such a burned mount can be made rapidly and scanned for species of interest. Living material from the same sample can then be examined and an attempt made to locate living cells matching the general size and shape of those you found in the burned mount.

Preparation of Permanent Mounts

Over the years many techniques have been employed in preparing diatoms for study. I will describe some of the more common ones. The technique chosen should fit both the requirements of the study attempted and the materials and facilities available to the researcher. I must caution you in advance that some of the techniques involve the use of chemicals that are highly dangerous if mishandled. Minimum precautions are listed as appropriate.

The techniques that follow are for general use in qualitative "survey" work. The methodology of quantitative study is beyond the scope of this book.

"Cleaning"

Identification of most species of diatoms requires specially prepared permanent mounts. The production of such mounts involves removing organic cell contents by incineration or chemical "cleaning" followed by mounting the cells in a mounting medium of high refractive index, for it is only when the mounting medium has a refractive index significantly higher than silica that the fine details on which species descriptions are based become visible.

Each of the following procedures has advantages and disadvantages. It is very important to pay attention to any *Caution* notes.

Removing organic matter from the diatom cells involves either oxidation or hydrolysis. For many years hydrolytic methods involving hot or cold acids were the standard. These methods require, however, a fully equipped chemical lab for safe use. I have decided to describe therefore only the oxidation methods. Oxidative cleaning may be done by incineration (the safest method) or by treatment with strong solutions of hydrogen peroxide. The incineration method is commonly called the "burned mount."

I recommend that you read all of the following procedures before choosing a method. *Please read all Cautions before attempting any Procedure!*

Incineration Methods (Burned Mounts)

Both methods use high temperature (provided by an oven or hot plate) to incinerate organic material in the sample. I am indebted to my father, John D. Dodd, for teaching me these techniques and especially the second one, which he has used over the years to produce beautiful slides from "impossible" samples.

Basic Incineration Method

Advantages: Allows preservation of lightly silicified cells such as those of *Rhizosolenia*; preserves colonial form; usually does not break cell walls apart into isolated valves (thus you will be able to examine complete frustules, which is important when one valve has a different pattern of markings from the other, as with *Achnanthes*);

requires very simple equipment and uses no hazardous chemicals.

Disadvantages: Cells are often clumped together, thus making identification difficult and interfering with some statistical sampling methods. Since frustules are not broken into component valves, cells with a large valve-to-valve (pervalvar) dimension will often lie girdle-side up, thus making identification difficult or impossible. Since it is impossible to remove all attached diatoms from coarse substrates, proportions of species in the finished slides may not be representative of those in nature. Since there is no good way to remove all coarse material, slides may appear dirty. Dissolved substances in the sample may form obscuring crystals.

When to Use: Use this method any time colonial form or lightly silicified valves must be observed. It gives best results with planktonic samples and poor results with soil samples. This, or the following procedure, is the method of choice when facilities for safe handling of hazardous chemicals are not available.

Materials: Hot plate (or oven), cover glasses (18 or 22 mm square, #1 or #2 glass, not plastic), clean (preferably unused) pipettes, scalpel or other instrument for scraping, dissecting needle, flat heavy-weight metal plate to support cover glasses during heating (a heavy-weight baking sheet has been used successfully for this purpose, but quarter-inch-thick aluminum or heavy copper is better), distilled water, clean vials or small bottles.

Cautions: Do not leave hot plates unattended. The metal plate used as a cover-glass support will remain hot enough to cause burns for some time after turning off the hot plate or removal from the oven. Do not use this method with samples that have been preserved in formaldehyde or alcohol. If these preservatives have been used, use the Modified Incineration Method.

Procedure

1. Place cover-glass support plate on hot plate or oven rack.

2. Arrange clean cover glasses on support plate in an orderly pattern and with plenty of space between adjacent glasses.

3. Sample preparation: a. For plankton: Mix thoroughly. If sample is dense, withdraw a portion with a pipette, place in a clean vial, and add distilled water until the sample is cloudy but not opaque.

b. For coarse substrates: Scrape the substrate with a scalpel and suspend the scrapings in distilled water in a vial. It may be necessary to use a glass rod to break up clumps of material. Add enough distilled water so that the suspension is cloudy but not opaque.

c. For coarse substrates that cannot be scraped: Wet the material with distilled water and squeeze it over a clean beaker. Alternatively, wrap it in cheesecloth, wet it, and squeeze over the beaker. Use the collected liquid as a subsample.

d. For bottom samples: Suspend sample in enough distilled water to make a suspension that is cloudy but not opaque. Samples with coarse sand should be mixed thoroughly, then allowed to settle briefly to allow the sand to drop. Be aware that some large diatoms may settle almost as fast as the sand.

e. For soils: Use the topmost layer of pieces of clay. Sandy soils or psammon samples with little fine material can be treated as is. In the case of silty soils, use less material than for sand. Suspend the subsample in enough distilled water to produce a suspension that is cloudy but not opaque. Soil preparations are unavoidably "dirty" but soil diatoms are interesting and worth the effort.

4. Withdraw a portion of the prepared sample with a clean pipette and transfer it to one of the cover glasses you placed on the support plate. Cover the glass evenly from edge to edge but do not overload it. (A slight bump may cause the water to overflow onto the support plate, thus ruining the preparation and perhaps adjacent ones as well!) Make a "map" to associate the position of a particular cover glass with a particular sample. In general, make up at least two cover glasses for each sample.

5. Allow the water on the cover glasses to evaporate. Do not use the hot plate or hot oven to speed drying. Doing so may cause spattering that will contaminate adjacent samples and in any case will usually result in formation of clumps of cells. Some researchers speed drying by arranging a gooseneck lamp above the samples. This usually does not cause problems. If you are using an oven, you may find that a very low heat level on the thermostat and an open door will speed drying without causing the water to boil and spatter.

6. When all samples are dry, increase the heat to 500°F or more. As incineration proceeds, the samples on the cover glasses will turn black and eventually gray as oxidation is completed.

7. Remove from heat, allow samples to cool, then proceed to the Slide-making procedure.

Modified Incineration Method

Advantages: Similar to those of the basic incineration technique with the added advantage of removing coarse materials and chemical preservatives. In addition it will reduce the concentration of naturally occurring solutes that might form obscuring crystals as the sample dries.

Disadvantages: Colonies and very delicate cells may be broken up, though to a lesser extent than with other methods. Some clumping of cells may occur, but to a lesser extent than with the basic method. It is not suitable for quantitative work since some of the sample will be lost and some steps may alter the proportions of species present in the prepared sample from what they were in nature.

When to Use: Use this method for any type of sample except pure plankton (which is best handled by the basic method unless it has been preserved with formaldehyde or alcohol or comes from waters with very high solute concentrations). It is particularly useful for moss samples and other vegetation samples with associated diatoms. It is the preferred method of pretreatment for samples preserved with formalin, formaldehyde, or alcohol and for those containing high concentrations of naturally occurring solutes. Use this method if laboratory facilities for handling hazardous chemicals are unavailable.

Materials: The materials for the basic method plus Pyrex (or similar borosilicate glass) beakers, beaker tongs, and cheesecloth (or other coarse-weave cloth).

Cautions: The same as for the basic method. If you are using the method to remove preservatives, be aware that some localities may have strict rules for the disposal of even small amounts of chemical wastes.

Procedure

This procedure replaces the sample preparation step (Step 3) of the Basic Incineration Method. All other steps follow the Basic Incineration Method.

1a. If you have a fairly clean sample, such as plankton, and need only to remove preservatives or to reduce hardness, place sample in a clean beaker, bring volume to 200 ml with distilled water, heat to boiling, allow to cool and settle, carefully pour off the water. Go to Step 6.

1b. If you have a sample with coarse particles, such as sand, place sample in a clean beaker, bring volume to 200 ml with distilled water, heat to boiling. Then go to Step 5.

1c. If you have a sample with various coarse substrates, start with Step 2.

2. Use a scalpel to scrape large surfaces, then place the scrapings and all other substrates in a clean beaker with around 100 ml of distilled water.

3. Boil the sample for 10–20 minutes, adding distilled water if necessary to keep the sample from boiling dry.

4. Add distilled water to bring the volume to around 200 ml. Mix well and strain the suspension through cheesecloth to remove the coarsest substrates. This step may result in loss of sample and/ or alteration of the numerical proportions of species since some cells may remain attached to the substrates you remove.

5. Mix the sample thoroughly. Let stand for 10 seconds to allow sand to settle, then carefully pour the rest of the sample into a clean beaker.

6. Add distilled water to fill the beaker. Stir thoroughly. Allow to settle for 4–6 hours to ensure that all diatoms have settled to the bottom.

7. Carefully pour off the water without disturbing the sediment. Refill the beaker with distilled water and resuspend the sediment. Let sample stand 4–6 hours to allow the suspended cells to settle again to the bottom. This step may be repeated as necessary to reduce the concentration of dissolved substances or preservatives.

8. Pour off the water without disturbing the sediment. Resuspend the sediment in a small amount of distilled water and use the resultant suspension as the prepared sample for the rest of the Basic Incineration Method. If the sample is very dense, you may reduce the density by suspending a portion of the sample in distilled water to produce a cloudy, but not opaque, suspension.

Hydrogen Peroxide Methods

Advantages: All of the hydrogen peroxide methods produce samples that are relatively free of organic matter. Since almost all organic substances are attacked by this reagent, attached cells are detached from their substrates and from each other. Hydrogen peroxide cleaning produces samples in which there is little clumping of cells and in which the numerical proportions of most species are usually not altered by the processing. Of the chemical cleaning methods, this is the safest for use in modestly equipped laboratories.

Disadvantages: Most, if not all, hydrogen peroxide methods will destroy lightly silicified cells such as those of *Rhizosolenia* and will break up all colonies that are held together by organic substances. Colonies produced by interlocking of cell-wall parts may or may not remain intact. The reagent itself is hazardous (see below under *Cautions*) and may not be available to members of the general public. Some of the hydrogen peroxide methods involve use of concentrated hydrochloric acid and potassium dichromate, which require special handling.

Materials: Pyrex or equivalent borosilicate glass beakers of 250– 500 ml size, wide mouth rather than tall style; distilled water; hydrogen peroxide, 27–30% reagent or technical grade (the technical grade is cheaper if you can find it and works just as well); potassium dichromate crystals (a few grams will be sufficient to treat a number of samples); hydrochloric acid in dropper bottle (concentration of acid is not critical, but should be as high as possible); cheesecloth; hot plate; glass stirring rods; chemical fume hood; eye protection (goggles or glasses designed for use in a chemical laboratory).

Cautions

1. Hydrogen Peroxide: The hydrogen peroxide used in these methods is not the weak solution used for medical purposes and as hair bleach but a 27–30% technical or reagent grade. *This reagent is an extremely strong oxidant* and can cause severe burns and eye damage. *Eye protection is mandatory.* Rubber gloves should be worn when handling. You must have a source of running water close at hand. If spilled on skin, the reagent should be washed off with running water for at least five minutes. *If this reagent contacts eyes, wash with running water for five minutes and seek medical attention at once!* Do not store this reagent near metallic substances such as zinc or near potassium dichromate. Do not store or use this reagent in metal containers. Do not use grades of this reagent in excess of 30% without assistance from a qualified chemist. Do not store in bottles with unvented caps.

While I have in the past used this reagent in a well-ventilated open room, I am now convinced that it should be used in a fume hood, especially when the method calls for boiling. In any case the vapors from the boiling liquid should never be inhaled or allowed to contact the eyes.

Samples preserved with formalin, formaldehyde, or alcohol should be pretreated via Steps 6 and 7 of the Modified Incineration Method to remove the preservative before treating with hydrogen peroxide, since *formalin and formaldehyde have been known to produce violent and unexpected reactions with peroxide* and high concentrations of *alcohol* pose a *fire hazard during steps involving heating!*

2. *Potassium Dichromate:* This reagent causes severe eye irritation and has been known to produce severe cases of dermatitis. It should therefore be considered a hazardous chemical and treated accordingly. Immediately wash the substance from skin. If eyes are involved, wash for at least five minutes with running water and seek medical attention at once. Follow all cautions on the reagent bottle. *Do not store this reagent near hydrogen peroxide!*

3. *Hydrochloric Acid:* Some steps involve treatment of the sample with concentrated hydrochloric acid. You should wear eye protection when adding acid to any sample. Do not add acid to hot samples. You should have baking soda (sodium bicarbonate) solution on hand to neutralize spills. If concentrated acid is spilled on skin, wash with running water at once. Use sodium bicarbonate solution after the initial rinse. If eyes are involved, instantly wash with running water and continue for at least five minutes. Seek medical attention!

Procedures

Initial Steps (All Methods)

Put about one gram of solid sample or 10–15 ml of suspended sample into a clean 250–500 ml beaker. Vegetation, such as mosses or pondweeds, should be cut up into small pieces to form a layer on the bottom of the beaker. (Remember that you should pretreat samples preserved in formaldehyde, formalin, or alcohol via the washing steps [Steps 6–7] of the Modified Incineration Method to remove the preservatives.) If a maximum separation of cells into isolated valves is desired, use the Catalyzed Method. If a gentler procedure is required to preserve delicate cells and/or reduce the amount of valve separation, use the Noncatalyzed Method.

Catalyzed Method

1. Bring sample volume to 20–25 ml with distilled water and add an equal volume of 27–30% hydrogen peroxide. Allow sample to stand for around five minutes. Typically very little activity will be noticed, but sometimes samples rich in iron or other metals will spontaneously begin to react. If this happens, allow the reaction to go to completion (be aware that the reaction mixture will become very hot), then go Step 4. If there is no obvious reaction, go to Step 2.

2. Add a few crystals of postassium dichromate ($K_2Cr_2O_7$). *Do not substitute potassium permanganate for potassium dichromate* as the permanganate may produce almost explosive reactions. With the addition of the potassium dichromate the reaction mixture will generally turn deep purple. Shortly, if there is sufficient organic matter, the sample will begin to boil. As the temperature rises, reaction velocity will increase. It may be necessary to use a stirring rod to keep the sample from boiling over. Some spattering may result, so it is best to keep all other samples covered or out of range. As I have said, this procedure is probably best carried out in a fume hood. Allow the reaction to go to completion (this is marked by a change in the color from purple to yellow or orange and an end to the vigorous boiling). Allow to cool, then go to Step 4. If no obvious reaction has begun within five minutes after adding potassium dichromate, go to Step 3.

3. If the reaction does not start spontaneously, heat the mixture *gently* on a hot plate until it does. Once it has gone to completion allow mixture to cool, then go to Step 4.

4a. If the sample is free of coarse substrates and sand, go to step 6.

4b. If the sample contains sand, add distilled water to restore the volume to 40–50 ml. Mix thoroughly then allow to settle *briefly* (around 10 seconds) to let the sand settle. Pour off the liquid into a clean beaker, and use this liquid for the rest of the procedure. Go to Step 5.

4c. If the sample contains other coarse substrates, strain through cheesecloth to remove them, and go to Step 5. *Note:* Straining may result in some loss of sample. *Caution:* Handle the used cheesecloth using the same precautions as for the reagents. Do not reuse cheesecloth. Dispose of it in the manner prescribed by your institution.

5. Restore volume to 40–50 ml with distilled water, resuspend the sample, and add an additional 20–25 ml of 27–30% hydrogen peroxide. Do not add additional potassium dichromate. Allow the sample to react with the peroxide as in Step 2. It is likely that you will have to heat the mixture to start the reaction. When the reaction is complete or if there is no reaction even after heating to boiling, allow mixture to cool and go to Step 6.

6. If carbonates are a problem, transfer the sample to a fume hood and add concentrated hydrochloric acid dropwise until no further gas evolves. Use no more acid than necessary. If you have ar-

rived at this step after treatment with potassium dichromate, you may find that the addition of acid will cause chromium compounds from the reaction mixture to form colored complexes (usually light blue or green). This is normal and will not interfere with further processing. If red iron oxides are present in the sample, continue adding hydrochloric acid until they dissolve (color will change from red to yellow).

7. Fill beaker with distilled water and go to the Washing Procedure.

Noncatalyzed Method

1. Place approximately 1 gram of solid sample or 15–20 ml of suspension in a clean 250 ml beaker. Cut vegetation such as mosses and pondweeds into small pieces.

2. Bring the volume to 50 ml with distilled water and add an equal volume of 27–30% hydrogen peroxide to the sample. Allow sample to stand for five minutes. It is not uncommon to find that iron or other metals in the sample may initiate a spontaneous reaction. If this happens, let the reaction to go to completion, allow mixture to cool, and go to Step 4. If no such reaction occurs, go to Step 3.

3. This step should be done in a fume hood. Transfer the beaker to a hot plate and heat to boiling. *Do not allow mixture to boil dry as explosion will result!* It is not necessary to boil the sample vigorously but only to keep it near the boiling point for around 10 minutes. Add distilled water as necessary to keep the mixture from boiling dry. At the end of the boiling period, remove sample from heat and allow to cool.

4. Remove remaining coarse materials from the sample as follows: If the coarse matter is sand or similar heavy matter, mix the sample thoroughly, allow to settle *briefly* (around 10 seconds) then pour off the liquid into a clean beaker. If the coarse matter does not drop after *brief* settling, remove it by straining through cheesecloth. *Caution:* Handle this step using the precautions appropriate for hydrogen peroxide. Do not reuse cheesecloth. *Note:* Straining through cheesecloth may result in some loss of sample.

5a. If the sample appears free of organic matter, go to Step 6.

5b. If there appears to be fine organic matter left after settling or straining, it may be necessary to repeat Steps 2 and 3. Allow sample to cool and go to Step 6.

6. If carbonates are a problem, add concentrated hydrochloric

acid dropwise until no further gas evolves. If red iron oxides are a problem, continue to add acid dropwise until the red oxides are replaced by the dissolved yellow oxides. 7. Fill beaker completely with distilled water and go to the Washing Procedure.

Washing Procedure (All Hydrogen Peroxide Methods)

Purpose: The object of washing is to remove all or most of the spent reagents and the reaction products leaving only the diatom cell walls.

Materials: Clean borosilicate glass beakers (250–500 ml Pyrex or equivalent), distilled water, glass stirring rods.

Cautions: The treated samples from the last stage of the cleaning processes contain fairly high levels of hazardous substances and should be handled accordingly.

Procedure

1. Allow the samples from the last stage of the chemical cleaning procedure to settle for 5–6 hours or preferably overnight.

2. Carefully pour off the liquid without disturbing the sediment. *Caution:* Handle this liquid using the same cautions as appropriate for the method in use for cleaning.

3. Refill the beaker with distilled water and resuspend the sediment using a glass stirring rod. (You may wish to transfer the sample to a clean beaker at this point.)

4. Allow the sample to settle for 4–6 hours or overnight.

5. Repeat Steps 2, 3, and 4. The number of times varies, but at least three repetitions are advisable. It is almost impossible to completely neutralize acid preparations by dilution alone, but certainly the pH should be increased to 4 at least.

6. After the last settling, pour off the water, resuspend the sediment in a small amount of water, and transfer it to a storage bottle. *Caution:* If you are using a hydrogen peroxide method that does not employ a catalyst, do not seal the storage bottle tightly, at least for a few weeks. It sometimes happens that some residual hydrogen peroxide may be present in the sample even after washing, and, if this decomposes during storage, pressures can develop that will shatter the bottle if tightly sealed. Using rubber stoppers rather than screw caps is an advisable safety measure.

7. Label the storage bottle and go to the procedure on Slide-making.

Slide-making

The making of permanent slides is necessary both to allow the observation of detail needed for identification and to provide a record of species found in a sample.

Materials

Glassware: Beakers (50–100 ml); microscope slides (standard 1 inch by 3 inch); cover glasses (#1 or #2 glass, not plastic, 18 or 22 mm square or round); clean pipettes (disposable Pasteur pipettes are best); glass stirring rods.

Hardware: Electric hot plate (the most suitable ones have a flat top with concealed heating coils and an adjustable thermostat); flat, heavy, metal plate (usually aluminum about one-quarter-inch thick and 8–10 inches square; if your hot plate has a flat surface, you can omit this extra metal plate and work directly on the surface of the hot plate); forceps (fine-pointed jeweler's forceps with straight or curved tips; these will be used for handling cover glasses); dissecting needles; scalpel or single-edged razor blades; diamond or carbide-tipped marking pencil.

Miscellaneous Materials: Low-lint tissue paper (laboratory wipers such as Kimwipes are best; if you must use facial tissue, use a cheap brand without perfume); slide labels (self-sticking preferred).

Reagents: Distilled water; 70–95% ethyl alcohol (denatured alcohol is satisfactory); xylene, (*Caution:* Xylene is a toxic substance. Do not inhale vapors. Avoid skin contact as much as possible. Use in fume hood or with adequate ventilation.)

Mounting Medium: For many years the preferred mounting medium has been the proprietary resin Hyrax, which has a refractive index in excess of 1.7. More recently another resin, Naphrax, which has a refractive index similar to that of Hyrax, has been adopted by some researchers. The ordinary resins, such as Canada balsam, that are used in general microtechnique are really not suitable for diatom work since they have refractive indices little greater than that of silica and provide poor resolution. However, if you cannot find or cannot afford Hyrax or Naphrax, you can still make useful permanent slides with ordinary mounting media if you can accept the limitation that the lower refractive index will place on resolution of closely spaced striae.

I recommend Hyrax as the medium of choice. It is easy to use, it is stable (preparations made with it thirty or forty years ago are still in good condition), and it is unsurpassed in refractive index (if you except the legendary claims for the poisonous and unstable realgar-based medium used experimentally at the end of the last century). Hyrax is very expensive, but a one-ounce bottle will make more slides than you can study carefully in a year.

Procedure

1a. If you have already prepared burned-mount cover glasses, begin with Step 12.

    1b. If you have a cleaned sample from one of the hydrogen peroxide procedures, begin with Step 2.

    2. The object of the first part of the procedure is to prepare a cover glass with an evenly distributed layer of diatoms.

    3. Wash as many cover glasses as you need in 70–95% ethyl alcohol and polish them with tissue paper to remove any residual film.

    4. Place the washed cover glasses in an orderly pattern on a flat metal plate or directly on the flat surface of a hot plate. If you are making slides from more than one sample, assign a number to each cover glass corresponding to the sample number. You should make a "map" of the surface of the plate on a piece of paper to help you remember which cover glasses come from which samples. Do not trust your memory here!

    5. Take each sample bottle and examine it before suspending the sediments. If there is a large amount of sediment, plan to make a dilution. If there is only a little sediment, plan to pour off (or pipette out) a considerable amount of water. There is no way to tell you exactly what "large amount" and "small amount" mean. The object is to produce a working sample that, when thoroughly mixed, has a faintly cloudy or light gray appearance. Mix a sample thoroughly by shaking or stirring. If the suspension is dark and opaque, it is too dense for slide-making. If it is almost as clear as pure water, it is too thin. Let the samples stand again until all sediments have settled, and you will be able to see what proportion of sediment to water produces dense, correct, and thin samples. Go to Step 6. Read 6a, 6b, and 6c. Choose whichever variant applies to your sample.

    6a. If the sample has little sediment, allow it to stand until it is obvious that no further sedimentation will occur. Next, withdraw almost all of the water with a clean Pasteur pipette (if you have steady hands you can pour off the water instead). Suspend the sediment in the remaining water. If the suspension appears too dense, add distilled water dropwise (mixing after each drop) until the sample is a light gray color. Go to Step 7.

    6b. If the sample is light gray when mixed, no dilution or concentration is needed. Go to Step 7.

    6c. If the sample appears very opaque when mixed, withdraw a subsample with a clean pipette and transfer it to a small beaker or clean sample bottle. Add distilled water dropwise (mixing after each drop) until the subsample is a light gray color. Go to Step 7.

    7. Mix the prepared sample thoroughly. If there is still some sand in the sample, allow the sand to settle out (do not wait more than a few seconds). Withdraw some of the sample with a clean pipette and discharge it onto one of the clean cover glasses on the plate. Do not overload the glass. It is better to leave some of the surface uncovered than to risk spilling the liquid onto the plate. If your prepared sample is on the dense side of correct, use less sample; if on the thin side, use as much as possible without risking a spill. Once the sample is on the cover glass, do not disturb it.

    8. Record the sample number on your "map."

    9. Repeat Steps 7–8 for each additional cover glass from the same sample, or Steps 5–8 for each additional sample. It is a good idea to make at least two slides from each sample. When you are doing the procedure for the first time, I would recommend working with several dilutions of one sample. Your concept of "cloudy" or "light gray" may be different from mine, and as a result you may produce cover glasses that are more or less dense than desired. During your first trials you should be seeking to calibrate your eye!

    10. When finished loading cover glasses from one sample, remove all used glassware from the work area before reaching for the next sample. This is one way of ensuring that you do not accidentally contaminate your samples by reusing dirty equipment. Some of my colleagues save pipettes by attaching the used pipette with a rubber band to the sample bottle in which it was used. Thus, if any further slides need to be made from that sample, the pipette can be used again.

    11. When all cover glasses have been loaded, allow them to dry at room temperature. Do not use the hot plate to speed drying since boiling will result in spattering liquid with the attendant possibility of contaminating adjacent samples. Even without boiling, forced drying tends to produce an uneven distribution of diatoms. If you must speed the drying, use the heat from a gooseneck lamp placed over the cover glasses.

12. When samples are air dry (or when incineration is complete if you have arrived at this step from Step 1a) you are ready to make the finished slides.

13a. If you have been working with a burned-mount technique and have just finished the incineration step, reduce the heat on the hot plate to around 300°F (low heat).

13b. If you are working with cover glasses that have been air-dried via Step 11 or if you are using burned-mount cover glasses that have cooled to room temperature, you will need to force-dry the glasses on the hot plate at around 300°F for a few minutes to drive off any residual moisture. Since the diatoms are fixed in place following the initial drying, forced drying in this step will not cause problems.

14. While the cover glasses are heating (or while the hot plate is cooling to 300°F if you used Step 13a), prepare slides as follows:

a. If you have a diamond or carbide glass marker, scratch an identifying number on each slide. The sample number is best.

b. Carefully clean the marked slide with alcohol (70–95% ethyl alcohol), and polish with tissue to remove any residue.

c. Place a drop of mounting medium in the center of the slide. Use only as much medium as you need. Once again, practice will show you the amount, but one drop is a good starting point. If you are using an ordinary histological mounting medium, go to Step 17. If you are using Hyrax or Naphrax, go to Step 15.

15. Remove the cover glasses from the hot plate with forceps. Place each cover glass, sample-side down, into the drop of mounting medium on a slide. There will usually be a slight bubbling in the medium when you do this. This is normal.

16. Set the mounting medium as follows (This procedure applies to both Hyrax and Naphrax. *Caution:* Hyrax can stand much more heat than Naphrax. Any heating of Naphrax should be very brief.):

Place the assembled slide and cover glass on the hot plate (which should still be at around 300°F). The medium should start to bubble very quickly. This is the solvent being driven off. Depending on the heat of the plate, this bubbling should subside in 10–30 seconds. (Watch the color of the medium. If it starts to darken noticeably, the heat is either too high or you have heated too long.) Remove the slide from heat, set it on a heat-resistant surface, and at once press the cover glass gently into the medium using the wooden handle of a dissecting needle. This should drive out the remaining bubbles. If you cannot remove all bubbles, reheat the slide briefly to melt the medium and try again. (If you still cannot remove the bubbles, you may be using too little medium or there may be sand or other coarse materials on the cover glass.) When the bubbles are removed, notice how much medium has spread beyond the edges of the cover glass. If there is a great deal of it, reduce the amount of medium you use on the next slide.

When the slide has cooled, test the medium at the edges of the cover glass with a dissecting needle, If it is soft and rubbery, you have not heated the slide long enough. Return the slide to the hot plate and heat again. If the excess medium on the cool slide is hard and turns to flakes or powder when you scratch it with the needle, it is correctly set.

Using a scalpel or razorblade scrape away excess medium from the edges and upper surface of the cover glass. Be sure not to break the cover glass loose from the slide. (If this happens, brief reheating will reset the glass.)

Using a piece of tissue paper with a small amount of xylene (see *Cautions* at beginning of Procedure), wipe any remaining medium from the upper surface of the cover glass. Do not use an excessive amount of xylene since this will soften the medium under the cover glass.

The slides are now ready to be labeled and examined with the microscope.

17. If you are using an ordinary histological medium, use the procedure recommended by the manufacturer. In general, the cover glasses should be force-dried as described in Step 13 but allowed to cool before inverting them into the drop of mounting medium. Press gently to drive out bubbles and excess medium; then allow the slide to dry for an appropriate amount of time at room temperature or on a warming tray. *Most media cannot stand the direct heating used with hyrax or naphrax.* In any case, the slides should be thoroughly dry before examination to avoid damage to the microscope objectives.

The Microscope

While much work, particularly in learning the genera, can be done with an ordinary classroom microscope, serious study requires the best optics funds will allow. The light microscope has a mechanical system for adjusting focus, supporting and manipulating slides, and ensuring the proper alignment of the optics. It has an illumination system with adjustments for the intensity and quality of light and the proper delivery of light to the optics. Finally, it has an optical system composed of the condenser, objectives, and oculars (eyepieces). The diatomist's microscope has special requirements for all three systems.

The supporting stand should be rugged. Focusing should allow smooth adjustment of both coarse and fine focus (the latter being the most important). A mechanical stage is a necessity since hand adjustment of the slide is almost impossible when using high magnifications.

The light source should be capable of providing even illumination of the field of view. It should be adjustable to allow for so-called Koehler illumination. There should be provision for use of color filters (many diatomists prefer a green filter). The light intensity should be variable since there are times when an intense light is needed for critical observations and some types of photography but many other times when a lower light level will be preferable for preventing eyestrain. Color temperature is not important for photography of "cleaned" diatoms since they have no true colors to preserve.

Most diatomists prefer ordinary bright-field optics. A few have found phase-contrast or interference-contrast useful. Almost no one uses dark-field. My preference is for bright-field, nonphase optics.

The heart of the diatomist's microscope is the "high-oil" objective lens with a magnification of $70-100 \times$ and a numerical aperture of 1.25 or higher. While color-correction is not really important, the highest numerical aperture (which *is* important) is found only in the semi-apochromatic ("fluorite") and apochromatic lenses. In addition to the "high-oil" objective, the microscope should have a $20-25 \times$ dry objective for scanning slides. Those who plan to do photography may find a medium-powered oil immersion lens ($40-54 \times$) a useful third objective. Those who wish to make extensive use of wet mounts may find a $40-43 \times$ dry objective valuable.

458

Little less important than the high-oil objective is the condenser. Only those condenser lenses with oilable top elements that match the numerical aperture of the high-oil objective will allow the system to achieve maximum performance, and then only if immersion oil is used on both the condenser top element and the objective. There is no point in spending a fortune on an oil immersion objective of high numerical aperture if you do not intend to match the condenser to it. In addition to its optics, the condenser must have an iris diaphragm to control the aperture.

Oculars should match the type of objective in use. For apochromatic and semi-apochromatic lenses this means the compensation-type (sometimes called compensating) ocular. Few people now are comfortable with monocular instruments. Binocular heads are standard, and photo-bincoular ("trinocular") heads may be needed if much photography is planned. The oculars should be of such magnifaction as to provide a system magnifcation of 900 to 1200×. While magnification above 1000 is largely "empty" in the sense that it does not show you more detail, it is often easier to count closely spaced striae at a higher magnification. I use magnifications as high as 2000× for this purpose.

Variable-magnification intermediate lenses or magnification-changers are sometimes installed between the ocular and the objective. These can provide high magnifications for counting of striae without the eyestrain associated with oculars of high magnification.

Measuring is critical in diatom work. One ocular must therefore be equipped with an ocular micrometer. This device is a clear disk with an engraved scale of 50–100 units that is so placed as to be in focus with the specimen. The scale units have no absolute dimension and so must be calibrated for each objective. Traditionally this is done by use of an absolute scale, called a stage micrometer. One other method is to mark a specimen and have it measured by someone with a reliably calibrated microscope. You can then observe this specimen with the various objectives of your own microscope and see how many units on the ocular micrometer are occupied by this specimen of known length.

There may be thousands of diatoms on a single slide. In order to re-locate a specimen of interest, you will need either to record its coordinates from the scales on the horizontal and vertical movements of the mechanical stage, or, better yet, mark a circle around it or a dot near it on the slide itself. This is difficult, since the immersion oil must be removed without moving the slide. Some manufacturers, notably Leitz, produce object markers that take the place of an objective lens. When the specimen is at the center of the field, the object marker is put in place, lowered until it touches the slide, and turned. A diamond point scratches a circle on the surface of the cover glass. One then can put additional marks in ink on the surface to aid in finding the circular scratch.

It is unfortunately the case that good microscopes have become so expensive that they are beyond even the budgets of many institutions. The high-resolution optics desirable in a diatomist's microscope are staggeringly expensive. If you are faced with budget limitations but still need a serviceable instrument, consider: 1) buying a used microscope made by a reputable manufacturer (a thirty-year old Zeiss, Leitz, or Bausch & Lomb can still be excellent if it was well cared-for); 2) loading superior optics onto a lower priced stand; or 3) refitting an old stand with new optics.

Also, it is not necessary to start with the "perfect" instrument. Much can be learned using lower-priced optics from good manufacturers. Leitz or Zeiss achromatic oil-immersion lenses are still very good, suffering only by comparison with their fluorite and apochromatic oil lenses. I do not, by the way, wish to imply that other manufacturers do not produce good optics. I have found Nikon lenses to be excellent and I have had good reports about those of other makers as well. I must caution you, however, that the so-called "student grade" lenses found on very low priced microscopes are indifferent at best and of no real use to the diatomist.

LITERATURE CITED

PUBLICATIONS REFERRING TO DIATOMS IN ILLINOIS

INDEX OF TAXA CONFIRMED, REPORTED, OR EXPECTED
IN ILLINOIS

LITERATURE CITED

A.N.S.P. File. A file of recently described species that is maintained by the Academy of Natural Sciences of Philadelphia.

Archibald, R. 1971. Diatoms from the Vaal Dam catchment area, Transvaal, South Africa. Botanica Marina 14:17–70.

A.S.A. *See* Schmidt et al. 1874–1959.

Begres, F. M. 1971. The diatoms of Clear Lake and Ventura Marsh, Iowa. Ph.D. diss. Iowa State University Library, Ames.

Bourelly, P. 1968. Les algues d'eau douce. Tome II. Les algues jaunes et brunes. Paris: N. Boubee & Cie. 438 pp.

Britton, M. E. 1944. A catalog of Illinois algae. Northwestern University Studies in the Biological Sciences and Medicine No. 2:1–177.

Cholnoky, B. J. 1956. Neue und seltene Diatomeen aus Afrika. II. Diatomeen aus dem Tugela-Gebiete in Natal. Oesterreichische Botanische Zeitschrift 103:53–97.

———. 1957. Neue und seltene Diatomeen aus Afrika. III. Diatomeen aus dem Tugela-Flusssystem, hauptsachlich aus den Drakensbergen in Natal. Oesterreichische Botanische Zeitschrift 104:25–99.

———. 1958. Beitrage zur Kenntnis der südafrikanischen Diatomeenflora. II. Einige Gewässer im Waterberg-Gebiet, Transvaal. Portugaliae Acta Biologia (Ser. B) 6(2):99–160.

———. 1960. Beitrage zur Kenntnis der Diatomeenflora von Natal (Sudafrika). Nova Hedwigia 2:1–128.

Cleve, P. T., & A. Grunow. 1880. Beitrage zur Kenntnis der Arctischen Diatomeen. Kongliga Svenska Vetenskaps-Akademiens Handlingar Bd. 17, No. 2. 121 pp. + 7 pls.

Cleve-Euler, A. 1953. Die Diatomeen von Schweden und Finnland. Teil III. Monoraphideae, Biraphideae 1. Kongliga Svenska Vetenskaps-Akademiens Handlingar, 4e Serien, 4(5):1–255.

———. 1955. Die Diatomeen von Schweden und Finnland. Teil IV. Biraphideae 2. Kongliga Svenska Vetenskaps-Akademiens Handlingar, 4e Serien, 5(4):1–232.

Drum, R. W., & H. S. Pankratz. 1965. Fine structure of an unusual cytoplasmic inclusion in the diatom genus *Rhopalodia*. Protoplasma 60(1):141–49.

Foged, N. 1976. Freshwater diatoms in Sri Lanka (Ceylon). Bibliotheca Phycologica Bd. 23:1–112.

Germain, H. 1964. *Navicula gothlandica* Grunow. Revue Algologique, n.s. T. VII, No. 2:196–201.

Granetti, B. 1968. Revisione critica di *Navicula nigrii*, *Navicula casertana*, *Pinnularia paserinii*, *Stauroneis verbania*, *Synedra juliana* De Notaris. Giornale Botanico Italiano 102:427–37.

Guermeur, P. 1954. Diatomées de l'Afrique occidentale française (Première liste: Sénégal). Institut Française d'Afrique Noire Catalogues no. XII. Dakar. 137 pp. + 24 pls.

Håkansson, H. 1976. Die Struktur und Taxonomie einiger *Stephanodiscus*-Arten aus eutrophen Seen Südschwedens. Botaniska Notiser 129 (1976):25–34.

Hohn, M. H., & J. Hellerman. 1963. The taxonomy and structure of diatom populations from three eastern North American rivers using three sampling methods. Transactions of the American Microscopical Society 82:250–329.

Huber-Pestalozzi, G. 1942. Das Phytoplankton des Süsswassers. *In* A. Thienemann, ed., Die Binnengewässer. Bd. 16, T. 2, 2. Hälfte. Stuttgart: Schweizerbart'sche Verlagsbuchhandlung. 182 pp.

Hustedt, F. 1930. Bacillariophyta (Diatomeae). *In* A. Pascher, ed., Die Süsswasser-Flora Mitteleuropas. Heft 10. Jena: Fischer. 466 pp.

———. 1930a. Die Kieselalgen. *In* L. Rabenhorst, ed., Kryptogamenflora von Deutschland, Österreich und der Schweiz. Bd. 7, T. 1. Leipzig: Geest und Portig. 311 pp.

———. 1931. Die Kieselalgen. *In* L. Rabenhorst, ed., Kryptogamenflora von Deutschland, Österreich und der Schweiz. Bd. 7, T. 2, Lief. 1: Leipzig: Geest und Portig. 176 pp.

———. 1932. Die Kieselalgen. *In* L. Rabenhorst, ed., Kryptogamenflora von Deutschland, Österreich und der Schweiz. Bd. 7, T. 2, Lief. 2. Leipzig: Geest und Portig. 143 pp.

———. 1937. Systematische und oekologische Untersuchungen über die Diatomeen-flora von Java, Bali und Sumatra nach dem Material der Deutschen Limnologischen Sunda-Expedition. Systematischer Teil. Archiv fur Hydrobiologie, Supplement-Band 15:131–77, 187–295, 393–506.

———. 1937a. Die Kieselalgen. *In* L. Rabenhorst, ed., Kryptogamenflora von Deutschland, Österreich und der Schweiz. Bd. 7, T. 2, Lief. 5. Leipzig: Geest und Portig. 161 pp.

———. 1942. Diatomeen aus der Umgebung von Abisko in Schwedisch-Lappland. Archiv für Hydrobiologie 39:82–174.

———. 1949. Susswasser-Diatomeen aus dem Albert-National Park in Belgisch-Kongo. *In* Exploration du Parc national Albert: Mission H. Damas (1935–36). Fasc. 8. Bruxelles: Marcel Hayez. 199 pp.

———. 1959. Die Kieselalgen. *In* L. Rabenhorst, ed., Kryptogamenflora von Deutschland, Österreich und der Schweiz. Bd. 7, T. 2, Lief. 6. Leipzig: Geest und Portig. 108 pp.

———. 1959a. Die Diatomeenflora des Neusiedler Sees im österreichischen Burgenland. Oesterreichische Botanische Zeitschrift 106: 390–430.

———. 1961. Die Kieselalgen. In L. Rabenhorst, ed., Kryptogamenflora von Deutschland, Österreich und der Schweiz. Bd. 7, T. 3, Lief. 1. Leipzig: Geest und Portig. 160 pp.

———. 1962. Die Kieselalgen. In L. Rabenhorst, ed., Kryptogamenflora von Deutschland, Österreich und der Schweiz. Bd. 7, T. 3. Lief. 2. Leipzig: Geest und Portig. 187 pp..

———. 1964. Die Kieselalgen. In L. Rabenhorst, ed., Kryptogamenflora von Deutschland, Österreich und der Schweiz. Bd. 7, T. 3, Lief. 3. Leipzig: Geest und Portig. 117 pp.

———. 1966. Die Kieselalgen. In L. Rabenhorst, ed., Kryptogamenflora von Deutschland, Österreich und der Schweiz. Bd. 7, T. 3, Lief. 4. Leipzig: Geest und Portig. 239 pp.

Kalinsky, R. 1973. An identification system for the genus Nitzschia Hassall (Bacillariophyta), with a nomenclatural revision. Ph.D. diss. Ohio State University, Columbus.

Koerner, H. 1970. Morphologie und Taxonomie der Diatomeengatung Asterionella. Nova Hedwigia, Bd. 20:447–724.

Krasske, G. 1951. Die Diatomeenflora der Acudas Nordostbrasiliens (Zur Kieselalgenflora Brasiliens II). Archiv für Hydrobiologie Bd. 49:639–53.

Lange-Bertalot, H. 1977. Eine Revision zur Taxonomie der Nitzschiae lanceolatae Grunow. Die "klassischen" bis 1930 beschriebenen Süsswasserarten Europas. Nova Hedwigia Bd. 28 (2 + 3):253–307.

———. 1978. Zur Systematik, Taxonomie und Ökologie des abwasserspezifisch wichtigen Formenkreises um "Nitzschia thermalis." Nova Hedwigia Bd. 30:635–52.

———. 1980. New species, combinations and synonyms in the genus Nitzschia. Bacillaria 3:41–77.

Lange-Bertalot, H., & R. Simonsen. 1978. A taxonomic revision of the Nitzschiae lanceolatae Grunow. 2. European and related extra-European fresh water and brackish water taxa. Bacillaria 1:11–111.

Lowe, R. 1972. Notes on Iowa diatoms X: New and rare diatoms from Iowa. Proceedings of the Iowa Academy of Science 79:66–69.

———. 1975. Comparative ultrastructure of the valves of some Cyclotella species (Bacillariophyceae). Journal of Phycology 11(4):415–24.

Lund, J. W. G. 1946. Observations on soil algae. I. The ecology, size and taxonomy of British soil diatoms. Pt. 2. New Phytologist 45:56–110.

Manguin, E. 1962. Contribution à la connaissance de la flore diatomique de la Nouvelle-Calédonie. Mémoires du Musée National d'Histoire Naturelle, n.s., t. 12, fasc. 1. 40 pp. + 8 pls.

———. 1964. Contribution à la connaissance des diatomées des Andes du Pérou. Mémoires du Musée National d'Histoire Naturelle, n.s., sér. B, Botanique, t. 12, fasc. 2. 98 pp. + 25 pls.

Mayer, A. 1928. Die bayerische Gomphonemen. Denkschriften der Königlichen Bayerischen Botanischen Gesellschaft in Regensburg Bd. 13, n.F. Bd. 11:83–128.

———. 1940. Die Diatomeenflora von Erlangen. Denkshriften der Königlichen Bayerischen Botanischen Gesellschaft in Regensburg Bd. 21, n.F. Bd. 15:113–225.

Mueller, O. 1911. Berichte ueber die botanischen Ergebnisse der Nyassa-See und Kinga-Gebirgs-Expedition. VIII. Bacillariaceen aus dem Nyassalande und einigen benachbarten Gebieten. 4e Folge. A. Engler Botanische Jahrbucher 45:69–122.

Patrick, R., & C. W. Reimer. 1966. The diatoms of the United States. Vol. 1. Philadelphia: Academy of Natural Sciences of Philadelphia. 688 pp.

———. 1975. The diatoms of the United States. Vol. 2, pt. 1. Philadelphia: Academy of Natural Sciences of Philadelphia. 213 pp.

Peragallo, H., & M. Peragallo. 1897–1908. Diatomées marines de France. Reprint. Amsterdam: Asher, 1965. 491 pp. + 137 pls.

Petersen, J. B. 1946. Algae collected by Eric Hulten on the Swedish Kamchatka Expedition, 1920–1922, especially from hot springs. Danske Videnskabernes Selskab Biologische Meddelser 20(1):1–122.

Reimer, C. W. 1966. Consideration of fifteen diatom taxa (Bacillariophyta) from the Savannah River, including seven described as new. Notulae Naturae 397:1–15.

———. 1970. Diatoms from Cayler Prairie. Nova Hedwigia Beiheft 31:235–49.

Schmidt, A., et al. 1874–1959. Atlas der Diatomaceen-Kunde. Leipzig: R. Reisland.

Schoeman, F. R. 1973. A systematical and ecological study of the diatom flora of Lesotho with special reference to the water quality. Pretoria: V & R. Printers.

Silva, P. C. 1962. Classification of algae. In R. A. Lewin, ed., Physiology and biochemistry of algae. New York: Academic Press.

Skvortzow, B. V. 1928. A contribution to the diatoms of Baikal Lake. Proceedings of the Sungaree River Biological Station 1(5):14.

———. 1937. Diatoms from Lake Michigan. I. American Midland Naturalist 18:652–58.

Stoermer, E. F. 1963. New taxa and new United States records of the diatom genus *Neidium* from West Lake Okoboji, Iowa. Notulae Naturae 358:1–9.

———. 1964. Notes on Iowa diatoms. VII. Rare and little known diatoms from Iowa. Proceedings of the Iowa Academy of Science 71:55–66.

———. 1967. Polymorphism in *Mastogloia*. Journal of Phycology 3:73–77.

Tiffany, L. H., & M. Britton. 1952. The algae of Illinois. Chicago: University of Chicago Press. 407 pp.

Van Heurck, H. 1880–83. Synopsis des diatomées de Belgique. Atlas. Anvers: Ducaju et cie. 132 pls.

Van Heurck Type Slides (A-VH). An exsiccata collection issued to accompany the Synopsis des diatomées de Belgique. One set of the slides is at the Academy of natural Sciences of Philadelphia and is designated A-VH. The numbers following this designation refer to the slide numbers within the set.

Wallace, J. H., & R. Patrick. 1950. A consideration of *Gomphonema parvulum* Kuetz. Butler University Botanical Studies 9:227–34.

Werner, D., ed. 1977. The biology of diatoms. Berkeley: University of California Press, 498 pp.

The following two references have been frequently used, but not cited directly:

Mills, F. W. 1933–35. An index to the genera and species of diatomaceae and their synonyms. Pts. I–XXI. London: Wheldon & Wesley. 1726 pp.

Van Landingham, S. L. 1967–79. Catalog of the fossil and recent genera and species of diatoms and their synonyms. 8 vols. Lehre, Federal Republic of Germany: J. Cramer. 4654 pp.

PUBLICATIONS REFERRING TO
DIATOMS IN ILLINOIS

Many of the publications in this list have not been cited directly in the text but may be of use to those readers who wish to examine the first reports of the diatoms in Illinois. I would especially like to call your attention to two works: *The Algae of Illinois* (Tiffany and Britton 1952) and *A Catalog of Illinois Algae* (Britton 1944) as being of major importance as references. Britton's *Catalog* has been my primary source of information on the synonymy of diatoms reported for the state prior to 1944. This work also provides a complete summary of the reported distribution of diatoms in Illinois prior to 1944. *The Algae of Illinois* remains a standard reference on Illinois algae.

Agersborg, H. P. K. 1930. Does high temperature in a frigid country limit diversification of the species? Transactions of the Illinois State Academy of Science 22:103–14.

Ahlstrom, E. H. 1936. The deep-water plankton of Lake Michigan. Transactions of the American Microscopical Society 55:286–99.

Babcock, H. H. 1872a. Chicago hydrant water. Grevillea 1:13.

———. 1872b. On the effect of the reversal of the current of the Chicago River on the hydrant water. The Lens 1:103–6.

Bailey, J. W. 1845. Notes on the infusoria of the Mississippi River. Proceedings of the Boston Natural History Society 2:33–35.

Baker, F. C. 1922. The molluscan fauna of the Big Vermillion River. Illinois Biological Monographis 7:7–126.

Briggs, S. A. 1872a. The diatomacceae of Lake Michigan. The Lens 1:41–44.

———. 1872b. *Rhizosolenia eriensis* H. L. Smith. Grevillea 1:1–14.

———. 1872c. *Rhizosolenia eriensis*. The Lens 1:15.

Britton, M. E. 1944. A catalog of Illinois algae. Northwestern University Studies in the Biological Sciences and Medicine, No. 2:1–177.

Daily, W. A. 1938. A quantitative study of the phytoplankton of Lake Michigan, collected in the vicinity of Evanston, Illinois. Butler University Botanical Studies 4:65–83.

Eddy, S. 1924. Presence of living organisms in lake ice. Transactions of the Illinois State Academy of Science 17:85–86.

———. 1925. Freshwater algal succession. Transactions of the American Microscopical Society 44:138–47.

———. 1927a. A study of algal distribution. Transactions of the American Microscopical Society 46:122–38.

———. 1927b. The plankton of Lake Michigan. Bulletin of the Illinois Natural History Survey 17:203–32.

———. 1931a. The plankton of the Sangamon River in the summer of 1929. Bulletin of the Illinois Natural History Survey 19:469–86.

———. 1931b. The plankton of some sink-hole ponds in southern Illinois. Bulletin of the Illinois Natural History Survey 19:449–67.

Eddy, S., & P. H. Simer. 1928. Notes on the food of the paddlefish and the plankton of its habitat. Transactions of the Illinois State Academy of Science 21:59–68.

Ehrenberg, C. G. 1856. Mikrogeologie. Das Erden und Felsen schaffende Wirken des unsichtbar kleinen selbstandigen Lebens auf der Erde. Fortetzung. Leipzig: Leopold Voss. 88 pp.

Forbes, S. A., & R. E. Richardson. 1913. Studies on the biology of the upper Illinois River. Bulletin of the Illinois State Laboratory of Natural History 9:481–574.

Galtsoff, P. S. 1924. Limnological observations in the upper Mississippi, 1921. Bulletin of the U.S. Bureau of Fisheries 39:347–438.

Grady, M. M. 1974. New algal records for Lake County, Illinois. Transactions of the Illinois State Academy of Science 67:318–22.

Jelliffe, S. E. 1893. The Chicago water supply in the World's Fair Grounds. American Monthly Microscopical Journal 14:310–11.

Kofoid, C. A. 1903. Plankton studies IV. The plankton of the Illinois River 1894–1899, with introductory notes upon the hydrography of the Illinois River and its basin. Part I. Quantitative investigations and general results. Bulletin of the Illinois State Laboratory of Natural History 6:95–635.

———. 1908. Plankton studies V. The plankton of the Illinois River, 1894–1899. Part II. Constituent organisms and their seasonal distribution. Bulletin of the Illinois State Laboratory of Natural History 8:3–361.

Lipsey, L. L. 1975. Notes on the diatom flora of Kane County, Illinois. I. The Fox River at Elgin. Transactions of the Illinois State Academy of Science 68:339–45.

———. 1976. Notes on the diatom flora of Lake and McHenry Counties, Illinois. Transactions of the Illinois State Academy of Science 69:283–91.

Purdy, W. C. 1930. A study of the pollution and natural purification of the Illinois River. II. The plankton and related organisms. Bulletin of the U.S. Public Health Service 198:1–212.

Skvortzow, B. V. 1937. Diatoms from Lake Michigan. I. American Midland Naturalist 18:652–58.

Smith, H. L. 1878. Description of new species of diatoms. American Quarterly Microscopical Journal 1:12–18.

Thomas, H. D., & H. H. Chase. 1887. Diatomaceae of Lake Michigan during the last sixteen years from the water supply of Chicago. Notarisia 2:328–30.

Tiffany, L. H., & M. Britton. 1952. The algae of Illinois. Chicago: University of Chicago Press. 407 pp.

Transeau, E. N. 1913a. Annotated list of the algae of eastern Illinois. Transactions of the Illinois State Academy of Science 6:69–89.

———. 1913b. The periodicity of algae in Illinois. Transactions of the American Microscopical Society 32:31–40.

———. 1916. The periodicity of freshwater algae. American Journal of Botany 3:121–33.

Welch, W. B. 1942. A study of the phytoplankton of Crab Orchard Lake. Transactions of the Illinois State Academy of Science 35:81–82.

Index of Taxa Confirmed, Reported, or Expected in Illinois

471

John Jeffrey Dodd received his B.S. degree in botany from Iowa State University, Ames, his M.S. degree in biology from Eastern Michigan University, Ypsilanti, and his Ph.D. degree in botany from Southern Illinois University, Carbondale. His articles on diatoms have appeared in *Proceedings of the Iowa Academy of Science*.

Robert H. Mohlenbrock, Distinguished Professor of Botany, Southern Illinois University, Carbondale, is the general editor and principal author of the Illustrated Flora of Illinois series, and the author of scores of articles and numerous other books.